AF805838

ADVANCES IN EXPERIMENTAL MEDICINE AND BIOLOGY

Editorial Board:

IRUN R. COHEN, *The Weizmann Institute of Science*
ABEL LAJTHA, *N. S. Kline Institute for Psychiatric Research*
JOHN D. LAMBRIS, *University of Pennsylvania*
RODOLFO PAOLETTI, *University of Milan*

For further volumes:
http://www.springer.com/series/5584

Heiko Enderling • Nava Almog • Lynn Hlatky
Editors

Systems Biology of Tumor Dormancy

Springer

Editors
Heiko Enderling
Center of Cancer Systems Biology
Steward Research & Specialty Projects Corp.
St. Elizabeth's Medical Center
Tufts University School of Medicine
Boston, MA, USA

Nava Almog
Center of Cancer Systems Biology
Steward Research & Specialty Projects Corp.
St. Elizabeth's Medical Center
Tufts University School of Medicine
Boston, MA, USA

Lynn Hlatky
Center of Cancer Systems Biology
Steward Research & Specialty Projects Corp.
St. Elizabeth's Medical Center
Tufts University School of Medicine
Boston, MA, USA

ISSN 0065-2598
ISBN 978-1-4614-1444-5 ISBN 978-1-4614-1445-2 (eBook)
DOI 10.1007/978-1-4614-1445-2
Springer New York Heidelberg Dordrecht London

Library of Congress Control Number: 2012947409

© Springer Science+Business Media New York 2013
This work is subject to copyright. All rights are reserved by the Publisher, whether the whole or part of the material is concerned, specifically the rights of translation, reprinting, reuse of illustrations, recitation, broadcasting, reproduction on microfilms or in any other physical way, and transmission or information storage and retrieval, electronic adaptation, computer software, or by similar or dissimilar methodology now known or hereafter developed. Exempted from this legal reservation are brief excerpts in connection with reviews or scholarly analysis or material supplied specifically for the purpose of being entered and executed on a computer system, for exclusive use by the purchaser of the work. Duplication of this publication or parts thereof is permitted only under the provisions of the Copyright Law of the Publisher's location, in its current version, and permission for use must always be obtained from Springer. Permissions for use may be obtained through RightsLink at the Copyright Clearance Center. Violations are liable to prosecution under the respective Copyright Law.
The use of general descriptive names, registered names, trademarks, service marks, etc. in this publication does not imply, even in the absence of a specific statement, that such names are exempt from the relevant protective laws and regulations and therefore free for general use.
While the advice and information in this book are believed to be true and accurate at the date of publication, neither the authors nor the editors nor the publisher can accept any legal responsibility for any errors or omissions that may be made. The publisher makes no warranty, express or implied, with respect to the material contained herein.

Printed on acid-free paper

Springer is part of Springer Science+Business Media (www.springer.com)

Foreword

Numerous organisms in nature have evolved defense behaviors to preserve themselves against predators. Ironically, many of these behaviors are seemingly at odds with the ultimate goal of survival. One such behavior, thanatosis (of Greek origin, meaning "putting to death"), is a defense mechanism in nature whereby an animal feigns death in order to avoid detection and possible death by a predator. This behavior is most commonly associated with the Virginia Opossum, where when threatened, it can switch on a near death appearance by "playing possum" and fooling would-be predators. This evolutionary trait holds insight for cancer research, because similar behaviors may be invoked by the cancer cells within the animal providing a defense for tumor cells.

Cancer is a complex disease and, by reputation and outcome, also an aggressive disease that can quickly overtake and kill its host. However, recent scientific advancements have shown that cancer is capable of a variety of growth patterns, from rapid replication and spreading to a more controlled dormant phenotype evading detection. Unfortunately, a dormant phenotype is by its very nature more difficult to detect and treat.

Despite amazing biomedical advances and billions of research dollars, cancer remains one of the most destructive and elusive diseases known to humankind. Statistically, cancer will be the cause of death for 25 % of the US population, and according to the World Health Organization, it will be the number one global killer this year. Ironically, part of the challenge is due to the success we have had in preventing and treating cancer and other acute diseases which increases survival, and subsequently, the at-risk population. Part of this can be explained by the increase in lifespan throughout the world and the knowledge that cancer is primarily a disease of the aged. In addition, the diversity of the disease and patient population suggests a multitude of etiologies and subsequent treatment strategies. The Cancer Genome Atlas (TCGA) has highlighted the complexity of cancer at the molecular level. Human behavior also plays a significant role especially in the prevention of cancer. Smoking and diet are the most common behaviors that continue to have an important impact on cancer incidence but remain difficult to alter.

With the knowledge that cancer incidence is increasing throughout the world, we must continue to advance detection and treatment of the disease. With encouraging and important exceptions, treatment remains somewhat generic and unchanged over the past decades. Surgery, which was pioneered over a century ago, remains the most recommended and successful treatment for solid tumors. Most chemotherapeutic strategies reflect more broad-based agents targeting fundamental cellular processes such as DNA replication. We currently know more about the physiology and biology of cancer than ever before, and we are beginning to use this knowledge for a more specialized approach to the prevention, detection, and treatment of cancer. Most indicative of this has been recent success of targeted therapies such as Herceptin or Gleevec, which are used to treat aggressive forms of breast cancer and leukemia, respectively. Unfortunately, these treatments, while promising, have pitfalls of patient selection and resistance development. Since cancer is most successfully treated at early or less aggressive stages of the disease, research into the growth kinetics of cancer continues to hold a great deal of promise for future advances against the disease.

Tumor dormancy is a critical stage in cancer development where cancer cells can remain occult and asymptomatic. Dormancy can occur at various stages of the cancer's progression from early stage development, as micro-metastasis, or as a residual disease following "successful" treatment. This last niche as residual disease is critical in long-term survival of the patient. While many questions remain unknown about tumor dormancy, we do know that the process, like so many in biology, involves multiple components and physical scales. At the cellular level, the cell cycle, senescence, and apoptosis are critical, while at the micro-environmental and organismal level, angiogenesis and the immune system are major players. The role that all of these components play in the initiation and cessation of dormancy remains a central question in cancer biology. Other questions exist as to the molecular and cellular markers of dormancy and how this phenotype is manifested in diverse tumor types under various conditions including current therapies. Obviously, answers to these questions will require a systems biology approach that can consider the variety of molecular and cellular components at work.

The National Cancer Institute (NCI) established the Integrative Cancer Biology Program in 2004 to study cancer biology from a systems biology perspective. The Center of Cancer Systems Biology at Steward St. Elizabeth's Medical Center is part of this effort and sponsored the first Annual Workshop on Systems Biology of Tumor Dormancy. The organizing committee included: Nava Almog, Heiko Enderling, Cassedra Enayo, Lynn Hlatky, Clare Lamont, and Melissa Klumpar. This international workshop brought together clinicians, biologists, mathematicians, and computer scientists to discuss the critical issues of tumor dormancy with emphasis on angiogenesis, the immune system, and cancer stem cells. The workshop included presentations by mathematicians Heiko Enderling, Kathleen Wilkie, and Philip Hahnfeldt, and biologists Nava Almog, Stefano Indraccolo, Tobias Schatton, Julio Aguirre-Ghiso, Bruno Quesnel, and Dean Felsher. Working groups held during the meeting allowed workshop participants to discuss current problems related to tumor dormancy and develop novel mathematical/computational models. Mathematicians, biologists, and clinicians in each working group engaged in interdisciplinary dialogues and model development. During the three-and-a-half day workshop, the modeling groups developed exciting new projects and laid the foundation for collaborations and joint manuscript submissions. The proceedings in this book reflect those presentations and discussions and in collection, represent an important reference for the state of science and hope in the field of tumor dormancy.

Tumor dormancy remains one of the least understood aspects of cancer biology. While its obvious phenotype represents a challenge in detection, elimination, and long-term survival, it also gives new hope in cancer treatment. If we can understand the mechanism of control of dormancy or gain new insights into the molecular and cellular controls of cancer growth or dormancy, then we have the potential to manipulate those processes for better therapies and outcomes. Knowledge gained from publications such as this will bring the field closer to practical approaches, knowing that while the tumor "can run, it can't hide." In the end, even the tricks of the opossum can be detected by the knowledgeable predator.

Dan Gallahan, Ph.D.
Deputy Director, Division of Cancer Biology
National Cancer Institute, National Institutes of Health
Bethesda, Maryland, USA

Introduction

Awareness of the existence and importance of tumor dormancy has come from a number of disparate clinical and translational directions, attesting to the broad applicability of this phenomenon. To do the topic justice, it is necessary to briefly recount the settings in which dormancy has been encountered, and most importantly, to assess what has been learned and what stands yet to be learned from those encounters, both in the clinic and in the laboratory.

Curiously, attention has turned to the subject of tumor dormancy amidst an intense clinical focus on the opposite phenomenon—advanced, "out-of-control" cancers. It is not lost on anyone in the field that, while very important therapeutic strides have been made against particular cancers, including blood-borne, germ-cell, and childhood cancers since the declared War on Cancer in 1971, the situation for adults manifesting most advanced epithelial cancers remains problematic. These high-profile refractory cancers, including those of the lung, breast, brain, pancreas, colon, and ovary, carry fearsome statistics and metastatic disease often foreshadows an inevitable course. Our mainstay strategies of direct tumor attack, employing a growing repertoire of chemotherapeutic and radiation protocols, often provide impressive initial responses, but over the long run frequently prove inadequate.

It therefore stands to reason attention is feverishly focused on finding new methods to detect cancer earlier while the condition remains treatable. Indeed, the search for cancer in asymptomatic people has taken on a life of its own, placing as much emphasis on discovering it in the seemingly healthy as treating it in the obviously sick. The battle has even pitted alternative methods of detection against one another. The National Lung Cancer Screening Trial (NLST) was conducted to resolve the issue of whether people at risk for lung cancer would benefit more from screening with the powerful low-dose spiral computed tomography (CT) than conventional chest X-ray. The trend was sufficient to end the trial early—a sizeable 20 % improvement in survival was noted when CT was employed. But there were some tradeoffs. With CT, only 3.6 % of lesions requiring clinical follow-up proved to be positive for cancer, while for X-rays, the rate was 52 % higher at 5.5 %.

It may at first seem paradoxical that a clear improvement in detection technology for cancer should also be yielding higher rates of false detection. One might argue that, as our ability to resolve increases, so should the accuracy of the claims surrounding what we are examining. But this would not take into account that with the power to resolve comes not only a better resolution of what was visible before, but also the ability to view what were previously undetected lesions, some of which may pose a distinctly different level of threat. This possibility was brought to the fore in a seminal study (Black and Welch, *NEJM*, 1993), which reported on histological findings from autopsies of adults dying of non-cancer causes. Similar to the NLST study, the limits of diagnostic capabilities using refined methods for gross visualization were tested. Surprisingly, for a range of cancer types, it was determined that the prevalence of microscopic detectable cancer far outweighs the actual macroscopic disease incidence—that virtually all of us by adulthood are cancer carriers, whether we manifest symptoms or not. Thus, by looking more closely for cancer disease in our quest to avoid its advanced refractory state, one is discovering that cancers commonly exhibit growth dynamics not characteristic of symptomatic disease. The picture emerging is that overtly transformed cancer cells commonly face cancer-host interaction bottlenecks that limit tumor growth before becoming overt disease.

One major realization of this altered dynamic is the state of tumor dormancy. Once thought an exceptional occurrence, dormancy is now appreciated to be a common stage in the course of many cancer types. The implications of this realization are nothing short of dramatic—extending in three major directions. The first is the epidemiologic notion of cancer risk, which if properly defined as the eventual experience of symptomatic disease, must now be conceptually disconnected from its current interpretation as the risk of creation of the first cancer cell. Secondly, we must reconsider the practical implications for whether to treat the ever-smaller tumors detectable by our improving technologies that may be dormant and therefore pose a much-reduced threat. Lastly, and perhaps most importantly, we must understand how intrinsic dormancy bottlenecks can effectively control cancer in ways we have not been able to match with our therapeutic anti-cancer armamentarium.

A proper accounting of dormancy in cancer progression would clearly improve risk estimation for symptomatic cancer presentation. Heretofore, classic thinking has maintained that stochastic DNA damaging events and gene mutations lead to eventual cell transformation and the first cancer cell, from which symptomatic cancer inevitably arises. The prevalence of tumor dormancy has removed the word "inevitably" from this statement, radically altering the classic risk models. Understanding the ramifications stands to better inform policy-making decisions, e.g., limits for exposures to carcinogens in the workplace and the environment. As cases in point, the Biological and Environmental Research Division of the Department of Energy (DOE) is charged with researching the cancer risk associated with nuclear waste cleanup, and more generally the hotly debated question of whether there exist low-dose limits to exposure below which there is no lasting damage. In addition, the Space Radiation Program Element of the Human Research Program at the National Aeronautics and Space Administration (NASA) is committed to estimating the excess radiogenic cancer risk for astronauts embarking on extended space missions. The matter of tumor dormancy is proving pivotal to both objectives.

The therapeutic question of treatment of slow-growing tumors, although not new, has also attracted attention in light of the question of "over-diagnosis" of cancers that would not progress. In contrast to "false positives," these represent the detection of histologically confirmed cancers, but cancers that are destined not to present as symptomatic disease over the person's lifetime. Contributing to this class are dormant and sufficiently slow-growing tumors—ironically the very types of tumors our early-detection technologies are best designed to detect. One recently published study of the subject involved 39,888 Norwegian women with diagnosed invasive breast cancer who had either participated in their new breast cancer mammography screening initiative, or not. What the investigators found was that, when tracking the number of detected cancers in the unscreened group, they never quite added up to the number detected by screening—the difference representing "pseudodisease," i.e., dormant or near-dormant tumors detected by screening that never would have advanced to routine clinical presentation over the patient's lifetime. They estimated that for every 2,500 people screened, one cancer death would be avoided, but six to ten individuals would undergo unnecessary treatment for a disease they were never destined to experience. More generally, the problem of overdiagnosis tends to exaggerate the success statistics for any screening study, as every treatment for screen-detected pseudodisease contributes a guaranteed "cure."

The most far-reaching implications of tumor dormancy, however, may well come from translational research. Looking forward, the phenomenon of tumor dormancy, or near dormancy, offers a unique opportunity to understand a natural means of modulating disease progression. Appreciating this, the Workshop on Tumor Dormancy held at the SEMC in Boston this last summer was focused on presenting for interactive discussion the various underpinnings and implications of this simple dynamic state. These settled into four broad contexts—the roles of (1) the immune response, (2) cancer stem cells (CSCs), (3) organ context, and (4) induction of angiogenesis.

The immune system was discussed for its rather complicated inclusion of tumor dormancy, sandwiched as it is as the second "E" (for "Equilibrium") between the earlier tumor attrition ("Elimination") phase and the final tumor release ("Escape") phase, known collectively as the three "E's" of immunoediting. The immune influence was portrayed as a contest of sorts between the tumor cells (prey) and the immune cells (predators), with the outcome being anything but intuitive. In line with recent studies, a biphasic immune response was noted. One surprising observation was that limited immunity may actually hasten escape from the equilibrium phase it helps establish and encourage cross-resistance to agents. In this way, tumor dormancy can actually limit the effectiveness of therapy.

By a quite distinct mechanism, CSCs, along with their non-stem counterparts, were proposed to play an analogous role in producing biphasic dependencies between cell targeting and overall population response. When CSC migration or non-stem cell killing is low, CSCs become encased in their own progeny, thwarting CSC expansion and thus population growth overall. However, when CSCs are able to occupy adjacent open areas of the tumor with the help of either higher cell migration or a higher attrition of the non-stem progeny, the tumor may more efficiently undergo "self-metastasis" at its periphery, thereby helping to expand the tumor as a whole. Other work demonstrated a more transcendent control, operating through the myc oncogene. The effect of its inactivation is to block self-renewal, tying this process again back to CSCs.

The role of context in controlling growth was seen also to extend beyond the stem and non-stem composition. Evidence exists that organ-specific molecular signaling can determine whether a metastatic lesion will expand or remain dormant. By examining the different signaling profiles at these sites, it has been possible to ascertain what may be dominant controlling factors. Key players prove to be stress-activated kinases, transcription factors, e.g., p53, and cell cycle inhibitors, e.g., p21.

Finally, a fourth major topic discussed was the role of tumor angiogenesis in defining the dormant state and its implications for tumor development. Once again cancer growth dynamics are seen to be controlled by a balance of opposing influences; either through balanced proliferation and cell death in the case of pre-vascular lesions that are not yet angiogenically competent, or a balance between pro- and anti-angiogenic factors emanating from the post-vascular tumor. Potential mechanisms governing dormancy control in these two cases were discussed; the former showing a novel influence of miRNAs, and the latter showing evidence of tumor exploitation of what are likely normal organogenic growth controls. In work that may be glimpsing a global influence of immunity, stem cells, and context in dormancy, a tumor model focusing on a stem-like ABCB5+ subpopulation of melanoma cells revealed simultaneous immune influence along with angiogenic control.

The take-home message from these seemingly disparate underpinnings of the dormancy state may well be the commonalities revealed. Tumor dormancy may generally be described as a balance between opposing forces, working through molecular, population, and inter-tissue levels. Most of the mechanistic drivers are proving not to be de novo creations, but mechanisms "borrowed" in a distorted way from normal tissue controls. This is providing an impetus for a new frontier in treatment approach—one that could conceivably limit progression of refractory cancers by employing existing natural control processes. "Putting the genie back in the bottle," if you will, a goal that has evaded tumor-directed attacks thus far, may well be achievable through exploitation of tumor dormancy—a dynamic which has already proven it can do just that.

Lynn Hlatky, Ph.D.
Director, Center of Cancer Systems Biology
Steward Research & Specialty Projects Corp.
St. Elizabeth's Medical Center
Tufts University School of Medicine
Boston, Massachusetts, USA

Preface

The concept of this book arose from the first in a series of annual workshops organized by the Center of Cancer Systems Biology at Steward St. Elizabeth's Medical Center, Tufts University School of Medicine, and supported by the National Cancer Institute's Integrative Cancer Biology Program. This inaugural workshop focused on Systems Biology of Tumor Dormancy and was held in Boston, Massachusetts in late July 2011. The goal of the workshop, and by extension, of this book, was to present research advances in the field of tumor dormancy from diverse experimental and clinical perspectives using biological, mathematical, and computational approaches.

As the editors, we are grateful to the team at the Center of Cancer Systems Biology who organized and hosted the workshop and would like to extend our appreciation to all workshop speakers and contributing authors who diligently worked on their respective chapters. We would also like to thank Melissa Klumpar and Brandy Weidow for their help in editing the chapters, and Melanie Tucker and Connie Walsh from Springer Publishing who guided us through this journey and kept us on course.

We hope that this book stimulates your interest in tumor dormancy, as well as in exploring interdisciplinary research techniques.

Enjoy.

Boston, MA, USA

Heiko Enderling
Nava Almog
Lynn Hlatky

Contents

Contributors

Julio A. Aguirre-Ghiso, Ph.D. Departments of Medicine and Otolaryngology, Tisch Cancer Institute, Black Family Stem Cell Institute, Mount Sinai School of Medicine, New York, NY, USA

Nava Almog, Ph.D. Center of Cancer Systems Biology, Steward Research & Specialty Projects Corp., St. Elizabeth's Medical Center, Tufts University School of Medicine, Cambridge St., Boston, MA USA

David I. Bellovin, Ph.D. Division of Oncology, Departments of Medicine and Pathology, Stanford University School of Medicine, Stanford, CA, USA

Khalid Boushaba, Ph.D. The Johns Hopkins Carey Business School, Baltimore, MD, USA

Paloma Bragado, Ph.D. Departments of Medicine and Otolaryngology, Tisch Cancer Institute, Black Family Stem Cell Institute, Mount Sinai School of Medicine, New York, NY, USA

Bikul Das, M.B.B.S., Ph.D. Division of Oncology, Departments of Medicine and Pathology, Stanford University School of Medicine, Stanford, CA, USA

Jayanta Debnath, M.D. Department of Pathology, Helen Diller Family Comprehensive Cancer Center, University of California San Francisco, San Francisco, CA, USA

Alberto D'Onofrio, Ph.D. Department of Experimental Oncology, Systems Biomedicine, European Institute of Oncology, Milan, Italy

Heiko Enderling, Ph.D. Center of Cancer Systems Biology, Steward Research and Specialty Projects Corp., St. Elizabeth's Medical Center, Tufts University School of Medicine Boston, MA, USA

Dean W. Felsher, M.D., Ph.D. Division of Oncology, Departments of Medicine and Pathology, Stanford University School of Medicine, Stanford, CA, USA

Daniel L. Gallahan, Ph.D. Division of Cancer Biology, National Cancer Institute, National Institutes of Health, Rockville, MD, USA

Philip Hahnfeldt, Ph.D. Center of Cancer Systems Biology, Steward Research & Specialty Projects Corp., St. Elizabeth's Medical Center, Tufts University School of Medicine, Boston, MA, USA

Leonid Hanin, Ph.D. Department of Mathematics, Idaho State University, 921 S. 8th Avenue, Stop 8085, Pocatello, ID USA

Stefano Indraccolo, M.D. Istituto Oncologico Veneto, IRCCS, via Gattamelata, Padua, Italy

Yangjin Kim, Ph.D. Department of Mathematics and Statistics, University of Michigan Dearborn, Dearborn, MI, USA

Sonja Kleffel, M.Sc. Harvard Skin Disease Research Center, Department of Dermatology, Brigham and Women's Hospital, Harvard Institutes of Medicine, Room 673B, 77 Avenue Louis Pasteur, Boston, MA, USA

Bruno Quesnel, M.D., Ph.D. Service des Maladies du Sang, Centre Hospitalier et Universitaire de Lille, Rue Polonovski, Lille, France

Tobias Schatton, Pharm.D., Ph.D. Harvard Skin Disease Research Center, Department of Dermatology, Brigham and Women's Hospital, Harvard Institutes of Medicine, Room 673B, 77 Avenue Louis Pasteur, Boston, MA, USA

Maria Soledad Sosa, Ph.D. Departments of Medicine and Otolaryngology, Tisch Cancer Institute, Black Family Stem Cell Institute, Mount Sinai School of Medicine, New York, NY, USA

Kathleen P. Wilkie, Ph.D. Center of Cancer Systems Biology, Steward Research & Specialty Projects Corp., St. Elizabeth's Medical Center, Tufts University School of Medicine, Boston, MA, USA

Part I
Angiogenesis

Chapter 1
Genes and Regulatory Pathways Involved in Persistence of Dormant Micro-tumors

Nava Almog

Abstract Micro-tumors can remain dormant for prolonged periods of time before they switch and enter the rapid growth phase. This initial stage in tumor progression is clearly understudied. In spite of high prevalence, significant clinical implications and increased interest by the research community, tumor dormancy is still poorly understood. The topic of tumor dormancy also suffers from a lack of definition and an agreed upon terminology to describe it. Additionally, the number of reproducible experimental models available for studying indolence of human micro-tumors is quite limited. Here, we describe the development of a general class of in vivo models of indolent human tumors and how these models can be used to elucidate molecular and cellular mechanisms involved in the regulation of dormancy. The models consist of human tumor cell lines that form microscopic cancerous lesions in mice. Although these lesions contain viable and fully malignant cancer cells, the tumors do not expand in size but remain occult for prolonged periods until they eventually spontaneously switch and become fast-growing tumors. Consistent with Judah Folkman's vision that tumors will remain occult and microscopic until they acquire the ability to recruit new and functional blood vessels, the dormancy period of the micro-tumors is associated with impaired angiogenic capacity. Such models can be used for dissecting the host and the tumor-derived regulatory mechanisms of tumor dormancy. Understanding the process by which dormant tumors can overcome growth constraints and emerge from dormancy, resuming size expansion, may provide insights into novel strategies to prolong the dormancy state or to block tumor formation in the early stages, before they are physically detected or become symptomatic.

N. Almog, PhD (✉)
Center of Cancer Systems Biology, Steward Research & Specialty Projects Corp., St. Elizabeth's Medical Center, Tufts University School of Medicine, 736 Cambridge St., Boston, MA 02135, USA
e-mail: Nava.almog@tufts.edu

H. Enderling et al. (eds.), *Systems Biology of Tumor Dormancy*, Advances in Experimental Medicine and Biology 734,
DOI 10.1007/978-1-4614-1445-2_1, © Springer Science+Business Media New York 2013

Keywords Micro-tumors • Angiogenesis • Occult cancer • Tumor progression • Microenvironment

Introduction

Tumor dormancy is a clinical phenomenon in which tumors do not expand in size over a prolonged period of time [1–6]. It has long been recognized as a significant problem in the management of cancer patients [3, 7–10]. Tumors can enter a latent phase during various stages in tumor progression including post-angiogenic stages of tumor progression [11]. However, in this chapter we only discuss dormancy of *microscopic tumors*, which are usually present with a maximal diameter of 1–2 mm. Here, dormant tumors are defined by their inability to expand beyond a microscopic size (see Fig. 1.1). Importantly, it is demonstrated that such small harmless lesions have the potential to switch to become fast growing, clinically apparent, and lethal. These microscopic cancerous lesions are observed as: one of the earliest stages in tumor development; as micro-metastasis in distant organs; and as minimal residual disease left after surgical removal or treatment of primary tumors [10, 12–15]. Indeed, the mortality of cancer patients is largely determined by the occurrence of metastases, which often are too small to be detected but eventually can lead to relapse [12, 16–19].

Although the concept of dormant tumors is more accepted and better studied when occurring as minimal residual disease, or as micro-metastases which remain

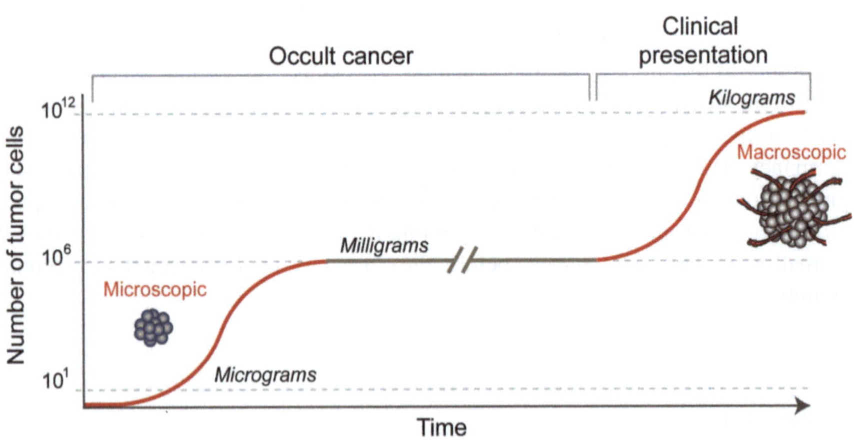

Fig. 1.1 Tumors often remain microscopic and clinically undetectable over long periods of time. This schematic of tumor growth can represent growth of primary or secondary cancers (metastases). Dormant micro-tumors can persist at a small steady size for years, remaining undetectable by commonly used imaging techniques

asymptomatic for decades, tumor dormancy is likely at least as prevalent as a stage of primary tumors. Microscopic and occult cancerous lesions are often found in otherwise healthy people [6, 20]. This implies that many people carry small and occult cancerous lesions without knowing it and that it is more common than frequently recognized.

In recent years, the field of tumor dormancy has been gaining significant attention: A number of reviews and essays on the topic have been published in leading journals such as *Nature* [6], *Nature Reviews Cancer* [2, 3], as well as *Nature Reviews Clinical Oncology* [4] and *Nature Medicine* [15], along with a specially dedicated issue of *APMIS* journal published in 2008. Moreover, dormancy has been the topic of dedicated sessions in several prominent cancer research conferences, and is currently a topic of considerable interest to NCI. Although an improved understanding of the manner by which tumors are induced to remain dormant would have important implications for cancer treatment and screening, and despite increased interest in the research community, to date, the dormant phase of tumor growth remains largely unexplored as a point of therapeutic intervention. The vast majority of cancer research is done on fast-growing and easily detectable tumors, which are more accessible for studies, such as signaling pathways investigations, drug response examinations, and biomarker analyses.

It is now clear that a number of biological processes can contribute to tumor dormancy, and they all support the role of the microenvironment, tumor stroma, and host response. These include tumor cell senescence, immune response of the host, hormonal control or block or insufficiency of tumor angiogenesis potential [1, 2, 5, 8, 17, 21–26]. Indeed, only in the last few years it has been fully appreciated that the tumor constitutes a highly integrated ecosystem in which different cellular populations depend upon each other. Clearly, dormancy of cancerous lesions depends on crucial signals from the microenvironment and the tumor stroma [27–29]. It is still to be determined, however, whether tumors attain dormancy through mechanisms involving extrinsic interactions (e.g., with the microenvironment) or from intrinsic properties of the cells.

Dormant Primary Tumors

The phenomenon of tumor dormancy can well explain the clinical phenomenon of minimal residual disease left after an apparent successful treatment of cancer and the very late relapses often seen in cancer. For example, it is well known that it can take years to decades following breast cancer treatment before local or distant recurrence becomes clinically detectable. The frequent recurrence of breast cancer strongly suggests that cancer seeds are left at the site of primary tumor growth or shed and seed in distant sites as dormant lesions. These lesions could eventually emerge from dormancy erupting into fast-growing tumors [17]. Moreover, patients can present with a metastasis, yet have "unknown primary tumors" which cannot be located [30].

Studying dormancy of *primary* tumors in clinical settings is extremely challenging. Evidence for the existence of primary micro-tumors in clinically healthy individuals comes primarily from histological studies that report a high prevalence of micro-tumors, even in young children (for a review, see [20]). However, even when such micro-tumors are detected in retrospective autopsy studies, it cannot be determined how long the lesions were present and whether they would remain occult or continue growing.

A feasible way to prove the existence of and to study the prevalence of primary dormant *microscopic* tumors is by comparing the frequency of micro-tumors found at autopsies with the frequency of detectable *macroscopic* tumors. When analyzing such data it is important to make sure that early precancerous lesions are excluded. It is also crucial to determine the prevalence of proliferating cells in the tumor, and exclude cases of very slow growing micro-tumors which have no or very few proliferating cells [31, 32]. While such studies are rare, many reports document the high prevalence of micro-tumors in retrospective autopsy studies.

One of the most striking observations of a high incidence of occult tumors was found in a systemic autopsy study of carcinoma of the thyroid. Although the frequency of detected occult papillary carcinoma in this study was 35.6% (at least one papillary carcinoma was found in 36% of the thyroids examined), the *estimated* frequency of such tumors, based on size and the sampling methodology, was over 100% (suggesting there could be more than one carcinoma in each thyroid examined) [33, 34]. Interestingly, such micro-tumors were found not only in older adults, but also in individuals younger than 40 years old [35]. Since clinically apparent thyroid cancer is found in less than 1% of the population [36], it can be concluded that the vast majority of micro papillary thyroid cancers remain occult. Similarly, it is estimated that over 33% of women aged 40–50 years old have clinically undetectable breast cancer [34], although only 1% or less of the population develops clinically detectable breast cancer [36, 37]. It is statistically evident therefore, that a large proportion of such micro-tumors *never* develop into clinical disease. In fact, tumor dormancy can last for a lifetime [5, 6].

Experimental In Vivo Models of Human Tumor Dormancy: How Can You Measure what You can't See?

Although tumor dormancy has been recognized as a clinical problem for many years, very few examples of spontaneous tumor dormancy have been documented in experimental animal models [20]. However, tumor dormancy in experimental animal models may be a frequent occurrence that goes unrecognized. The fact that it is not commonly seen in the laboratory is most probably due to the fact that the majority of researchers select for rapid and consistent tumor growth.

To better understand the pathogenesis and underlying regulatory mechanisms of dormancy in human tumors, we previously established in vivo xenograft models of human tumor dormancy that include breast cancer, glioblastoma, osteosarcoma, and liposarcoma. In these models, human tumor cell lines are injected into immunocompromised (SCID) mice and form microscopic dormant tumors. We showed that in these models, tumor dormancy is associated with impaired angiogenic potential. To date, these are the only available in vivo models we are aware of, in which human tumor cell lines derived from malignant cancers form dormant and occult tumors when injected into mice and then spontaneously emerge from dormancy into rapid growth. All of these models were generated from commercially available human tumor cell lines and did not include any artificial genetic modification. These models were generated using two discrete approaches, developed in the Folkman laboratory. Both approaches are based on recognizing the heterogeneity of tumor cell populations in fully malignant tumors.

Achilles et al. described the first approach in 2001 [38]. The angiogenic heterogeneity in a human liposarcoma was studied by the generation of single-cell derived clones from a liposarcoma cell line (Fig. 1.2). While this parental liposarcoma cell line, as well as a majority of the single-cell derived clones, generated fast-growing and highly angiogenic tumors when injected into mice, other clones generated non-angiogenic microscopic tumors that did not grow and instead remained occult over 100 days after cell inoculation. This was the first direct proof that an angiogenic tumor can contain subpopulations of tumor cells with little or no angiogenic activity. These cells, when expanded in culture and injected into mice, will form small, avascular tumors at the site of injection. The angiogenic capacity of tumor cells can be therefore, correlated with the growth rate of the tumors they can generate.

This work implies that non-angiogenic tumor cells can "hitchhike" in tumors that contain angiogenic cells and suggests that the growth rate of a tumor will rely on the total angiogenic output of all the tumor cell subpopulations. This is consistent with the "hot spots" often observed in histological analysis of tumor vascular density [39].

Two of the single-cell derived clones that were generated in the Achilles studies were used for our studies: Clone 9 which generates fast-growing liposarcomas and Clone 4 which generates dormant, non-angiogenic liposarcomas (Fig. 1.3). Although we have used only two of these clones, this method of isolating cells that form dormant tumors is applicable to other tumor types and tumor cell lines (data not shown). However, biochemical markers of tumor dormancy could make this approach much easier and cost-effective.

The second model approach was developed from the observation that many human tumor cell lines do not "take," or do not form aggressive tumors, when injected into immune-deficient mice, coupled with Dr. Folkman's hypothesis that such cell lines might actually "take" and generate dormant tumors that remain microscopic and occult for long periods of time. Indeed, a number of such cell lines were shown to form microscopic and avascular tumors at site of injection [30]. Some of the dormant tumors generated by this means eventually switch and "escape" from dormancy to form aggressive tumors (Fig. 1.2b).

For our studies, three human cell lines from different tumor types were chosen based on their "no-take phenotype." These include breast carcinoma, glioblastoma, and osteosarcoma. Similar to the dormant clones of the liposarcoma, when injected into SCID mice, these cell lines generated microscopic tumors that remained occult

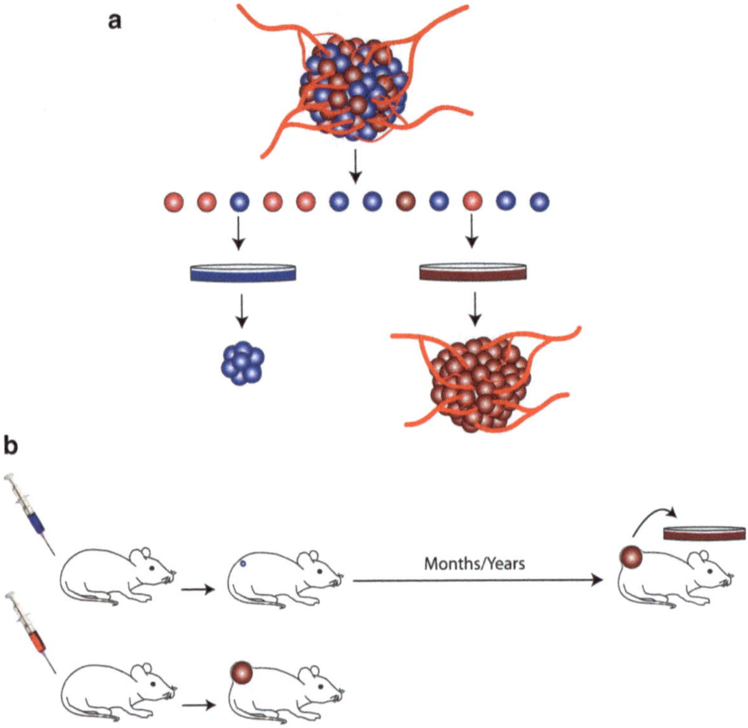

Fig. 1.2 Summary of the two approaches to generate pairs of cell lines that form either dormant or fast-growing tumors. (**a**) Isolation of cellular subpopulations that form dormant or fast-growing tumors. Single-cell derived clones are prepared from a heterogeneous cancer cell population (such as a human tumor cell line) that forms fast-growing tumors (shown in figure as a heterogeneous tumor cell population of *red* and *blue cells*). Tumors generated from these different clones have a spectrum of growth rates. While majority of the single-cell derived clones will generate fast-growing tumors (*red* tumor cells in figure), a percentage of such clones will generate dormant, microscopic tumors (*blue* tumor cells in figure). Screening for clones that form either dormant or fast-growing tumors requires in vivo tumor growth monitoring. (**b**) Isolation of cells from tumors that had spontaneously escaped from dormancy. Human tumor cell lines that are known to have a "no take" phenotype are injected into mice (*blue* tumor cell suspension in figure). These cells form microscopic tumors that remain occult for long periods of time until some of them spontaneously switch to rapid growth (*red* tumor in figure). These tumors that escaped dormancy are used to generate new tumor cell lines (*red* tumor cell in culture dish in figure). Cell lines from tumors that switched (*red* tumor cell suspension in figure) form fast-growing tumors immediately after injection into new mice

for over 100 days [40]. Following prolonged periods of dormancy, however, some of the tumors spontaneously emerged from dormancy and formed fast-growing and aggressive tumors at the site of injection. Cells from these "switched" tumors were cultured, confirmed to be of a human origin and maintained as clones in tissue culture conditions. When these cells, cultured from dormant tumors that had switched to fast growing were re-injected into mice, fast-growing tumors were observed soon after tumor cell injection. This implies that cells from tumors that emerged from

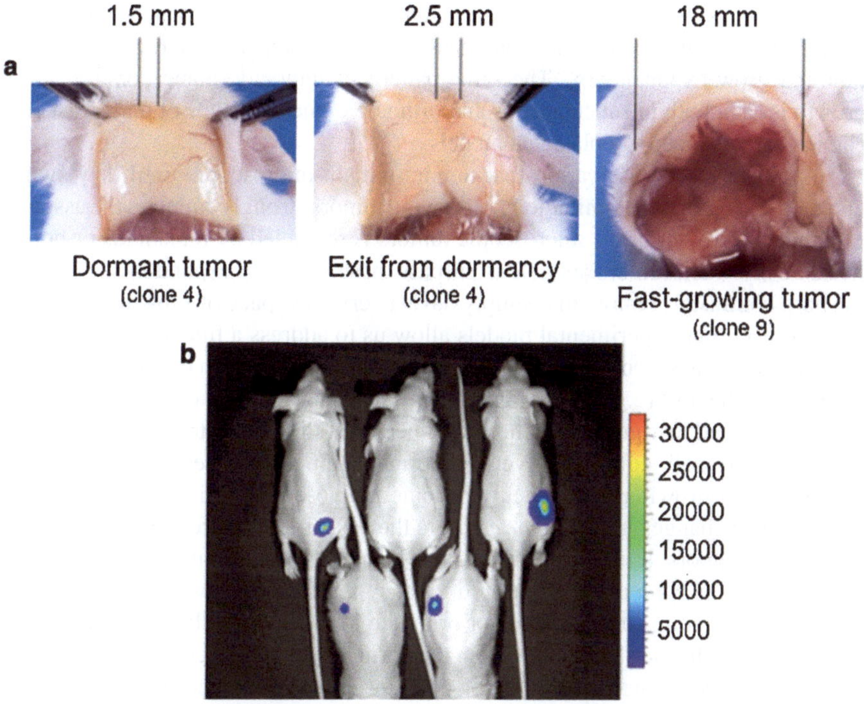

Fig. 1.3 Representative images of human liposarcoma grown in mice. (**a**) 37 days after subcutaneous injection of human SW872 liposarcoma Clone 4 cells into SCID mice, small tumors can be detected only after flipping the skin. Rarely, a more vascularized tumor with a diameter over 2 mm can be observed. Such tumors might be during the initiation of the "switch" from dormancy. In sharp contrast, at that same time point, tumors generated from Clone 9 of the human SW872 liposarcoma are considerably larger, easily detected by gross examination, and highly vascularized. (**b**) Persistence of dormant tumors generated from Clone 4 of the human SW872 liposarcoma can be detected by bioluminescence. Luminescence from tumor cells (that were labeled with luciferase before injection) indicates the presence of viable and metabolically active cells at the site of injection. 80 days after subcutaneous injection, the tumors can be detected by bioluminescence although they are not detected by gross examination

dormancy had acquired *stable intrinsic changes* that confer the tumor growth ability beyond the limiting diameter of a few millimeters.

For each tumor type (glioblastoma, osteosarcoma, and breast carcinoma), we currently have a pair of clones: One clone that generates dormant tumors (the parental cell line) and one clone that generates fast-growing tumors (established from the tumors that escaped from dormancy). Together with the pair of dormant and fast-growing liposarcoma (described above), we have a panel of pairs of cell lines from four different tumor types that each share a common genetic background but differ in their in vivo tumor growth patterns. In all these tumor models, the dormant tumors remain occult at the site of injection for prolonged periods of time until they eventually switch to rapid growth. Once these tumors pass the dormancy threshold, they grow at kinetics similar to the fast-growing and angiogenic tumors. On the other hand, the fast-growing tumors initiate rapid mass expansion soon after the tumor cell injection and grow exponentially. The same pattern of tumor dormancy or fast tumor growth is seen both in subcutaneous and orthotopic injection sites of our breast cancer, liposarcoma, and glioblastoma models.

Importantly, the observation that tumors remain microscopic until they switch and then grow at a pace similar to fast-growing tumors strongly supports the assumption that the growth of the microscopic tumors is restricted by thresholds or bottlenecks. Only when tumors are able to surpass these, can they expand in size. This is in sharp contrast to tumors that simply have a very slow pace of growth.

Moreover, our experimental models allow us to address a fundamental question in tumor dormancy: Do the elements necessary for the induction of dormancy originate within the host (e.g., tumor microenvironment) or within the tumor cells? Both the dormant and fast-growing tumors are injected at the same sites and are grown in identical "stromal" conditions, yet the dormant tumors will remain microscopic, while the fast-growing ones quickly become macroscopic. This strongly suggests that intrinsic changes in the *tumor cells* are the basis for the differential growth patterns of the tumors. It is also clear that intense intercellular communication with the tumor stroma plays a critical role in dormancy regulation. However, it seems that the signals dictating stromal behavior originate in the tumor cells. Importantly, the selection for cells that "switched" from the dormancy period is evident only in vivo, since prolonged growth of the tumor cells in culture does not affect the growth kinetics of the tumors generated from them (Almog, unpublished). Clearly, the selection for cells that can generate fast-growing tumors is derived from microenvironment pressure and signaling communication with the host.

The unique advantage of the experimental system we developed is the unlimited source of cells that will form dormant tumors (which are otherwise rarely obtained from in vivo tumors), together with counterpart tumor cells that will form fast-growing tumors, both derived from the same parental tumor cell population. This enables detailed and extensive molecular and cellular analyses. Indeed, we are currently using these models to study common pathways that are uniquely expressed in dormant tumors of various tumor types. Understanding the underlying mechanisms of tumor dormancy could have significant implications in the prevention and treatment of cancer: The human tumor cell lines that form dormant tumors in mice can be used not only as models for dormant primary tumors, but also as possible models for the clinical observations of very late cancer recurrences.

Angiogenesis Regulation of Tumor Dormancy

Dr. Judah Folkman was the first to suggest the fundamental relationship between angiogenic potential and the ability of a tumor to grow malignantly, and that dormancy can be associated with lack of angiogenesis [22, 41]. By now, it is well established that tumor growth beyond the size of 1–2 mm is angiogenesis-dependent

[42–45], and several experimental models of angiogenesis-related dormancy have been reported. Evidence of this was first observed using tumor implants in rabbit eyes [22] in which the same tumor remained dormant and avascular when implanted in the anterior chamber, but grew progressively when implanted in the iris. Holmgren et al. described another example of spontaneous dormancy. They observed micrometastases that remained occult as a result of systemic inhibition of angiogenesis mediated by the primary tumor [23].

In a spontaneous tumor model (RIP1-Tag2) in transgenic mice, tumors arise in the pancreatic islets as a result of the expression of the simian virus 40T antigen (Tag) oncogene. After 13 weeks, only 4% of tumors are angiogenic and contain evidence of neovascularization, whereas the remaining 96% stay microscopic and nonangiogenic. The spontaneous progression of nonangiogenic lesions to the angiogenic phenotype in this model was termed the *angiogenic switch* [46]. Although this name implies a short-acting "on-off" switch, the transition of a non-angiogenic avascular cancerous lesion to a highly angiogenic and fast-growing tumor encompasses a series of steps [47]. The successful culmination of this continuously productive process is the development of fully functional (although possibly abnormal) vessels capable of sustaining sufficient blood flow to support tumor mass expansion.

In our experimental models, tumor dormancy is clearly associated with impaired angiogenic potential. While no major cellular differences can be observed in vitro between cell lines that form dormant or fast-growing tumors, including morphology, proliferation, migration, and colony formation in soft agar, the tumor growth patterns in vivo are strikingly different [40, 49]. Similar to dormant tumors generated from other cell lines [30], dormant tumors generated from all of our models (SW872 liposarcoma, MDA-MB-436 breast carcinoma, T98G glioblastoma, and KHOS-24 osteosarcoma) have a high prevalence of proliferating cells. Tumor mass does not expand due to the high rate of apoptosis of tumor cells, which balances their proliferative capacity.

Noticeably, in contrast to fast-growing tumors that are highly vascularized, dormant tumors are mostly avascular. In most cases, vasculature can be observed only on the periphery of dormant tumors. In immunohistochemistry analysis and staining of endothelial cells, large vessels with open lumens are frequently seen in fast-growing tumors, while in dormant tumors, rare aggregates of endothelial cells are observed [40, 49]. In a detailed examination of tumor vasculature in liposarcomas by confocal analysis, a typical tumor vasculature comprised of interconnected and tortuous vessels is observed in the fast-growing tumors, whereas the vessels observed on the periphery of dormant tumors appear as nonfunctional tubes with aberrant morphology and many blunt ends [1].

Furthermore, when the relative area of endothelial cells in dormant tumors was followed over time, a decrease in microvessel density (MVD) was observed between days 14 and 60 after cell injection. A sharp increase in MVD was associated with the transition of tumors from dormancy to rapid growth and mass expansion [49]. This suggests that tumor dormancy is associated not just with impaired angiogenic capacity, but also with *inhibition* of angiogenesis. The inhibition is terminated following the induction of the angiogenic switch.

A significant and consistent difference between cells that form dormant tumors and those that form fast-growing tumors in our models is the secretion of the angiogenic inhibitor, thrombospondin-1 (TSP-1) [40, 49]. When in vitro secretion of pro- and anti-angiogenic factors from cells that form dormant tumors was compared with those that form fast-growing tumors, the dormant tumor-forming cells, regardless of tumor type, secreted relatively high levels of TSP-1. Other angiognesis inhibitors might also play a role in dormancy regulation, but these have yet to be determined.

It should be noted that once these dormant tumors undergo the angiogenic switch and initiate growth and expansion of mass, the tumor growth kinetics are similar to those of their paired rapidly growing angiogenic tumors. This further supports the concept that the fundamental mechanism underlying tumor dormancy in these models is impaired angiogenic capacity, rather than a decreased proliferation rate.

In summary, in our experimental models, blockage of tumor progression and persistence of micro-tumors is associated with the inability of the tumor cells to sustain the induction of functional new capillary blood vessels. This implies that not only the onset, but also the extent of angiogenesis is a critical determinant of tumor progression and growth.

Molecular Signature of Tumor Dormancy

The fact that tumor cells undergo genetic alterations during the switch from dormancy to rapid growth prompted us to identify the genetic profiles of indolent tumors. For this purpose, we utilized our experimental model of paired dormant and fast-growing tumors originating from the same parental cell lines. We ran genome-wide expression profiling assays to determine the consensus signature across our human tumor dormancy models.

We identified several genes that were differentially expressed between our dormant and fast-growing tumors, regardless of tumor type, and characterized common tumor dormancy-associated genes [50]. Around 700 genes were significantly differentially regulated in the same pattern (either induced or suppressed) in all four dormant and fast-growing tumor models examined. A number of these dormancy-associated genes had previously been shown to be involved, or associated with, tumor angiogenesis and tumor progression.

The molecular process most differentially expressed between dormant and fast-growing tumors was the *regulation of angiogenesis. Thrombospondin*, a known angiogenesis inhibitor [43], *angiomotin*, a mediator of the angiogenesis inhibitor angiostatin [51], and *tropomyosin*, a suggested mediator of the anti-angiogenic activity of endostatin [52], were shown to be upregulated in all of the dormant tumor cells examined. Dormant tumors also expressed TGFbeta2, which was previously shown to inhibit FGF-2-induced corneal endothelial cell proliferation [53] and to modulate extracellular matrix component expression [54]. In addition, dormant tumors induced the expression of proline-4-hydroxylase, which was previously shown to upregulate levels of several angiogenesis inhibitors [55]. Interestingly, although all dormant tumors had elevated levels of EphA5 and the histone H2BK, the RNA levels of these two genes in dormant glioblastoma was particularly and dramatically elevated [50]. The expression of EphA5 was shown to correlate with disease stage in glioma patients and to have distinct patterns of expression in plasma of both control and tumor-bearing mice. EphA5 plasma levels decrease in tumor-bearing mice even when the tumors are still microscopic in size.

On the other hand, tissue inhibitor of metalloproteinases (TIMP-3), shown to control tube stabilization and tube morphogenesis [56], CD73/Ecto-5′-Nucleotidase, shown to prevent vascular leakage [57]; and endothelial cell specific molecule-1 (ESM-1), whose expression is increased by VEGF and FGF-2 [58], were all shown to be upregulated in angiogenic fast-growing tumors. RNA levels of EGFR (EGF receptor 1) and PI3k, well-known players in tumor progression, were also elevated in angiogenic fast-growing tumors. Interestingly, two members of the insulin-like growth factor pathways have opposite patterns of expression. While the insulin-like growth factor receptor I (IGFR1) RNA is elevated in fast-growing angiogenic tumors, the RNA levels of the insulin-like growth factor binding protein 5 (IGFBP5) are elevated in the dormant non-angiogenic tumors.

In summary, the specific expression profiles of tumor dormancy-associated genes are associated with the conversion of dormant tumors into fast-growing angiogenic tumors and can be considered as molecular characteristics of indolent tumors.

Future Direction: What Needs to be Changed?

By the time a tumor is diagnosed, it is very likely that it has been growing undetected for a number of years [5]. The recent focus in translational cancer research on developing biomarkers for early stages in tumor development may enable the diagnosis of cancer well before a patient becomes symptomatic or before the anatomical location of a tumor is detectable. The relatively long period in which tumors remain occult and asymptomatic could provide critical time for assessment of efficacy of several treatment methods directed to eliminate such micro-tumors [59–61]. The development of novel anti-angiogenic drugs with minimal toxicity, together with the implication of "metronomic therapy" in which low doses of currently used chemotherapy drugs are used [62–64], both offer the possibility of a long-term prophylactic treatment of occult cancer to create a manageable chronic disease. Based on these ideas, it is reasonable to foresee that tumor dormancy could be prolonged therapeutically. Still, better models for dormancy of micro-tumors and improved methods for determining which ones will progress to emerge from dormancy are crucially needed.

The experimental models that we have developed reliably recapitulate aspects of dormancy found in human tumors. While it is clear that intrinsic genetic determinants in the cancer cells dictate whether these cells will form dormant or fast-growing tumors, it is the pressure from the microenvironment that *selects* for clones that have switched to become angiogenic. Still, further analysis of master regulators of tumor dormancy are needed to better control this phase in tumor progression.

It is becoming well accepted that microRNAs (miRNAs) play a critical role in cancer [65]. MiRNAs repress expression of target genes and can act as either oncogenes or tumor suppressors in tumor development. It is estimated that one miRNA could regulate gene expression of multiple target genes and therefore act as a *master regulator* of gene expression [66]. Importantly, disregulation of miRNAs is well known to correlate with various diseases, including cancer [67]. Therefore, it is possible that dormant tumors might have a unique miRNA expression profile that can be distinguished from that in fast-growing tumors. Identifying tumor dormancy-specific miRNAs could serve as a major contribution to our understanding of molecular mechanisms underlying tumor dormancy and potentially as novel biomarkers that would indicate the presence or state of an otherwise occult tumor.

A promising opportunity for cancer research is the recent shift from an almost exclusive focus on macroscopic and aggressive symptomatic tumors [5, 12], to inclusion of current investigations into the *regulators* of the *early stages* in micro-tumor (primary and metastases) growth. Such a shift, however, requires a fundamental change in thinking by both basic biology and clinical oncology researchers. Active dialogue and interaction with investigators across disciplines would facilitate novel approaches to studying this biologically complex topic. Chances for successful treatment of cancer significantly increase with early detection. Innovative tumor systems designed to model dormancy, can aid in the identification of biomarkers of early tumors and facilitate an understanding of the mechanisms which lead tumors to exit the dormancy stage and to initiate rapid tumor growth. Such pre-clinical models can also serve to provide molecular insight for the design of combination therapies that block the mechanisms allowing escape from dormancy, with the goal of better controlling the progression stage of cancer thereby reducing cancer-associated mortality and morbidity.

Acknowledgments The author wishes to thank Melissa Klumpar for critical reading and editing of the manuscript and to Clare Lamont for help with figure preparation. This research was supported by the Office of Science (BER), U.S. Department of Energy, under Award Number DE-SC0002606 and by the National Cancer Institute under Award Number U54CA149233 (both to L. Hlatky). The content is solely the responsibility of the authors and does not necessarily represent the official views of the National Cancer Institute or the National Institutes of Health.

References

1. Almog N (2010) Molecular mechanisms underlying tumor dormancy. Cancer Lett 294(2):139–146
2. Aguirre-Ghiso JA (2007) Models, mechanisms and clinical evidence for cancer dormancy. Nat Rev Cancer 7(11):834–846
3. Goss PE, Chambers AF (2010) Does tumour dormancy offer a therapeutic target? Nat Rev Cancer 10(12):871–877
4. Pantel K, Alix-Panabieres C, Riethdorf S (2009) Cancer micrometastases. Nat Rev Clin Oncol 6(6):339–351
5. Uhr JW, Scheuermann RH, Street NE, Vitetta ES (1997) Cancer dormancy: opportunities for new therapeutic approaches. Nat Med 3(5):505–509

6. Folkman J, Kalluri R (2004) Cancer without disease. Nature 427(6977):787
7. Demicheli R, Retsky MW, Hrushesky WJ, Baum M (2007) Tumor dormancy and surgery-driven interruption of dormancy in breast cancer: learning from failures. Nat Clin Pract Oncol 4(12):699–710
8. Hart IR (1999) Perspective: tumour spread—the problems of latency. J Pathol 187(1):91–94
9. Quesnel B (2008) Dormant tumor cells as a therapeutic target? Cancer Lett 267(1):10–17
10. Uhr JW, Pantel K (2011) Controversies in clinical cancer dormancy. Proc Natl Acad Sci USA 108(30):12396–12400
11. Hahnfeldt P, Panigrahy D, Folkman J, Hlatky L (1999) Tumor development under angiogenic signaling: a dynamical theory of tumor growth, treatment response, and postvascular dormancy. Cancer Res 59(19):4770–4775
12. Hedley BD, Chambers AF (2009) Tumor dormancy and metastasis. Adv Cancer Res 102:67–101
13. Weinberg RA (2008) The many faces of tumor dormancy. APMIS 116(7–8):548–551
14. Wikman H, Vessella R, Pantel K (2008) Cancer micrometastasis and tumour dormancy. APMIS 116(7–8):754–770
15. Bissell MJ, Hines WC (2011) Why don't we get more cancer? A proposed role of the microenvironment in restraining cancer progression. Nat Med 17(3):320–329
16. Riethdorf S, Wikman H, Pantel K (2008) Review: biological relevance of disseminated tumor cells in cancer patients. Int J Cancer 123(9):1991–2006
17. Brackstone M, Townson JL, Chambers AF (2007) Tumour dormancy in breast cancer: an update. Breast Cancer Res 9(3):208
18. Demicheli R, Terenziani M, Valagussa P, Moliterni A, Zambetti M, Bonadonna G (1994) Local recurrences following mastectomy: support for the concept of tumor dormancy. J Natl Cancer Inst 86(1):45–48
19. Goss P, Allan AL, Rodenhiser DI, Foster PJ, Chambers AF (2008) New clinical and experimental approaches for studying tumor dormancy: does tumor dormancy offer a therapeutic target? APMIS 116(7–8):552–568
20. Alsabti EA (1979) Tumor dormancy: a review. J Cancer Res Clin Oncol 95(3):209–220
21. Naumov GN, Folkman J, Straume O (2009) Tumor dormancy due to failure of angiogenesis: role of the microenvironment. Clin Exp Metastasis 26(1):51–60
22. Gimbrone MA Jr, Leapman SB, Cotran RS, Folkman J (1972) Tumor dormancy in vivo by prevention of neovascularization. J Exp Med 136(2):261–276
23. Holmgren L, O'Reilly MS, Folkman J (1995) Dormancy of micrometastases: balanced proliferation and apoptosis in the presence of angiogenesis suppression. Nat Med 1(2):149–153
24. Willis L, Alarcon T, Elia G, Jones JL, Wright NA, Tomlinson IP et al (2010) Breast cancer dormancy can be maintained by small numbers of micrometastases. Cancer Res 70(11):4310–4317
25. Marches R, Scheuermann R, Uhr J (2006) Cancer dormancy: from mice to man. Cell Cycle 5(16):1772–1778
26. Koebel CM, Vermi W, Swann JB, Zerafa N, Rodig SJ, Old LJ et al (2007) Adaptive immunity maintains occult cancer in an equilibrium state. Nature 450(7171):903–907
27. Aguirre-Ghiso JA (2006) The problem of cancer dormancy: understanding the basic mechanisms and identifying therapeutic opportunities. Cell Cycle 5(16):1740–1743
28. Barkan D, Green JE, Chambers AF (2010) Extracellular matrix: a gatekeeper in the transition from dormancy to metastatic growth. Eur J Cancer 46(7):1181–1188
29. Kenny PA, Bissell MJ (2003) Tumor reversion: correction of malignant behavior by microenvironmental cues. Int J Cancer 107(5):688–695
30. Udagawa T, Fernandez A, Achilles EG, Folkman J, D'Amato RJ (2002) Persistence of microscopic human cancers in mice: alterations in the angiogenic balance accompanies loss of tumor dormancy. FASEB J 16(11):1361–1370
31. Klein CA (2011) Framework models of tumor dormancy from patient-derived observations. Curr Opin Genet Dev 21(1):42–49

32. Welch HG, Black WC (2010) Overdiagnosis in cancer. J Natl Cancer Inst 102(9):605–613
33. Harach HR, Franssila KO, Wasenius VM (1985) Occult papillary carcinoma of the thyroid. A "normal" finding in Finland. A systematic autopsy study. Cancer 56(3):531–538
34. Black WC, Welch HG (1993) Advances in diagnostic imaging and overestimations of disease prevalence and the benefits of therapy. N Engl J Med 328(17):1237–1243
35. Franssila KO, Harach HR (1986) Occult papillary carcinoma of the thyroid in children and young adults. A systemic autopsy study in Finland. Cancer 58(3):715–719
36. Feldman AR, Kessler L, Myers MH, Naughton MD (1986) The prevalence of cancer. Estimates based on the Connecticut Tumor Registry. N Engl J Med 315(22):1394–1397
37. Udagawa T (2008) Tumor dormancy of primary and secondary cancers. APMIS 116(7–8) :615–628
38. Achilles EG, Fernandez A, Allred EN, Kisker O, Udagawa T, Beecken WD et al (2001) Heterogeneity of angiogenic activity in a human liposarcoma: a proposed mechanism for "no take" of human tumors in mice. J Natl Cancer Inst 93(14):1075–1081
39. Hlatky L, Hahnfeldt P, Folkman J (2002) Clinical application of antiangiogenic therapy: microvessel density, what it does and doesn't tell us. J Natl Cancer Inst 94(12):883–893
40. Naumov GN, Bender E, Zurakowski D, Kang SY, Sampson D, Flynn E et al (2006) A model of human tumor dormancy: an angiogenic switch from the nonangiogenic phenotype. J Natl Cancer Inst 98(5):316–325
41. Folkman J (1971) Tumor angiogenesis: therapeutic implications. N Engl J Med 285(21): 1182–1186
42. Folkman J (1992) The role of angiogenesis in tumor growth. Semin Cancer Biol 3(2):65–71
43. Folkman J (1995) Angiogenesis in cancer, vascular, rheumatoid and other disease. Nat Med 1(1):27–31
44. Moserle L, Amadori A, Indraccolo S (2009) The angiogenic switch: implications in the regulation of tumor dormancy. Curr Mol Med 9(8):935–941
45. Naumov GN, Akslen LA, Folkman J (2006) Role of angiogenesis in human tumor dormancy: animal models of the angiogenic switch. Cell Cycle 5(16):1779–1787
46. Hanahan D, Folkman J (1996) Patterns and emerging mechanisms of the angiogenic switch during tumorigenesis. Cell 86(3):353–364
47. Abdollahi A, Schwager C, Kleeff J, Esposito I, Domhan S, Peschke P et al (2007) Transcriptional network governing the angiogenic switch in human pancreatic cancer. Proc Natl Acad Sci USA 104(31):12890–12895
48. Abdollahi A, Folkman J (2010) Evading tumor evasion: current concepts and perspectives of anti-angiogenic cancer therapy. Drug Resist Updat 13(1–2):16–28
49. Almog N, Henke V, Flores L, Hlatky L, Kung AL, Wright RD et al (2006) Prolonged dormancy of human liposarcoma is associated with impaired tumor angiogenesis. FASEB J 20(7):947–949
50. Almog N, Ma L, Raychowdhury R, Schwager C, Erber R, Short S et al (2009) Transcriptional switch of dormant tumors to fast-growing angiogenic phenotype. Cancer Res 69(3):836–844
51. Troyanovsky B, Levchenko T, Mansson G, Matvijenko O, Holmgren L (2001) Angiomotin: an angiostatin binding protein that regulates endothelial cell migration and tube formation. J Cell Biol 152(6):1247–1254
52. MacDonald NJ, Shivers WY, Narum DL, Plum SM, Wingard JN, Fuhrmann SR et al (2001) Endostatin binds tropomyosin. A potential modulator of the antitumor activity of endostatin. J Biol Chem 276(27):25190–25196
53. Lu J, Lu Z, Reinach P, Zhang J, Dai W, Lu L et al (200) TGF-beta2 inhibits AKT activation and FGF-2-induced corneal endothelial cell proliferation. Exp Cell Res 312(18):3631–3640
54. Fuchshofer R, Birke M, Welge-Lussen U, Kook D, Lutjen-Drecoll E (2005) Transforming growth factor-beta 2 modulated extracellular matrix component expression in cultured human optic nerve head astrocytes. Invest Ophthalmol Vis Sci 46(2):568–578
55. Teodoro JG, Parker AE, Zhu X, Green MR (2006) p53-mediated inhibition of angiogenesis through up-regulation of a collagen prolyl hydroxylase. Science 313(5789):968–971

56. Saunders WB, Bohnsack BL, Faske JB, Anthis NJ, Bayless KJ, Hirschi KK et al (2006) Coregulation of vascular tube stabilization by endothelial cell TIMP-2 and pericyte TIMP-3. J Cell Biol 175(1):179–191
57. Thompson LF, Eltzschig HK, Ibla JC, Van De Wiele CJ, Resta R, Morote-Garcia JC et al (2004) Crucial role for ecto-5′-nucleotidase (CD73) in vascular leakage during hypoxia. J Exp Med 200(11):1395–1405
58. Sarrazin S, Adam E, Lyon M, Depontieu F, Motte V, Landolfi C et al (2006) Endocan or endothelial cell specific molecule-1 (ESM-1): a potential novel endothelial cell marker and a new target for cancer therapy. Biochim Biophys Acta 1765(1):25–37
59. Almog N, Klement GL (2010) Platelet proteome and tumor dormancy: can platelets content serve as predictive biomarkers for exit of tumors from dormancy? Cancers 2(2):842–858
60. Cervi D, Yip TT, Bhattacharya N, Podust VN, Peterson J, Abou-Slaybi A et al (2008) Platelet-associated PF-4 as a biomarker of early tumor growth. Blood 111(3):1201–1207
61. Klement GL, Yip TT, Cassiola F, Kikuchi L, Cervi D, Podust V et al (2009) Platelets actively sequester angiogenesis regulators. Blood 113(12):2835–2842
62. Hahnfeldt P, Folkman J, Hlatky L (2003) Minimizing long-term tumor burden: the logic for metronomic chemotherapeutic dosing and its antiangiogenic basis. J Theor Biol 220(4):545–554
63. Kerbel RS, Klement G, Pritchard KI, Kamen B (2002) Continuous low-dose anti-angiogenic/metronomic chemotherapy: from the research laboratory into the oncology clinic. Ann Oncol 13(1):12–15
64. Browder T, Butterfield CE, Kraling BM, Shi B, Marshall B, O'Reilly MS et al (2000) Antiangiogenic scheduling of chemotherapy improves efficacy against experimental drug-resistant cancer. Cancer Res 60(7):1878–1886
65. Kasinski AL, Slack FJ (2011) MicroRNAs en route to the clinic: progress in validating and targeting microRNAs for cancer therapy. Nat Rev Cancer 11(12):849–864
66. Bartel DP (2004) MicroRNAs: genomics, biogenesis, mechanism, and function. Cell 116(2):281–297
67. Zhang B, Pan X, Cobb GP, Anderson TA (2007) microRNAs as oncogenes and tumor suppressors. Dev Biol 302(1):1–12

Chapter 2
The Host Support Niche as a Control Point for Tumor Dormancy: Implications for Tumor Development and Beyond

Philip Hahnfeldt

Abstract An increasingly appreciated focus of carcinogenesis research is on mechanisms governing tumor growth after the fact of cancer cell creation. Of particular interest are dynamical interactions between tumor and host cell populations that can themselves strongly impact the fate of established cancer lesions. Regardless of tumor type, all cancers face the common problem of having to breach the barrier of angiogenic competency in order to advance from a microscopic lesion to symptomatic disease. If pre-angiogenic tumor cells are held in dormancy due to cell cycle arrest, this will postpone the need to traverse this higher-level barrier. On the other hand, the barrier itself may prove limiting to a tumor at its diffusion-limited size, creating a population-level dormancy characterized by balanced proliferation and cell death. In both cases of dormancy, the "angiogenic switch" has not yet occurred. We here describe and mathematically quantify an underappreciated third dormancy state defined by an angiogenic balance following the angiogenic switch. In this state we term "post-vascular dormancy," a tumor has attained angiogenic competency, but again demonstrates balanced proliferation and cell death because ambient pro- and anti-angiogenic influences are offsetting. Interestingly, autopsies have shown virtually all of us carry latent tumors in pre- or post-vascular states, many of which lie under the threshold of routine clinical detection. We show how, in the post-vascular case, tumor latency can arise from an elaborate mechanism of self-controlled growth, mediated through the tumor–vascular interaction. Underlying this observation is the finding that a tumor produces both angiogenesis stimulators and inhibitors, with the latter having greater influence, both locally and systemically, as the tumor grows—a mechanism we hypothesize is an aberrant co-option of normal organogenic regulation.

P. Hahnfeldt, PhD (✉)
Center of Cancer Systems Biology, Steward Research and Specialty Projects Corp., St. Elizabeth's Medical Center Tufts University School of Medicine, 736 Cambridge St., Boston, MA 02135, USA
e-mail: Philip.Hahnfeldt@tufts.edu

H. Enderling et al. (eds.), *Systems Biology of Tumor Dormancy*, Advances in Experimental Medicine and Biology 734, DOI 10.1007/978-1-4614-1445-2_2, © Springer Science+Business Media New York 2013

That a tumor can limit its own growth raises the prospect that chronic therapies aimed at suppressing this tumor–host dynamic may compare favorably to current strategies which often yield favorable short-term responses but fail to deliver long-term tumor suppression.

Keywords Angiogenic balance • Tumor–host dynamics • Organogenic regulation • Post-vascular dormancy • Carrying capacity • Logistic growth • Stable disease

Introduction

Against the backdrop of intra- and inter-tumor heterogeneity and genomic instability that are hallmarks of carcinogenesis [1–3], uncovering unifying principles of action that underlie the complex nature of cancer would seem improbable. Yet, there is abundant evidence that tumors are exquisitely dependent on their host environment to manifest the malignant phenotype. The classic experiments of Illmensee and Mintz [4] showed that teratocarcinoma tumor cells can phenotypically revert to contribute to normal mouse development when inserted into blastocysts. More recently, Bhowmick et al. [5] showed that the loss of TGF-beta responsiveness in adjacent fibroblasts can result in prostate neoplasia and invasive squamous cell carcinoma of the forestomach. These events show, respectively, that an aberrant, unstable genome is not inconsistent with controlled phenotypic behavior, while conversely cells with no prior overt oncogenic alteration to their genomes can go on to exhibit malignancy. These and other findings [6, 7] demonstrate an overriding ability of the tumor niche to control tumor development. Perhaps the best evidence for the potent role of the tumor/microenvironment dynamic in controlling growth of established tumors comes from the finding that nearly all adults harbor indolent tumor lesions [8, 9]. As the sizes of many of these tumors lie at or somewhat beyond what diffusion of nutrients could accommodate, both pre- and post-vascular forms are likely represented, undergoing balanced proliferation and cell death [10–13] as they await an environmental alteration favoring net growth. Such an alteration could come from increased angiogenic induction by the tumor. But host response is not simply one of accommodating pro-growth tumor cues. Instead, the tumor plays an elaborate role in influencing whether the niche permits or denies tumor growth. Kaplan et al. [14] showed that a certain class of vascular endothelial growth factor receptor-1-positive hematopoietic progenitors must first be recruited by the primary tumor to initiate the "pre-metastatic niche" before a tumor metastasis can seed there and develop. Previously, the revelation that tumors produce both stimulators and inhibitors of angiogenesis [15, 16] had already pointed to a potentially complicated tumor control dynamic. Exploring this possibility, we have shown that tumor growth can be controlled rheostatically by shifting the balance between tumor-derived angiogenesis stimulators and inhibitors [17]. Moreover, as the inhibition is shown to be systemic in scope and would eventually dominate for large enough tumors, we have gone on to propose the tumor-metastasis system should also exhibit asymptotic self-control. This organized growth has profound implications for the nature of oncogenesis and treatment. Among these is the notion that chronic therapy designed to maintain a cancer as "stable disease" indefinitely might have decided advantage as a therapeutic recourse in those instances where strategies with eradicative intent have typically fallen short of their mark.

The Bottlenecks of Tumor Dormancy

Histologically confirmed tumors may fail to progress beyond a certain size for a number of reasons. The first to be proposed is the "dormant cancer cell," a hypothesis advanced by Willis [18] to explain discrepancies between natural progression and observed tumor latency, i.e., the time to recurrence following surgery. The effect has variously been attributed to extended mitotic arrest of tumor-propagating cancer stem cells and their progeny (Fig. 2.1a) [19, 20] and immune response [21, 22]. But interruptions to growth may also occur at the population level. As mentioned, a major barrier to tumor development can be the natural impediment of diffusion-limited nutrient availability. For non-angiogenic tumors, this causes them to become dormant at less than a millimeter or so in size, as cells must lie within ~200 μm or so of the nearest capillary to be adequately nourished. One way this might be realized is through the balanced creation and death of cancer stem cell progeny (Fig. 2.1b). In addition, for those somewhat larger tumors that already evidence a vascular contribution, we propose that a limitation can be the failure to stimulate the additional vascularization required for continued expansion (Fig. 2.1c). Both types of dormancy would be classified as population-level because the cells themselves are quite active, proliferating and dying at balanced rates [10–13]. Apparently, tumors held in one or the other form of population-level dormancy are much more prevalent than overt clinical cancers [8]. Further, it is clear many of these dormant cancers will not progress sufficiently rapidly to ever pose a disease threat [8, 9]. Indeed, in the Mayo lung cancer screening trial, consisting of chest X-ray and sputum cytology testing [23], chest X-ray and sputum testing over 6 years detected 143 lung cancers (90 specifically by screening plus 53 in connection with clinical procedure) compared to just 87 in the control group. Ten more cancers were detected over the next 5 years in the control group, leaving a difference of 46 cancers that persisted over a 16-year follow-up [24]. Similar disparities between prevalence and clinical incidence have been observed in breast cancer. In a consecutive autopsy study of 110 women [25], histologically confirmed breast cancers were detected in 39% of women in the age 40–50 cohort, while only 1% of women in this age group actually suffer from the disease. These results underscore the power of the tumor–host interaction to impede the course of even confirmed cancers.

At the same time, there is no guarantee a given lesion will remain nonthreatening. Almog et al. [26] have shown that a non-angiogenic clone of human liposarcoma can escape dormancy after an extended period (~130 days). In their islet cell RIP-Tag model, Hanahan and Folkman [27] also found a certain percentage of non-angiogenic tumors would spontaneously convert, i.e., "switch," after 13 weeks. Considering stages of advancing pancreatic disease, we showed at the molecular

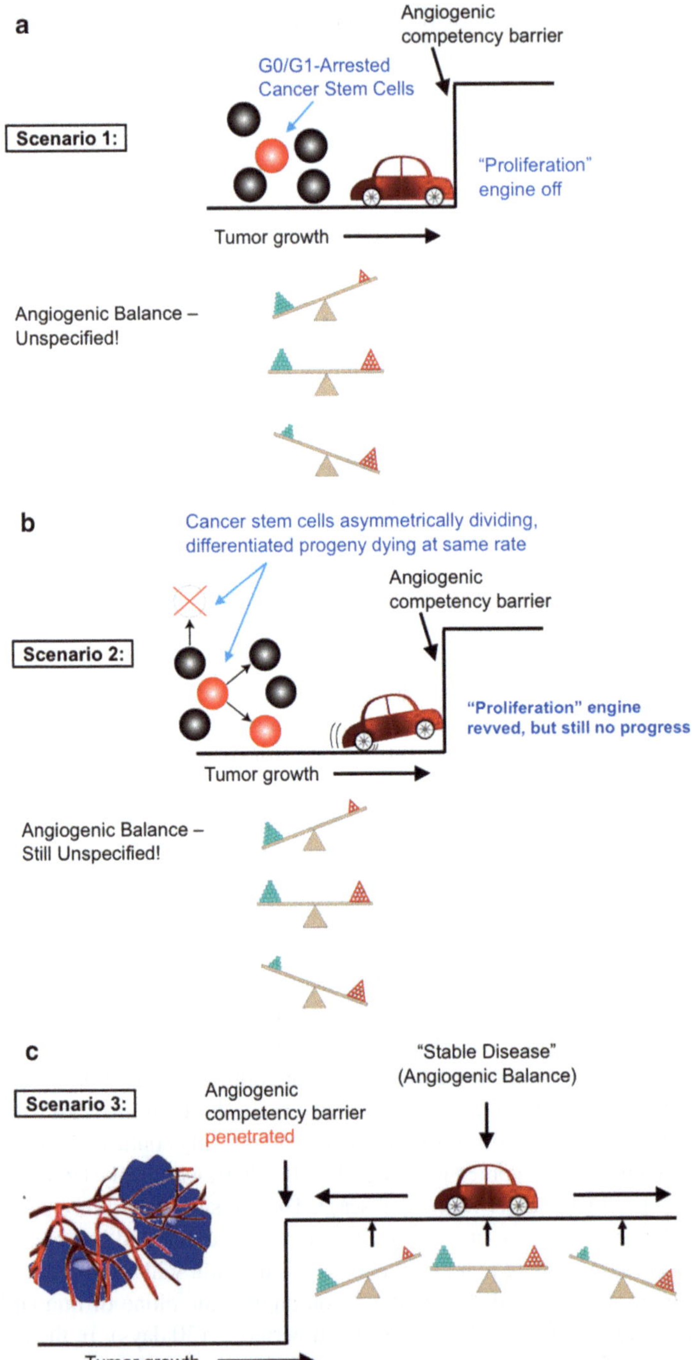

Fig. 2.1 (**a**–**c**) Three scenarios for tumor dormancy. In *Scenario 1*, cancer cells are arrested in G0/G1. As the population cannot proliferate, the barrier of angiogenic competency is not tested (the "car" is off). Simple diffusion remains adequate to sustain the population. The population is considered

level that as disease transits from chronic pancreatitis to pancreatic cancer to metastatic disease, there is a progressive upregulation of the large class of pro-angiogenic genes, with a simultaneous downregulation of a class of anti-angiogenic genes [28]. The work demonstrated that the angiogenic switch may be triggered over the course of an *en masse* pro-angiogenic regulation by the tumor of a large network of normal genes associated with vascular control. The extent of genetic involvement in the switch would be consistent with a finding by Indraccolo et al. [29], who showed that a more intense pro-angiogenic stimulus may be required for a pre-vascular lesion to breach the diffusion limitation barrier than is required to drive a post-angiogenic lesion after the fact (Figs. 2.1b,c). If so, this does not appear to be an obstacle that limits growth of subsequent metastases. The frequent observation of explosive growth of pre-vascular metastases after excision of the primary [15, 30, 31] is comparatively easy to reconcile with the idea that the loss of primary tumor-derived angiogenesis inhibitors is sufficient to release the metastases from dormancy. In this scenario, no second burst of angiogenic activity at the pre-vascular metastatic sites is required. On the other hand, it may be argued that the metastatic sites have already been pre-conditioned so as to obviate the need for the burst [14]. A second possibility for why a pre-vascular lesion may need more angiogenic stimulus is because it is nowhere near the tipping point of angiogenic balance in the first place, so that induction of angiogenesis may require more stimulatory upregulation than would otherwise be required for growth after a switch. In any event, escape from pre-vascular dormancy arguably constitutes a major step in cancer progression. At the same time, there is no reason to think the original notion of angiogenic balance in the post-switch context has now become irrelevant—in fact, the meaning becomes more important than ever.

The Post-vascular Dormant Tumor

Inherent in the notion of a post-vascular dormant tumor is the notion of an ongoing, balanced expression of angiogenesis stimulators and inhibitors in the tumor milieu. Putting this another way, a dormant tumor that has switched at the level of a pre-vascular lesion could in principle be restored to a dormant status as a larger, post-vascular

Fig. 2.1 (continued) pre-vascularly dormant at the cell level. In terms of an angiogenic balance, just where the population is relative to the tipping point is unspecified. In *Scenario 2*, there is no cell cycle arrest, but cells are proliferating and dying as the population attempts to breach the angiogenic competency barrier (the "car" is running and pushing against the barrier, but cannot surmount it). The population cannot grow, and so is considered pre vascularly dormant at the population level. Again, while the population awaits angiogenic competency, it still remains unspecified how close it is to actually crossing the switch. In *Scenario 3*, the population has previously attained angiogenic competency and breached the angiogenic switch barrier. Because it is now equipped with vasculature and in a position to readily grow or regress in response to either a net pro- or an anti-angiogenic stimulus (the "car" is free to move forward or backward in response to any forward or reverse throttle), a state of neither growth nor shrinkage (stable disease) implies an angiogenic balance lesion through a suitable re-balancing of factors in the niche (Fig. 2.1c). If this were to happen spontaneously, or if the tumor/niche system were to hover at the point of angiogenic balance for prolonged periods of time, larger but dormant or very slowly growing variants of many clinical tumors should be observable. Such is the case. In the breast cancer microlegal autopsy study of 110 women aged 20–54 years [25], 20% had occult tumors, with some reaching >5 mm in size, well beyond the size where vascularization would be required. In a directed study of growth rates of 147 breast cancers based on mammography scans, von Fournier et al. [32] found a remarkable distribution of doubling times, ranging from 44 to 1,869 days. Tumors at the higher end would be considered dormant by most standards. Moreover, at detection, the mean diameter of the 147 tumors was 17 mm (95% confidence limits: 15–18 mm), so the dormant and slow-growing lesions observed were certainly post-vascular. These data suggest that a tumor that escapes pre-vascular dormancy by undergoing an angiogenic switch [10, 26, 27, 33] may potentially remain near or be restored to a dormant or near-dormant state, even as a post-vascular lesion. This is important from the standpoint of competing risks, in the sense that the growth may be slow enough for the tumor to no longer pose a risk of compromising morbidity or survival in that person's lifetime (Fig. 2.2) [9].

Nevertheless, the potential for macroscopic, post-vascular dormancy remains to be recognized in the general literature. This is especially curious given that the concept of angiogenic balance [34] has often been used to characterize pre-vascular dormancy (e.g., [13]), a state where the proximity to the precise tipping point of angiogenic induction is quite unclear (Fig. 2.1a, b). In point of fact, the concept may better apply to what Kerbel and Folkman [35] have termed "stable disease." This is a form of clinically evident dormancy that, because it already possesses a vascular component poised to expand or shrink in response to even slight shifts in the angiogenic state of the microenvironment, is more arguably in a state of angiogenic balance (Fig. 2.1c).

A Dynamic Carrying Capacity Representation for Angiogenesis-Dependent Tumor Growth

In a typical ecological system consisting of a population and a defined level of environmental support, various quantitative relationships, including the Verhulst equation and its logistic variants, have been used to describe population growth over time [36]. These generally show the population size $V(t)$ to be exponential initially (reflecting abundant support vs. demand), but later to exhibit growth deceleration in the form of an asymptotic approach to some final value K. K reflects the maximum population size the environment can support, otherwise known as the carrying capacity of the environment. Models generally presume the level of support K is fixed, with some functional deviations to adjust for technical features, e.g., the point of inflection (the point $(t_0, V(t_0))$ where $d^2V(t)/dt^2=0$) or the rate at which the population

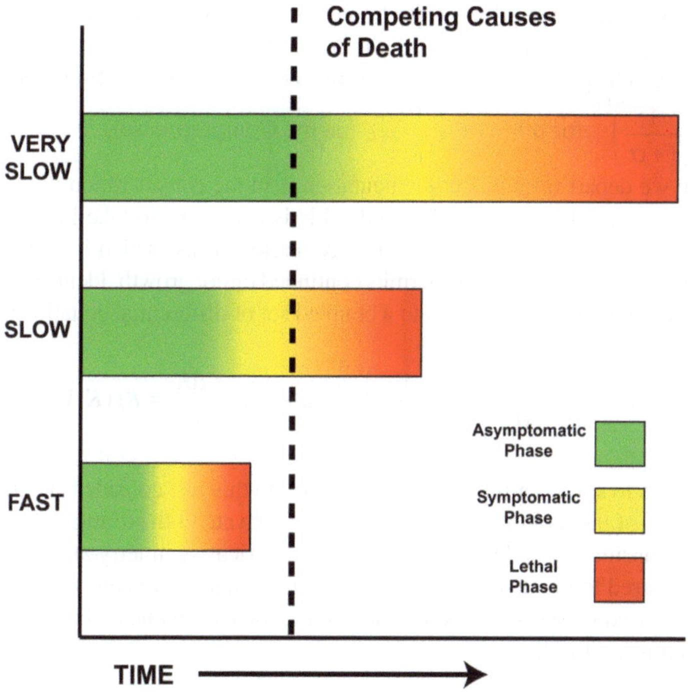

Fig. 2.2 The threat of cancer is determined in the progression phase. Even when the steps leading to cancer are completed and a tumor is growing, whether the cancer will ever present as symptomatic disease is determined by its rate of progression. If the tumor is growing quickly, symptoms followed by potential lethality is a strong possibility. If instead the tumor is growing more slowly, it may reach the point of clinical detection, but may not become life-threatening over the normal lifetime of the patient. On the other hand, if the tumor is growing very slowly, such as in the case of dormancy or near dormancy, the tumor may not even be diagnosed. In this case, the patient effectively had no disease, even though from a genetic standpoint a cancer was created. Autopsy data suggest most of us are in this category, carrying latent disease that will not present itself clinically

approaches its carrying capacity. These principles have proven to be robust enough to find application to tumor growth [37].

The generalized logistic equation for tumor size $V(t)$ can be expressed in the form:

$$\frac{\mathrm{d}V}{\mathrm{d}t} = \frac{\lambda}{\alpha} V\left(1 - \left(\frac{V}{K}\right)^{\alpha}\right) \quad \text{for } \alpha \neq 0, \lambda > 0, t \geq 0, V(t) < K. \tag{2.1}$$

In the limit $\alpha \to 0$, this reduces to the Gompertz equation:

$$\frac{\mathrm{d}V}{\mathrm{d}t} = -\lambda V \log\left(\frac{V}{K}\right). \tag{2.2}$$

The growth rate overall is proportional to λ, and α controls the rate at which tumor size approaches the asymptotic limit, i.e., the steepness of the ascending part of the curve. The parameter α also determines the point of inflection, which occurs at $\frac{V}{K}=\left(\frac{1}{1+\alpha}\right)^{1/\alpha}$ for $\alpha > -1$ (or $\frac{V}{K}=\mathrm{e}^{-1}$ in the Gompertz case).

Where we depart from this implementation is in the consideration of the carrying capacity K as variable, rather than fixed. This is necessary to take into account the well-established ability of a tumor to induce angiogenesis, which increases nutrient delivery to the tumor and thereby permits continued tumor growth. Identifying induced vascularization with K, then, leads to a coupled set of differential equations:

$$\text{(a)}\ \frac{\mathrm{d}V}{\mathrm{d}t}=\frac{\lambda}{\alpha}V\left(1-\left(\frac{V}{K(t)}\right)^{\alpha}\right) \quad \text{and} \quad \text{(b)}\ \frac{\mathrm{d}K}{\mathrm{d}t}=Kf(K,V) \tag{2.3}$$

It remains to solve for the K dependence. To do this, we consider the quantitative implications of the fact that tumors produce or activate both stimulators and inhibitors of its vascular niche [15, 16]. Assuming spherical symmetry and a growth rate small compared to the distribution of angiogenesis agent, we can write a diffusion-consumption equation for the concentration m of angiogenesis factor within and outside a tumor of radius r_0.

$$D^2\nabla^2 m - cm + s = 0 \quad \text{or} \quad \frac{\mathrm{d}^2 m}{\mathrm{d}r^2}+2\frac{\mathrm{d}m}{\mathrm{d}r}-\frac{cm}{D^2}+\frac{s}{D^2}=0. \tag{2.4}$$

where D^2 is the diffusion of agent, c is its clearance rate, and s is its rate of production (we will assume $s=s_0$ inside the tumor and $s=0$ outside).

By making the substitutions

$$\text{(a)}\ r = xD/\sqrt{c} \quad \text{and} \quad \text{(b)}\ m = y/\sqrt{x}+s/c \tag{2.5}$$

the equation reduces to the modified Bessel function of order ½

$$x^2\frac{\mathrm{d}^2 y}{\mathrm{d}x^2}+x\frac{\mathrm{d}y}{\mathrm{d}x}-\left(x^2+\frac{1}{4}\right)=0, \tag{2.6}$$

which has two linearly independent solutions. If we choose the first to be finite at $x=0$, and the other to be finite as $x\to\infty$, they become:

$$y_1=\frac{\sinh(x)}{\sqrt{x}} \quad \text{and} \quad y_2=\frac{\exp(-x)}{\sqrt{x}} \tag{2.7}$$

From Eq. (2.5) and assumptions about s, these expressions immediately yield the agent concentrations inside and outside the tumor

$$m_{\text{inside}}=\frac{A}{r}\sinh\left(\frac{r\sqrt{c}}{D}\right)+\frac{s_0}{c} \quad \text{and} \quad m_{\text{outside}}=\frac{B}{r}\exp\left(\frac{-r\sqrt{c}}{D}\right) \tag{2.8a}$$

where A and B are constants. The constants are solved for by having the two solutions and their derivatives be continuous across the tumor boundary at $r=r_0$. We obtain

$$\mathrm{A} = \frac{-s_0 D}{c\sqrt{c}}\left(1+\frac{r_0\sqrt{c}}{D}\right)\exp\left(\frac{-r_0\sqrt{c}}{D}\right) \quad \text{and}$$
$$\mathrm{B} = \frac{s_0 D}{c\sqrt{c}}\left(\frac{r_0\sqrt{c}}{D}\cosh\left(\frac{r_0\sqrt{c}}{D}\right)-\sinh\left(\frac{r_0\sqrt{c}}{D}\right)\right). \tag{2.8b}$$

To assess the action of angiogenesis stimulators and inhibitors, we now take into account the relatively slow clearance of endogenous inhibitors and the relatively fast clearance of endogenous stimulators [15, 16, 38, 39].

We consider the two limiting cases $c \ll D^2/r_0^2$ and $c \gg D^2/r_0^2$ and values of r near or inside the tumor.

$$c \ll D^2/r_0^2 : m_{\text{inside}} \approx \frac{s_0}{6D^2}(3r_0^2 - r^2) \quad \text{and} \quad m_{\text{outside}} \approx \frac{s_0 r_0^3}{3D^2 r} \tag{2.9a}$$

$$c \gg D^2/r_0^2 : m_{\text{inside}} \approx \frac{s_0}{c}\left(1-\frac{r_0}{2r}\exp\left(\frac{-(r_0-r)\sqrt{c}}{D}\right)\right) \approx \frac{s_0}{c} \quad \text{and}$$
$$m_{\text{outside}} \approx \frac{s_0 r_0}{2cr}\exp\left(\frac{-(r-r_0)\sqrt{c}}{D}\right) \approx 0 \tag{2.9b}$$

The concentration profiles are shown in Fig. 2.3.

It is clear from Case 1 that at any location $r_0 f$ near or within the growing tumor (i.e., for $0 \leq f < \approx 1$), the concentration of inhibitor is increasing as the square of the tumor radius, or as $V^{2/3}$. The inhibitor term in the growth rate $f(K, V)$ of the carrying capacity K would therefore be expected to be proportional to $V^{2/3}$. By contrast, stimulator concentration does not increase at all anywhere, so can be said formally to increase as V^0. It could also be said to increase as V/K, which also has zero net volume scale. We decided on the latter because tumors were thought not to display transient oscillation in tumor size for smaller tumors, as the use of V^0 predicts theoretically, although in retrospect, it appears some oscillations in tumor size for small tumors do take place. In any event, the choice of V/K is not expected to alter the results dramatically.

The form for the carrying capacity growth rate $f(K, V)$, then, becomes $aV/K - bV^{2/3}$. Eq. (2.3b) becomes

$$\frac{\mathrm{d}K}{\mathrm{d}t} = aV - bKV^{2/3}. \tag{2.10}$$

As a confirmation of these results, we previously performed animal experiments where we implanted Lewis lung tumors subcutaneously in the flanks of C57Bl/6

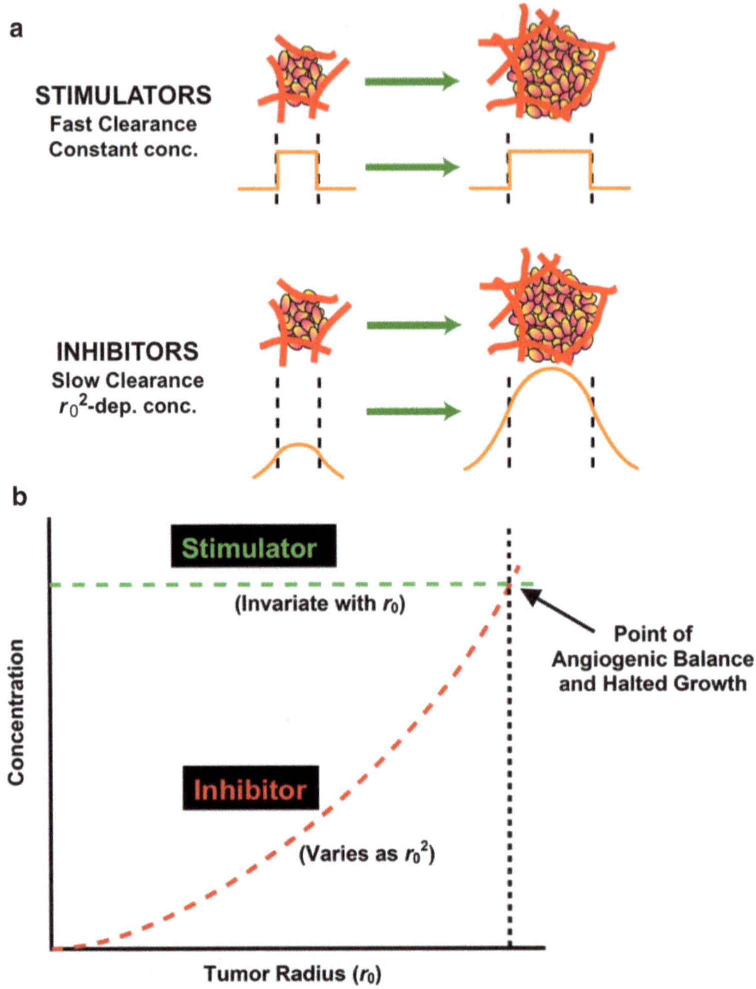

Fig. 2.3 Tumor growth is dictated by the pharmacokinetics of tumor-derived angiogenic stimulation and inhibition. The observation that a tumor produces both angiogenesis stimulators and inhibitors, and that the inhibitors are cleared much more slowly, has important implications for tumor growth. (**a**) Mathematical analysis shows that, in the extreme case of very fast clearance of stimulators and very low clearance of inhibitors, stimulators maintain a constant concentration within the tumor independent of its size. Stimulator concentration away from the tumor is negligible (rectangular function shown in yellow). Meanwhile, inhibitor concentration tends to grow at a rate proportional to the square of the tumor radius r_0 everywhere (bell-shaped function in yellow grows in height as r_0). (**b**) The faster accumulation of inhibitor everywhere assures that inhibitor will eventually overtake stimulator, causing the tumor to become dormant (stable disease). While this is a predicted outcome for any tumor, whether this happens before it becomes symptomatic or life-threatening will depend on individual patient circumstance. In any case, shifting the angiogenic state of the tumor environment towards inhibition, e.g., with therapeutic intervention, could cause stable disease to occur at a point consistent with host viability. Such a strategy of chronic tumor maintenance may comprise a favorable alternative to eradicative-intent treatments in some situations

mice [17]. This was the model used by O'Reilly et al. [15] in their study of angiostatin, which was actually first isolated from this tumor. We observed tumor growth in a normal setting and in settings where angiogenesis inhibitors (angiostatin, endostatin, and an exogenous inhibitor TNP-470) were introduced exogenously by various schedules. The goal was to see if the model as derived could explain tumor growth and in particular, the effect of angiogenesis stimulation and suppression by the tumor on its own growth. For the purpose of modeling the inhibitor injections, we appended a term –d*Ke*(*t*) to Eq. 2.10 to account for background antiangiogenic drug administration.

$$\frac{\mathrm{d}K}{\mathrm{d}t} = aV - bKV^{2/3} - dKe(t), \text{ where } e(t) = \int_0^t r(t')\exp(-c_{\text{inh}}(t-t'))dt' \quad (2.11)$$

Here, e(*t*) is the concentration of injected inhibitor, $r(t')$ is the rate of injection of inhibitor (in practice, nearly an impulse function), and c_{inh} is the clearance rate of the injected inhibitor. We assumed basic exponential clearance pharmacokinetics.

We fit the model to control growth of tumors, solving for the tumor–host parameters λ, α, a, and b (finding α to be about zero, giving us a Gompertz form for Eq. 2.5a) then tested its predictive power for cases where the systemic environment was artificially made angio-inhibitory by injection, supplementing any tumor-derived angiogenesis inhibition. Tumor responses under the three antiangiogenic agents were used to calculate the inhibitor effectiveness coefficient d and the clearance rate c_{inh} for each agent in the equation for e(*t*). We were able to confirm that clearance rates for the administered inhibitors angiostatin and endostatin were indeed quite long, supporting the same assumption made for the action of the endogenous inhibitor activated by the Lewis lung tumor (angiostatin). Our data also predicted *de novo* that TNP-470 should be quite effective in terms of its suppression per unit concentration per unit time, but that its effectiveness is likely limited by a relatively fast clearance rate compared to the other inhibitors examined. This is supported by direct pharmacokinetic analysis [40]. The behavior of K/V with time is shown in Fig. 2.4a. Of note, K is not simply advancing marginally ahead of V, as might be expected if it is just accommodating the growing nutritional needs of the tumor, nor is the ratio monotonically decreasing, as would be expected if the growth were in accordance with Gompertz or any conventional logistic form with a fixed carrying capacity. Instead, an entirely new form is revealed, defined by an active advancement of carrying capacity well ahead of growth early on, followed later by an equally active curtailment of carrying capacity, and thus tumor growth.

The Mathematics of Control of Distant Metastases

A basis for thinking about the tumor-metastasis system is laid out in general terms by the finding that tumors are capable of producing stimulators and inhibitors of angiogenesis, with the inhibitors generally having longer half-lives. That this may translate to distant control of a metastatic site is suggested by the solutions to our

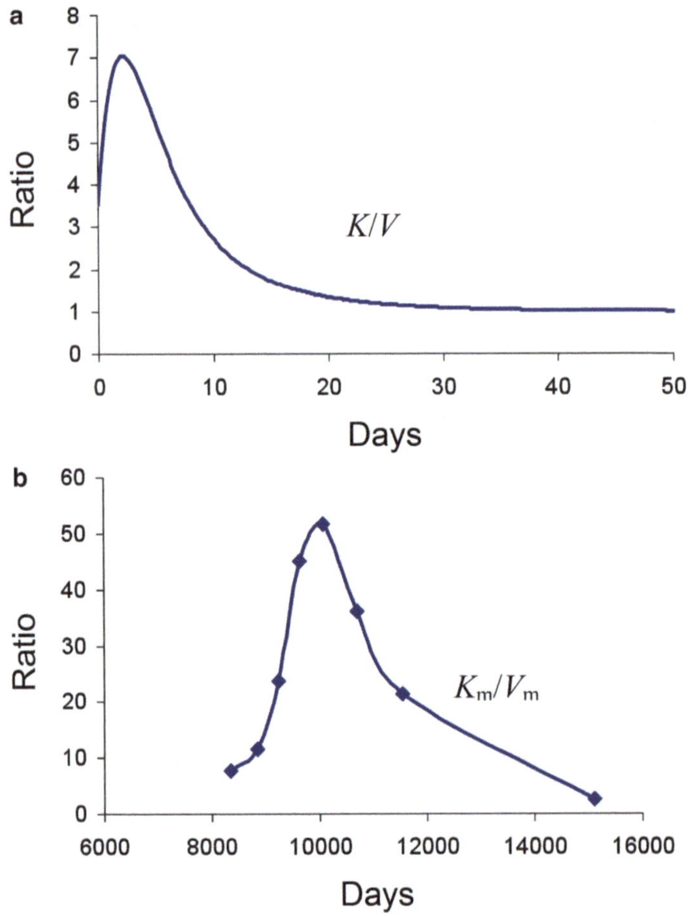

Fig. 2.4 The dynamic carrying capacity model for the progression-level bottleneck of angiogenesis we previously derived based on murine studies (**a**) was demonstrated to be predictive of angiogenesis-dependent breast tumor growth humans (**b**). In the human data, growing tumors call for proportionally quickly rising support/tumor ratios K_m/V_m early on (7.62–45.1), then a collapsing ratio later on (45.1–2.5). This reveals an active tumor–host dynamic that is inconsistent with simple vascular induction due to nutrient demand. It is also inconsistent with conventional Gompertz or logistic growth dynamics. That the basic dynamic applies to both mice and humans points to a conserved mechanism that may have an origin in organogenesis

original dynamic carrying capacity construct for the distribution of angiogenesis factor within and outside the tumor. It is seen that the inhibitor can accumulate at some distance from the tumor, while stimulator does not accumulate, despite tumor size. Of course, we used extremes of clearance rate to model this point. In fact, while inhibitors do generally have more persistence, there will be a variation in clearance rates, with some of the stimulators on the more slowly clearing end of the spectrum perhaps having an influence comparable to that of inhibitors that happen to be on the more quickly clearing end. One would have to consider the details of each instance. For now, we do know there is some connection between the growth of metastasis and signals from the primary. There are a number of reports citing increased vascular density in metastases after removal of the primary [41] and increased detection of metastases overall [42].

Human Cancers

To test how our "dynamic carrying capacity" interpretation of angiogenic control of tumor growth carries over to the human circumstance, we have begun to study the growth of 420 nondormant, untreated breast tumors as part of a study conducted at the University of Heidelberg [37]. As the data for each patient i come in the form of volume scans V_{ij} corresponding to scans $j = 1, \ldots, n_i$, with the first scan being assigned a time $\tau_{i1} = 0$ days, the objective was to merge the scan data across patients to permit a global analysis of tumor/vascular development to compare to that performed with the mouse data. To do this, we again applied the general logistic growth relationship Eq. (2.1), but to first derive a sense of absolute times to associate with the scans, we regressed the data onto the solution to Eq. (2.1) expressed as

$$\lambda t = f(K, V_0, a) \quad \text{for} \quad \alpha \neq 0, \lambda > 0, t \geq 0, V(t) < K. \tag{2.12}$$

Here, V_0 is the tumor size at tumor age $t=0$, assumed to be one cell, i.e., 1.0×10^{-6} mm^3. The objective of the regression was to find formal expressions for the overall λ and time offsets $T_i(K, \alpha)$ for the scan data (τ_{ij}, V_{ij}) of Patient i that would translate that patient's scan times τ_{ij} into estimated absolute tumor ages t_{ij} ($= \tau_{ij} + T_i(K, \alpha)$) based on the collective behavior of the intra-patient tumor volume measurements. We then reinserted these formal expressions back into the solution of Eq. (2.1), now written more conventionally using $\ln(V(t)/V_0)$ as the dependent variable:

$$\ln(V(t)/V_0) = g(K, V_0, \alpha, t). \tag{2.13}$$

Finally, we used Eq. (2.13) to perform another regression of the data ($t_{ij}(K, \alpha)$, V_{ij}) to find the best fit with respect to K and α, thereby solving explicitly for K, α, and so the t_{ij} themselves. With the tumor data points thus rendered in the form (t_{ij}, V_{ij}) for the jth scan of Patient i, we then refit to Eq. (2.13) subsets of these data lying in the volume cohorts $0 < V \leq V_m$, for $V_1 = 100$ mm^3, $V_2 = 250$ mm^3, $V_3 = 500$ mm^3, $V_4 = 1{,}000$ mm^3, $V_5 = 2{,}000$ mm^3, $V_6 = 5{,}000$ mm^3, $V_7 = 15{,}000$ mm^3, and $V_8 = 215{,}600$ mm^3 to find analogous parameter values K_m, α_m, and λ_m corresponding to each of these cohorts. These fits included a final one to the full cohort ($V_8 = 215{,}600$ mm^3), which contained all tumor measurements (the largest tumor size measured was 215,600 mm^3). This last fit was not redundant to the original that allowed us to solve for K, α, and t_{ij} in the first place, because the full cohort fit was done with explicit knowledge of the t_{ij}, just as with the other cohort fits. Times corresponding to the arbitrary V_m values were inferred from the full-cohort fit to

Eq. (2.13), and plotted against the ratio of K_m/V_m surmised from each of the cohort fits $m = 1, \ldots, 8$. The overall intent of this cohort analysis was to crudely trace out the dynamic behavior of K/V as a function of tumor age by independent methods to compare with that of the animal model.

What we found is shown in Fig. 2.4b. Of note, both in this case (Fig. 2.4b) and in the mouse (Fig. 2.4a), there is a slightly left-skewed curve describing how the ratio of carrying capacity (vascular support) K leaps well ahead of tumor size V initially, then almost as rapidly descends so as to cap off the potential tumor size. The human data thus corroborates the finding in mice; that tumor support is not passively controlled (e.g., by need for O_2), as originally believed, but is dynamically reset by the tumor throughout growth, up to a point where a tumor will start to curtail its own growth (approach a dormant state) through active capping of endothelial support. Whether it can attain this state soon enough to be host-viable will depend on the precise case-specific details of the tumor–niche interaction, as modified by therapeutic intervention. The potential significance of this reciprocal tumor–host dynamic is also glimpsed in work by Kaplan et al. [14], where it is shown that homing of VEGFR1-positive hematopoietic progenitors to would-be sites of tumor metastasis ("pre-metastatic niches"), although a tumor-directed event, is a precondition for eventual tumor metastasis to those sites. If the pre-metastatic conditioning program is blocked, metastases do not occur. Suggesting a generalization beyond cancer, Lammert et al. [43] and Yoshitomi and Zaret [44] described a control in line with our proposed "endothelio-centric" paradigm [17] for organ development, in that endothelial growth at the organ site is seen to precede and permit growth of the parenchyma, which in turn controls growth of the supporting vasculature. As further support for our hypothesized connection between angiogenesis and organogenesis, Greene et al. [45] tracked liver regrowth after partial hepatectomy, and showed that the regenerating organ plateaus to a final size that varies with the angiogenic status of the host. Mice given angiogenesis stimulators during regeneration developed larger-than-normal livers, while those given angiogenesis inhibitors had smaller final liver sizes.

Conclusion

The notion that cancer often occurs as "stable disease," a state of post-vascular dormancy viewable clinically, has vital implications for antiangiogenic therapy, and in particular how we interpret therapeutic progress towards overcoming the original angiogenic switch. Once it is recognized that achieving an angiogenic balance, and thus "stable disease," may be achievable with a macroscopic tumor without shrinking it to microscopic size, this opens the door for new thinking about what constitutes successful response. Current notions of "complete or partial response," referring to total or partial tumor shrinkage as measures of the effectiveness of classic maximum tolerated dosing (MTD) regimens, would give way to "failure of the tumor to progress," the likely hallmark of successful antiangiogenic therapy.

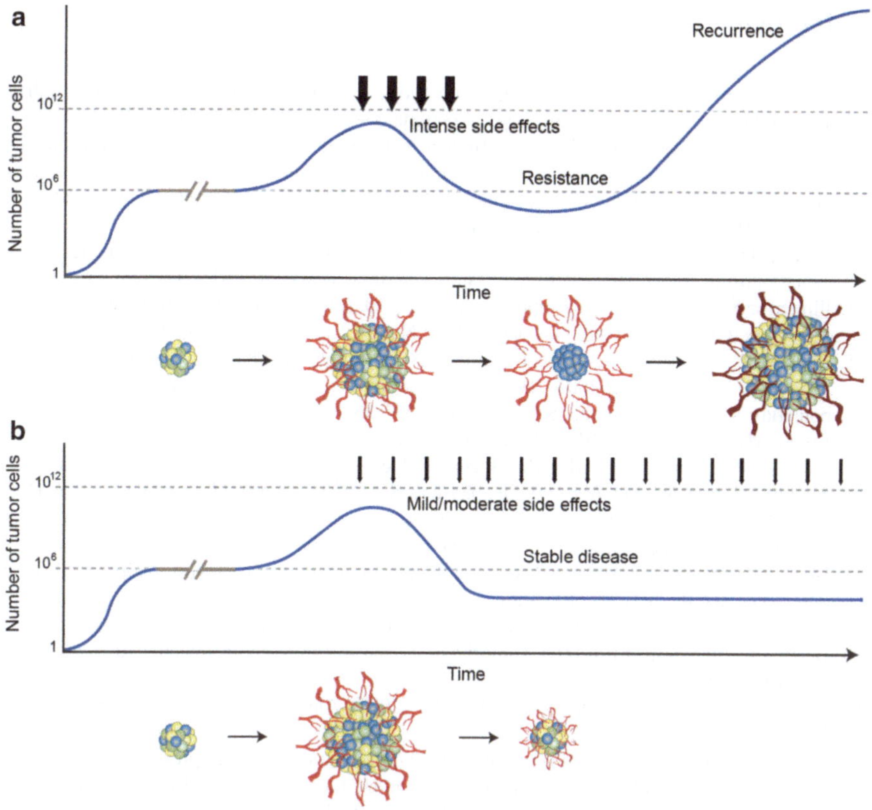

Fig. 2.5 Two strategies for tumor therapy compared. Maximum tolerated dosing (MTD) strategies and targeted strategies aimed at total tumor eradication hold the promise of cure. For some cancers, this is a course with a substantial success rate, albeit often with undesirable side effects. For most cancers, however, good responses are often followed by resistance and tumor regrowth, assisted in rebound by the host support left behind (**a**). Part of the reason may be that the tumor and its vascular niche are not being co-suppressed as a unit, which may be better accomplished using more chronic treatments that include anti-angiogenics. While cures are no longer the goal, stable disease consistent with excellent quality of life may be achievable (**b**)

New therapeutic designs might entail altering the self-imposed theoretical dormancy "set point" of the tumor downward to a level consistent with symptom-free disease over the lifetime of the patient (Fig. 2.5). In this way, the achievement of a tolerable equilibrium between tumor and host could stand quite favorably against eradicative-intent strategies whose very aggressiveness may often be the instrument of their own defeat.

Acknowledgments This project was supported by the National Cancer Institute under Award Number U54CA149233 (to L. Hlatky). The content is solely the responsibility of the author and does not necessarily represent the official views of the National Cancer Institute or the National Institutes of Health. Figure 2.5 graphics due to Clare Lamont.

References

1. Cheng L, Bostwick DG, Li G, Wang Q, Hu N, Vortmeyer AO, Zhuang Z (1999) Allelic imbalance in the clonal evolution of prostate carcinoma. Cancer 85(9):2017–2022
2. Stilgenbauer S, Sander S, Bullinger L, Benner A, Leupolt E, Winkler D, Kröber A, Kienle D, Lichter P, Döhner H (2007) Clonal evolution in chronic lymphocytic leukemia: acquisition of high-risk genomic aberrations associated with unmutated VH, resistance to therapy, and short survival. Haematologica 92(9):1242–1245
3. Losi L, Baisse B, Bouzourene H, Benhattar J (2005) Evolution of intratumoral genetic heterogeneity during colorectal cancer progression. Carcinogenesis 26(5):916–922
4. Illmensee K, Mintz B (1976) Totipotency and normal differentiation of single teratocarcinoma cells cloned by injection into blastocysts. Proc Natl Acad Sci USA 73(2):549–553
5. Bhowmick NA, Chytil A, Plieth D, Gorska AE, Dumont N, Shappell S, Washington MK, Neilson EG, Moses HL (2004) TGF-beta signaling in fibroblasts modulates the oncogenic potential of adjacent epithelia. Science 303(5659):848–851
6. Bissell MJ, Radisky D (2001) Putting tumours in context. Nat Rev Cancer 1(1):46–54
7. Nguyen DH, Oketch-Rabah HA, Illa-Bochaca I, Geyer FC, Reis-Filho JS, Mao JH, Ravani SA, Zavadil J, Borowsky AD, Jerry DJ, Dunphy KA, Seo JH, Haslam S, Medina D, Barcellos-Hoff MH (2011) Radiation acts on the microenvironment to affect breast carcinogenesis by distinct mechanisms that decrease cancer latency and affect tumor type. Cancer Cell 19(5):640–651
8. Black WC, Welch HG (1993) Advances in diagnostic imaging and overestimations of disease prevalence and the benefits of therapy. N Engl J Med 328(17):1237–1243
9. Welch HG, Black WC (2010) Overdiagnosis in cancer. J Natl Cancer Inst 102(9):605–613
10. Bergers G, Benjamin LE (2003) Tumorigenesis and the angiogenic switch. Nat Rev Cancer 3(6):401–410
11. Davis DW, Herbst RS, Abbruzzese JL (eds) (2007) Antiangiogenic cancer therapy. Boca Raton, CRC
12. Cao Y, O'Reilly MS, Marshall B, Flynn E, Ji RW, Folkman J (1998) Expression of angiostatin cDNA in a murine fibrosarcoma suppresses primary tumor growth and produces long-term dormancy of metastases. J Clin Invest 101(5):1055–1063
13. Udagawa T, Fernandez A, Achilles EG, Folkman J, D'Amato RJ (2002) Persistence of microscopic human cancers in mice: alterations in the angiogenic balance accompanies loss of tumor dormancy. FASEB J 16(11):1361–1370
14. Kaplan RN, Riba RD, Zacharoulis S, Bramley AH, Vincent L, Costa C, MacDonald DD, Jin DK, Shido K, Kerns SA, Zhu Z, Hicklin D, Wu Y, Port JL, Altorki N, Port ER, Ruggero D, Shmelkov SV, Jensen KK, Rafii S, Lyden D (2005) VEGFR1-positive haematopoietic bone marrow progenitors initiate the pre-metastatic niche. Nature 438(7069):820–827
15. O'Reilly MS, Holmgren L, Shing Y, Chen C, Rosenthal RA, Moses M, Lane WS, Cao Y, Sage EH, Folkman J (1994) Angiostatin: a novel angiogenesis inhibitor that mediates the suppression of metastases by a Lewis lung carcinoma. Cell 79(2):315–328
16. O'Reilly MS, Boehm T, Shing Y, Fukai N, Vasios G, Lane WS, Flynn E, Birkhead JR, Olsen BR, Folkman J (1997) Endostatin: an endogenous inhibitor of angiogenesis and tumor growth. Cell 88(2):277–285
17. Hahnfeldt P, Panigrahy D, Folkman J, Hlatky L (1999) Tumor development under angiogenic signaling: a dynamical theory of tumor growth, treatment response, and postvascular dormancy. Cancer Res 59(19):4770–4775
18. Willis RA (1948) Pathology of Tumours. Butterworth, London
19. Hadfield G (1954) The dormant cancer cell. Br Med J 2(4888):607–610
20. Guba M, Cernaianu G, Koehl G, Geissler EK, Jauch KW, Anthuber M, Falk W, Steinbauer M (2001) A primary tumor promotes dormancy of solitary tumor cells before inhibiting angiogenesis. Cancer Res 61(14):5575–5579
21. Wheelock EF, Weinhold KJ, Levich J (1981) The tumor dormant state. Adv Cancer Res 34:107–140

22. Meltzer A (1990) Dormancy and breast cancer. J Surg Oncol 43(3):181–188
23. Fontana RS, Sanderson DR, Woolner LB, Taylor WF, Miller WE, Muhm JR, Bernatz PE, Payne WS, Pairolero PC, Bergstralh EJ (1991) Screening for lung cancer. A critique of the Mayo Lung Project. Cancer 67(4 suppl):1155–1164
24. Marcus PM, Bergstralh EJ, Zweig MH, Harris A, Offord KP, Fontana RS (2006) Extended lung cancer incidence follow-up in the Mayo Lung Project and overdiagnosis. J Natl Cancer Inst 98(11):748–756
25. Nielsen M, Thomsen JL, Primdahl S, Dyreborg U, Andersen JA (1987) Breast cancer and atypia among young and middle-aged women: a study of 110 medicolegal autopsies. Br J Cancer 56(6):814–819
26. Almog N, Henke V, Flores L, Hlatky L, Kung AL, Wright RD, Berger R, Hutchinson L, Naumov GN, Bender E, Akslen LA, Achilles EG, Folkman J (2006) Prolonged dormancy of human liposarcoma is associated with impaired tumor angiogenesis. FASEB J 20(7):947–949
27. Hanahan D, Folkman J (1996) Patterns and emerging mechanisms of the angiogenic switch during tumorigenesis. Cell 86(3):353–364
28. Abdollahi A, Schwager C, Kleeff J, Esposito I, Domhan S, Peschke P, Hauser K, Hahnfeldt P, Hlatky L, Debus J, Peters JM, Friess H, Folkman J, Huber PE (2007) Transcriptional network governing the angiogenic switch in human pancreatic cancer. Proc Natl Acad Sci USA 104(31):12890–12895
29. Indraccolo S, Favaro E, Amadori A (2006) Dormant tumors awaken by a short-term angiogenic burst: the spike hypothesis. Cell Cycle 5(16):1751–1755
30. Coffey JC, Wang JH, Smith MJ, Boucher-Hayes D, Cotter TG, Redmond HP (2003) Excisional surgery for cancer cure: therapy at a cost. Lancet Oncol 4(12):760–768
31. Baum M, Demicheli R, Hrushesky W, Retsky M (2005) Does surgery unfavourably perturb the "natural history" of early breast cancer by accelerating the appearance of distant metastases? Eur J Cancer 41(4):508–515
32. von Fournier D, Weber E, Hoeffken W, Bauer M, Kubli F, Barth V (1980) Growth rate of 147 mammary carcinomas. Cancer 45(8):2198–2207
33. Abdollahi A, Folkman J (2010) Evading tumor evasion: current concepts and perspectives of anti-angiogenic cancer therapy. Drug Resist Updat 13(1–2):16–28
34. Rastinejad F, Polverini PJ, Bouck NP (1989) Regulation of the activity of a new inhibitor of angiogenesis by a cancer suppressor gene. Cell 56(3):345–355
35. Kerbel R, Folkman J (2002) Clinical translation of angiogenesis inhibitors. Nat Rev Cancer 2(10):727–739
36. Tsoularis A (2001) Analysis of logistic growth models. Res Lett Inf Math Sci 2:23–46
37. Spratt JS, Meyer JS, Spratt JA (1996) Rates of growth of human neoplasms: Part II. J Surg Oncol 61(1):68–83
38. Zetter BR (1998) Angiogenesis and tumor metastasis. Annu Rev Med 49:407–424
39. Ribatti D (2008) Judah Folkman, a pioneer in the study of angiogenesis. Angiogenesis 11(1):3–10
40. Figg WD, Pluda JM, Lush RM, Saville MW, Wyvill K, Reed E, Yarchoan R (1997) The pharmacokinetics of TNP-470, a new angiogenesis inhibitor. Pharmacotherapy 17(1):91–97
41. Peeters CF, Westphal JR, de Waal RM, Ruiter DJ, Wobbes T, Ruers TJ (2004) Vascular density in colorectal liver metastases increases after removal of the primary tumor in human cancer patients. Int J Cancer 112(4):554–559
42. van der Bij GJ, Oosterling SJ, Beelen RH, Meijer S, Coffey JC, van Egmond M (2009) The perioperative period is an underutilized window of therapeutic opportunity in patients with colorectal cancer. Ann Surg 249(5):727–734
43. Lammert E, Cleaver O, Melton D (2003) Role of endothelial cells in early pancreas and liver development. Mech Dev 120(1):59–64
44. Yoshitomi H, Zaret KS (2004) Endothelial cell interactions initiate dorsal pancreas development by selectively inducing the transcription factor Ptf1a. Development 131(4):807–817
45. Greene AK, Wiener S, Puder M, Yoshida A, Shi B, Perez-Atayde AR, Efstathiou JA, Holmgren L, Adamis AP, Rupnick M, Folkman J, O'Reilly MS (2003) Endothelial-directed hepatic regeneration after partial hepatectomy. Ann Surg 237(4):530–535

Chapter 3
Insights into the Regulation of Tumor Dormancy by Angiogenesis in Experimental Tumors

Stefano Indraccolo

Abstract While it is well established that an angiogenic switch marks escape from tumor dormancy in xenograft models, the molecular pathways involved in the control of tumor cell proliferation or survival by angiogenesis remain substantially uncharted. We recently demonstrated that signals stemming from angiogenic endothelial cells (EC) regulate the behavior of dormant cancer cells. Specifically, we observed that the Notch ligand Dll4, induced by angiogenic factors in EC, triggers Notch3 activation in neighboring tumor cells and promotes a tumorigenic phenotype. Evidence that Notch signaling is involved in tumor dormancy was further strengthened by the observation that MKP-1 levels—a broadly expressed phosphatase—are controlled by Notch3 by regulation of protein ubiquitination and stability. Notch3 and MKP-1 levels are consistently low in dormant tumors, and this is accompanied by relatively high levels of phosphorylated p38, a canonical MKP-1 target previously associated with maintenance of tumor dormancy. These results elucidate a novel angiogenesis-driven mechanism involving the Notch and MAPK pathways that controls tumor dormancy. More in general, angiogenic EC could form part of the vascular niche, a specialized microenvironment which appears to regulate metastatic outgrowth and future studies are needed to clarify the contribution of EC in the regulation of cancer stem cell behavior in the niche.

The notion that EC could communicate signals to tumor cells raises questions about the possibility of achieving tumor dormancy by counteracting angiogenesis. In experimental tumors, anti-VEGF drugs typically prune the newly formed vasculature, thus reducing microvessel density, blood flow, and perfusion. These drugs eventually increase hypoxia and cause tumor necrosis but dormancy is rarely observed. Our group recently reported that anti-VEGF therapy causes a dramatic depletion of glucose and an exhaustion of ATP levels in tumors. Moreover, we found that the central metabolic checkpoint LKB1/AMPK—a cellular sensor of ATP levels

S. Indraccolo, MD (✉)
Istituto Oncologico Veneto—IRCCS,
via Gattamelata, 64, 35128 Padua, Italy
e-mail: stefano.indraccolo@unipd.it

H. Enderling et al. (eds.), *Systems Biology of Tumor Dormancy*,
Advances in Experimental Medicine and Biology 734,
DOI 10.1007/978-1-4614-1445-2_3, © Springer Science+Business Media New York 2013

that supports cell viability in response to energy stress—is activated by anti-VEGF therapy in experimental tumors and it has a key role in induction of sustained tumor regression. These functional links between activation of the LKB1/AMPK by anti-angiogenic therapy and tumor dormancy suggest a role for metabolism in the regulation of this phenomenon.

Keywords Angiogenesis • Notch • Dll4 • MKP-1 • VEGF • LKB1 • AMPK

Aim of the Chapter

Tumor dormancy is a condition defined by the presence of fully transformed cells which do not cause symptomatic cancer [1]. Existence of this phenomenon in patients is supported by various observations, including very late relapses of cancer (especially, melanoma, breast, and renal cancer) [2], the development of cancer of donor origin following organ transplantation [3], and the common detection of disseminated tumor cells (DTC) in patients lacking bone marrow metastasis [4]. However, progress in decoding the molecular mechanisms behind clinical tumor dormancy has been very low, in part due to the difficulties of obtaining samples of dormant tumor cells from patients.

In preclinical models, dormant tumor cells exist either as solitary cells interspersed into organs or, alternatively, as non-angiogenic subcutaneous tumors or micrometastases. Cancer cells found in the context of spontaneous or experimental quiescent metastasis are generally in proliferation-arrested or low-proliferation conditions [5–9]. In contrast, in angiogenesis-dependent models of tumor dormancy, a balance of proliferation and apoptosis is generally observed [10–14]. The mechanisms involved in the regulation of the behavior of single, quiescent cells and micrometastatic foci are likely different. This chapter describes the signaling pathways which might be triggered by the angiogenic switch or by anti-angiogenic therapies in tumors and discuss possible implications for tumor dormancy.

The Angiogenic Switch and Tumor Dormancy

Failure to induce an angiogenic response has been considered of paramount importance in the establishment of the dormant state of tumors which have passed the pre-angiogenic stage [15]. The escape of tumors from dormancy is considered to depend, at least in some experimental systems, on the "angiogenic switch," a discrete event that can be triggered by various signals including genetic mutations, hypoxia and other metabolic stress, mechanical stress, and the immune/inflammatory response [16]. In some cases, angiogenic profiling of early and late cancer samples has shown that the latter often acquire expression of a broader angiogenic pattern [17], thus supporting the hypothesis that tumors abandon dormancy due to increased angiogenic output. As human tumors often contain cell populations heterogeneous in angiogenic activity [13], spontaneous progression of non-angiogenic lesions to the angiogenic phenotype could occur following a selection process. In this regard, the notion that dormant tumors often contain cycling cells may in part explain their evolution to aggressive tumors; cell proliferation is likely to contribute by favoring the emergence of angiogenic clones or tumor cells otherwise able to escape from starvation imposed by insufficient nutrients supply. Alternatively, micrometastases may escape dormancy due to lack of a systemic control, such as the disappearance of circulating angiogenesis inhibitors (e.g., angiostatin or endostatin) [18, 19] or loss of control by the adaptive immunity [20, 21].

Although the importance of the angiogenic switch during tumor progression is firmly established, it is assumed that a sustained production of angiogenic factors is required for tumor growth. In different tumor models, these factors can be released by (1) the tumor cells themselves, (2) tumor-infiltrating host inflammatory cells, or (3) resident fibroblasts. Dormancy could theoretically arise from interference with any of these sources of angiogenic factors, depending on the model considered, as extensively reviewed elsewhere [22–26]. Hypoxia and necrosis are often found in solid tumors and could contribute to the angiogenic switch both by up-regulating pro-angiogenic factors such as VEGF and CXCL12 through induction of hypoxia inducible factor 1α(alpha) HIF-1α(alpha) [27], or by recruitment and activation of pro-angiogenic myeloid cells through necrosis-mediated mechanisms [28]. It should be considered that the glycolytic phenotype of cancer cells is directly associated with the extent of necrosis in some tumor models [29], thus suggesting that certain metabolic features of cancer cells (i.e., the Warburg effect) could indirectly control the angiogenic switch.

Transgenic mouse models which develop spontaneous tumors via well-defined stages provide a valuable resource to investigate several aspects of the angiogenic switch. In several of these models, indeed, the development of foci of transformation is strictly associated with the induction of a neo-vasculature [30–34]. VEGF is often involved in the angiogenic switch, as observed for instance in the TRAMP model of prostatic cancer, where its expression marks the transition from pre-angiogenic to angiogenic PIN lesions [33]. In this and other models, up-regulation of pro-angiogenic molecules appears to be partly driven by hypoxia through HIF-1α(alpha) accumulation, which occurs as a consequence of increased cell proliferation [33]. In other transgenic models, such as the Rip Tag model of pancreatic cancer and the K14-HPV16 model of cutaneous squamous cell carcinoma, hematopoietic cells infiltrating the tumors release MMP-9, a metalloproteinase that in turn makes extracellular matrix-bound VEGF biologically active and brings about the angiogenic switch [35–38].

In apparent contrast with the dominant view that tumor growth demands sustained angiogenesis, a few years ago we observed that a short-term angiogenic burst may also suffice to break dormancy. Poorly angiogenic human T acute lymphoblastic leukemia cells (T-ALL) failed to form tumors in NOD/SCID mice; microscopic or small, dormant tumors containing viable cancer cells were found to persist in vivo in the injection sites. Our key finding was that the local co-injection of angiogenic third-party cells or a single injection of the angiogenic factor bFGF interrupted the state of cell dormancy by providing a temporarily limited and spatially confined angiogenic burst [12]. Once angiogenesis was switched on, and endothelial cells (EC) had escaped from their quiescent state, low level production of angiogenic factors by the tumor cells sufficed to allow progressive tumor growth. Overall, these findings were summarized in a hypothesis that we termed "spike hypothesis" [12]. We hypothesized that an induction threshold has to be overcome to start the process of neoangiogenesis in the tumor microenvironment. This threshold may be physically represented by the amount of angiogenic factors required to turn quiescent EC into proliferating cells that is conceivably higher than levels required to keep them proliferating, which define the maintenance threshold [12]. In this regard, it could be predicted that a short-term exogenous "spike" of angiogenic factors within the tumor microenvironment could suffice to start the process of angiogenesis and tumor growth. As an implication of this hypothesis, large tumors may not necessarily be formed by highly angiogenic cells, assuming that they passed through an angiogenic phase during their early phases of development. What could bring about the "spike" in quiescent tumors? In our opinion, inflammation could represent a relatively common event capable of delivering a timely regulated and localized angiogenic switch. In this respect, although frank inflammation is not often encountered, a sort of smouldering inflammatory reaction, mainly sustained by myeloid cells of the monocyte/macrophage lineage, is commonly observed in tumors, including clinical samples (for a review see references [39, 40]), and there is sound evidence of its contribution to tumor angiogenesis [41, 42]. Since in our experimental model transient angiogenesis triggered the outgrowth of a population of subcutaneously injected tumor cells [43], one cannot conclude whether or not this mechanism might contribute to reactivation of isolated dormant cells. However, Husemann et al. recently showed that quiescent DTC from transgenic NeuT mice can be recruited into proliferation following bone marrow transplantation into irradiated non-transgenic recipients [44]. Speculatively, poorly characterized events related to irradiation (such as inflammation) might have contributed to unmask the tumorigenic potential of DTC.

Angiogenesis Triggers Both Positive and Negative Regulatory Signals in Tumor Cells: Implications for the Homeostasis of Cancer Stem Cells

Recent studies have shown that a specialized microenvironment, termed niche, could be required to support cell subpopulations endowed with stem cell potential [45]. The principal function of the niche is not merely to provide a docking site, but also to dynamically modulate stem cell function, normally maintaining stem cells in a quiescent state and activating their proliferation under conditions of physiologic challenge [46].

Several reports highlight the importance of EC for the promotion of stem cell survival. Glioblastoma stem cells, for example, appear to depend on a vascular niche for their survival. The first observation supporting this hypothesis was provided by the study of Bao et al., who showed that high-level production of VEGF by stem cell-like glioma cells might contribute to their tumorigenic potential [47]. Calabrese and colleagues, however, were the first to provide conclusive evidence that stem cells from various brain tumors are maintained within vascular niches that mimic the neural stem cell niche [48]. Importantly, by co-transplanting brain tumor cells and EC into immunodeficient mice, it was found that EC-derived factors promote tumor formation in the brain. These results were supported by a subsequent study which showed that C6 rat glioma cells with stem cell features are also maintained by factors secreted by EC [49]. More recently, the existence of a vascular niche relevant for the establishment of liver metastasis has been validated in colon cancer models [50].

Tumor cells can also program the niche at distant sites of colonization, as elegantly shown by Lyden and co-workers [51], who analyzed the cellular events leading to the formation of the "pre-metastatic niche" in distant organs. By flow cytometry and immunofluorescence studies, these authors showed that hematopoietic progenitor cells (HPC) labeled with green fluorescent protein arrive and form clusters of cells in the tissue parenchyma before establishment of metastasis. These HPC express the VEGF receptor 1 (VEGFR1) along with other hematopoietic markers, including CD34, CD11b, c-kit, and Sca-1. A critical anchor for HPC is fibronectin, which is newly synthesized by resident fibroblasts and fibroblast-like cells and interacts with the integrin VLA-4, which is expressed by VEGFR1$^+$ HPC, thus allowing their adherence and initiation of cellular cluster formation [52]. This pre-metastatic niche assembled by VEGFR1$^+$ HPC, EC, fibronectin, and other components, such as tenascin C (TNC) [53], provides a marked support to metastasis growth, and its disruption has been hypothesized to predispose to tumor dormancy [52]. These seminal studies highlight the existence of a specialized microenvironment, termed "niche," and the metastasis-promoting role of soluble factors produced by EC or HPC or by ECM components.

Our group has recently investigated the possible involvement of a cell-to-cell interaction in the escape from tumor dormancy. Using a model of angiogenesis-dependent dormancy of T-ALL cells [12], we found that bFGF induces expression by the EC and other stromal cells of Dll4, a ligand of the family of Notch known to regulate Notch signaling during angiogenesis [54]. Dll4 appears to interact with Notch 3, which is abundantly expressed by leukemia cells, and activates Notch 3 signaling in the T-ALL cells, which protects them from apoptosis and initiates progressive tumor growth [55]. Other Notch paralogs could also be activated by ligands expressed by angiogenic EC in other tumor types.

One of the mechanisms downstream of Notch3 in this tumor model involves MKP-1—a broadly expressed phosphatase—whose levels are controlled by Notch3 by regulation of protein ubiquitination and stability [56]. Notch3 and MKP-1 levels are consistently low in dormant tumors, and this is accompanied by relatively high levels of phosphorylated p38, a canonical MKP-1 target previously associated with

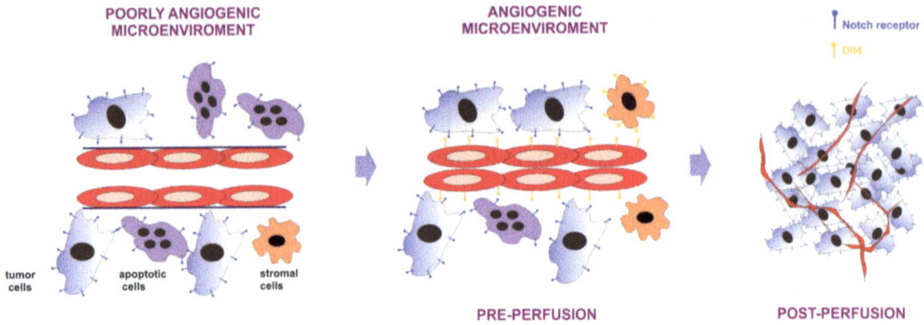

Fig. 3.1 Positive signals stemming from angiogenic blood vessels regulate tumor growth. *Left panel*: cancer cells incapable of inducing the angiogenic switch remain dormant in tissues due to a balance between proliferation and apoptosis. These cells receive limited oxygen and nutrient supply from the existing vasculature. Notch receptors are expressed on tumor cells but ligands (Dll4) expression in the tumor microenvironment is low. *Mid panel*: an exogenous angiogenic switch, for instance due to inflammation and recruitment of angiogenesis-promoting haematopoietic cells, such as macrophages, activates EC proliferation and migration and stimulates EC expression of the Notch ligand Dll4. Re-modeling of existing vessels allows for cell-to-cell contacts between tumor and EC followed by triggering of Notch signaling in tumor cells. Augmented Notch activation conveys a survival signal into tumor cells which can favor proliferation over apoptosis. Escape from tumor dormancy begins in a pre-perfusion phase. *Right panel*: upon completion of the angiogenic process, functional new blood vessels are in place and feed tumor cells thus allowing further expansion of the tumor mass

maintenance of tumor dormancy [57–59]. Interruption of dormancy by angiogenic factors is accompanied by increased MKP-1 expression and reduction of phosphorylated p38/ERK levels. MKP-1 seems indeed to be essential in this tumor model, as its attenuation by shRNA maintains dormancy [56]. On the other hand, MKP-1 overexpression in dormant tumor cells does not lead to progressive tumor growth (unpublished data), thus indicating that MKP-1 does not fully replace Notch signaling in this model.

These findings suggest that EC embedded in tissues undergoing angiogenesis may communicate activation signals to tumor cells mediated by the Notch-Dll4 molecular interaction, which contributes to the switch towards an aggressive phenotype (Fig. 3.1). Intriguingly, Dll4 expression by the EC is an early event which precedes perfusion of the newly formed blood vessels; later on, angiogenesis supports further tumor growth by supplying oxygen and nutrients to tumor cells, according to its canonical function (Fig. 3.1).

As Notch receptors and ligands are commonly expressed in solid tumors [60], we speculate that regulation of Notch activation by the tumor microenvironment could occur quite frequently, although consequences on tumor growth could largely vary depending on the cellular context considered and the intensity/duration of the signals. In line with this hypothesis, Sansone et al. recently described Jagged-1-dependent activation of Notch3 signaling in breast cancer cells and its implications in the regulation of stem cell features [61], and there is increasing evidence of the importance of this triggering mechanism in colon cancer [50, 62]. Notably, new findings indicate that TNC, a component of the extracellular matrix whose expression is elevated during inflammation, can be downstream of Notch in certain tumor cells [63] and further stimulates Notch signaling, hypothetically by acting as non-canonical Notch ligand [53]. This feed-forward loop further highlights the complexity of regulation of Notch signaling by the tumor microenvironment. It remains to be determined whether pathway-activating signals within the metastatic niche are delivered selectively to specific subsets of tumor cells, such as cancer stem cells.

Ultimately, this novel role of angiogenesis in tumors is not unexpected. EC were indeed known to enhance neurogenesis and to promote formation of pancreatic and liver tissue independently of their ability to form vasculature [64–66]. Moreover, angiogenesis has been hypothesized to control tissue mass in the prostate [67]. In liver, activation of VEGFR1 on sinusoidal EC results in the paracrine release of hepatocyte growth factor, IL-6, and other cytokines that stimulate the proliferation of hepatocytes [68]. In liver, it is possible that anti-angiogenic drugs like sorafenib—which block VEGFR1 signaling and are used to treat hepatocellular carcinoma [69]—might interfere with EC-tumor cell communication, although this has not been experimentally addressed so far.

It is also important to stress that EC can also send negative signals to tumor cells. In this respect, it was previously reported that interaction of KAI1 on tumor cells with DARC on EC can suppress the metastatic properties of prostate cancer cells [70]. Adhesion of $KAI1^+$ tumor cells to $DARC^+$ EC inhibited cell proliferation, reduced cell survival, and induced senescence markers p21 and TBX2 in the tumor cells. The translational relevance of this observation for patients is, however, still unknown.

In conclusion, there is some evidence that EC may regulate proliferation of normal or transformed cells, either through secreted signaling molecules or intercellular contacts.

Counteracting the Angiogenic Switch with Anti-VEGF Drugs: Implications for Tumor Dormancy

The notion that angiogenic EC may signal to tumor cells has provoking theoretical implications for anti-angiogenic therapy. Will regression of the vasculature down-regulate key oncogenic pathways in tumor cells and promote tumor dormancy? Conceivably, EC-mediated activation of oncogenic signaling pathways (such as Notch) will be critical only for the subset of tumors which depend on paracrine or third-party stimulation (Fig. 3.2). A clinical setting where this mechanism could be important is when anti-angiogenic therapy is administered to patients with microscopic metastases, following surgery of the primary tumor.

Alternatively, one could expect that anti-angiogenic therapy causes cancer cell death—due to oxygen and nutrients starvation—thus perturbing the homeostasis of quiescent cancer stem cells. This scenario—quite possible in established tumors—would ultimately favor relapse, as opposed to dormancy (Fig. 3.2).

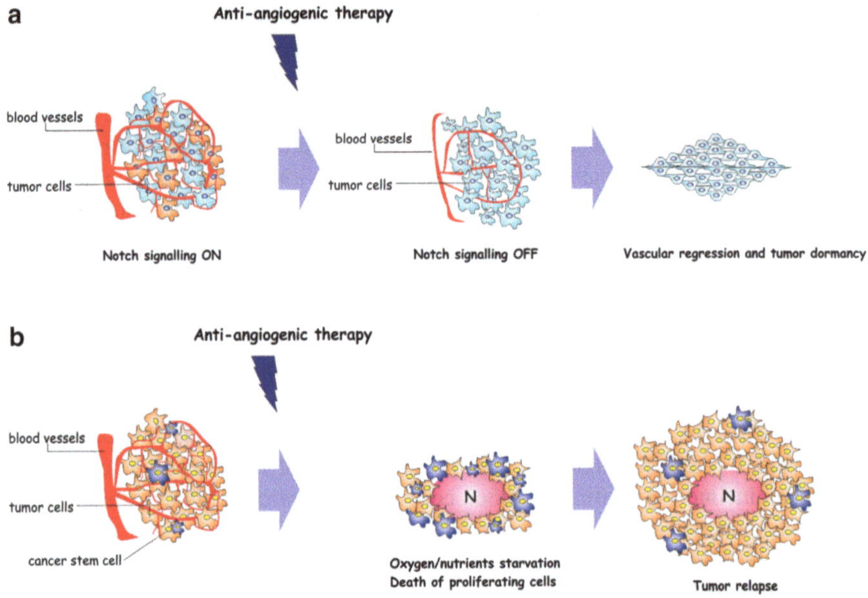

Fig. 3.2 Anti-angiogenic therapy and tumor dormancy: two alternative scenarios can be envisaged. (**a**) Vessel regression interrupts signaling between EC and tumor cells, thereby lowering activation of key oncogenic pathways in cancer cells, such as Notch, and leading to tumor dormancy. (**b**) Anti-angiogenic therapy kills a large majority of metabolically active tumor cells, due to oxygen and nutrients starvation. However, quiescent cancer stem cells survive and are recruited into cell division, due to perturbation of homeostatic mechanisms. Proliferation of the surviving cells is high and tumor relapse eventually occurs due to acquired resistance mechanisms

If we consider clinical studies, tumor dormancy was not observed in patients treated with first generation anti-angiogenic drugs such as sunitinib or bevacizumab, although prolonged disease stabilization was occasionally reported. Remarkably, radiologic results have been quite heterogeneous. In some cases, tumors respond to anti-angiogenic therapy by decreasing tumor volume by more than 33%, qualifying it for a partial response according to RECIST criteria (response evaluation criteria in solid tumors). In most patients, however, significant changes in tumor density with no decrease in tumor dimensions are observed. This is often associated with central tumor cavitation and necrosis [71], an observation which suggests that VEGF blockade may perturb the energy balance in cancer cells.

Stimulated by these clinical observations, in a recent study we investigated how metabolic parameters contribute to determine the pathologic response to VEGF blockade in tumor xenografts [29]. A landmark observation of our study was that the level of "glucose addiction" of tumor cells dictates the amount of necrosis caused by angiogenesis inhibition. This was explained by the fact that VEGF blockade acutely perturbs glucose levels in tumor xenografts, as shown by bioluminescence metabolic imaging. Although it is well recognized that very low glucose concentrations can be present in large solid tumors, due to their compromised vascular functions [72, 73], with regard to the effects of anti-angiogenic drugs previous studies mainly characterized hypoxia changes in tumors, assuming that glucose would remain available due to its high diffusion capacity in tissues. Our results indicate that this assumption should be revised. Notably, glucose uptake was maintained following anti-angiogenic therapy, as shown by fluorodeoxyglucose PET imaging, indicating that delivery of glucose through the vasculature is not compromised despite a substantial decrease in microvessel density [29]. Similar observations were reported in patients with rectal cancer after bevacizumab monotherapy [74]. So it appears that glucose catabolism is very high after anti-angiogenic therapy—accounting for very low steady-state levels—whereas glucose uptake is high. These results are partially explained by HIF-1α(alpha) accumulation in treated tumors.

A related study investigated by Magnetic Resonance Spectroscopy (MRS) metabolic changes in glioblastoma following anti-VEGF treatment. The authors observed a tendency toward accumulation of lactate, alanine, choline, myo-inositol, creatine, taurine, and mobile lipids together with induction of HIF-1α(alpha) and activation of the phosphatidyl-inositol-3-kinase pathway [75]. This combination of metabolic changes has previously been associated with increased hypoxia in human brain tumor spectra [76] and partially overlaps with our findings in ovarian cancer xenografts [29]. In patients, radiologic techniques including perfusion CT and dynamic contrast-enhanced MRI have shown suppression of tumor vascular permeability induced by anti-angiogenic agents, which transiently ameliorates hypoxia and improves delivery of chemotherapy [77]. In any case, following this initial normalization window, hypoxia and nutrients insufficiency increase, as found in animal models.

The marked metabolic changes caused by anti-angiogenic therapy might be predicted to activate AMP-activated protein kinase (AMPK), a central metabolic sensor found in all eukaryote systems that governs glucose and lipid metabolism in response to alterations in nutrients supply and intracellular energy levels, as well as cell polarity, cell proliferation, and gene expression regulation [78, 79]. AMPK is a heterotrimer that consists of a catalytic subunit, AMPKα(alpha), and two regulatory subunits, AMPKβ(Beta) and AMPKγ(gamma). There are two distinct isoforms of the AMPKα(alpha) subunits designated AMPKα(alpha)1 and AMPKα(alpha)2 which differ in their tissue specificity, subcellular localization, and mechanism of activation. Mainly, this protein kinase is activated in response to an increase in the AMP/ATP ratio within the cell and it is phosphorylated at Thr-172 in the catalytic subunit by upstream kinases including LKB1 or calmodulin-dependent protein kinase kinase beta [CAMKKβ(Beta)] [80, 81]. In addition, AMPK can also be activated by a variety of pharmacological agents, such as metformin and AICAR [82]. AMPK activation reprograms cellular metabolism and enforces metabolic checkpoints to stop proliferation by acting on mTOR complex 1 [mTORC1, p53, and other molecules [83]]. In particular, AMPK acts to restore cellular energy balance by promoting ATP generating processes, such as fatty acid beta oxidation, and simultaneously by inhibiting ATP consuming processes, such as fatty acid synthesis, gluconeogenesis, and protein synthesis. This is initially achieved by direct phosphorylation of some key metabolic enzymes (such as Acetil-CoA carboxylase,

ACC) and subsequently by modulation of gene expression [84]. Several recent studies in cell culture models and in vivo have shown that growth of tumor cell lines was inhibited by AMPK activation, highlighting this kinase as cancer relevant "druggable" target, particularly in combination with chemotherapy [85, 86].

Since AMPK is activated when intracellular levels of ATP decline and intracellular levels of AMP increase, as often happens during nutrient starvation and hypoxia, a certain level of AMPK activation is commonly seen in solid tumors [87], and we also observed AMPK activation in the peri-necrotic areas of our control xenografts [29]. Moreover, we found that anti-angiogenic therapy increased AMPK activation levels in tumors, probably as a consequence of the dramatic glucose depletion and ATP level exhaustion, as demonstrated by immunohistochemistry analysis of pAMPK and pACC levels in tumor xenografts treated with anti-VEGF [29]. Our results are supported by a clinical study that showed that bevacizumab increased total AMPK and pAMPK levels in renal cell carcinoma patients [88].

AMPK-deficient tumor cells are hypersensitive to energy stress-inducing agents [89, 90]. It is thus possible that defects of AMPK activation may in part account for the reduced survival of certain tumor cells under glucose starvation and/or hypoxia in vitro or anti-angiogenic therapy in vivo. Indeed, in our study highly glycolytic cells that failed to activate AMPK developed large necrotic areas after short-term anti-VEGF therapy. Intriguingly, attenuation of AMPKα(alpha)2 in poorly glycolytic cells compromised their survival under glucose deprivation in vitro, caused a metabolic switch (i.e., increased glycolysis), and increased necrosis in xenografts following anti-angiogenic therapy [29], thus supporting this hypothesis.

How does this impact on the outcome of anti-angiogenic therapy? In the context of established tumors, we may predict that tumors lacking LKB1 or AMPK continue to proliferate faster than their AMPK-containing counterparts, but then succumb to necrosis owing to the inevitable energy shortage. Necrosis is rarely complete in tumors, even following treatment with vascular damaging agents which cause acute and profound perturbations of oxygen and nutrients supply [41]. The few tumor cells surviving these treatments, generally located within the viable rim of the tumor, account for an expected tumor relapse. Necrosis is also a very potent inducer of macrophage recruitment and activation [26], and it could be foreseen that these hematopoietic cells actively contribute to the angiogenic switch, thereby facilitating engraftment of tumors bearing mutations in the LKB1/AMPK pathway (Fig. 3.3). In contrast, tumors with a functional LKB1/AMPK pathway and integrity of downstream effectors involved in control of cell proliferation, such as p53, might be driven into a quiescent state (Fig. 3.3). Indeed, preliminary results from my laboratory using IGROV-1 xenografts—a prototype of poorly glycolytic tumors with wild-type p53 and LKB1/AMPK pathways [91]—indicate that prolonged anti-VEGF treatment causes marked tumor regression not followed by tumor relapse upon discontinuation of anti-angiogenic therapy (unpublished data). Although further studies are needed to investigate proliferation and apoptosis in the regressed tumors, our findings provide initial support of the hypothesis illustrated in Fig. 3.3. Moreover, reintroduction of functional LKB1 in cancer cells lacking endogenous LKB1 activity greatly impaired

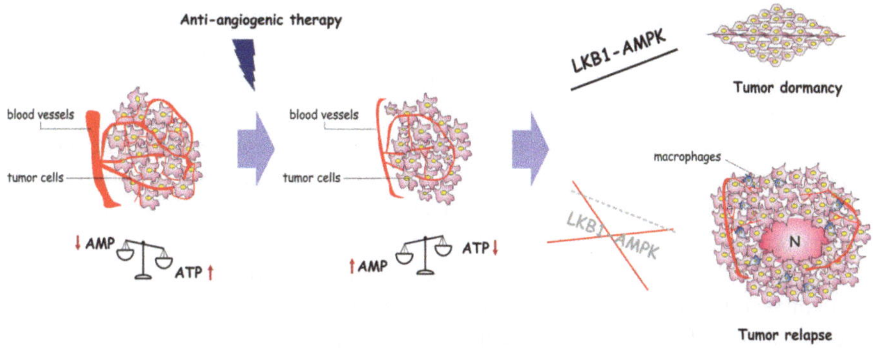

Fig. 3.3 Metabolic perturbations caused by anti-angiogenic therapy: predictions for tumor dormancy. As recently shown in preclinical models [29], glucose and ATP steady-state levels are dramatically reduced by anti-VEGF therapy. The LKB1/AMPK pathway is activated: tumor cells slow down anabolic processes and cell proliferation, eventually entering quiescence. However, if LKB1/AMPK or downstream components are disabled, metabolic processes are not down-regulated and cell proliferation is not compromised. Tumors initially undergo partial necrosis due to energy imbalance, followed by infiltration of macrophages and tumor relapse. *N* necrotic area. From Zulato E. et al., Metabolic effects of anti-angiogenic therapy in tumors. Biochimie Apr 2012;94(4):925–931. Modified with permission from Elsevier Ltd

subcutaneous tumor engraftment in SCID mice, possibly leading to a dormant state (unpublished data). In this experimental setting, it is conceivable that limitations on glucose and oxygen diffusion imposed by the initial lack of angiogenesis caused activation of an AMPK-mediated metabolic growth checkpoint and growth inhibition in LKB1-proficient cells. Altogether, these ongoing studies suggest a role for the LKB1/AMPK pathway in angiogenesis-dependent types of tumor dormancy.

Conclusions

Lack of angiogenic potential can explain certain types of tumor dormancy, conceivably due to limiting oxygen and nutrients supply. I propose here that angiogenesis could play a wider role in the regulation of this phenomenon than currently held. Indeed, there is increasing evidence that tumor cells, and in selected instances cancer stem cells, need to interact with a vascular niche to fully express their oncogenic potential. I hypothesize that angiogenesis could be involved in the establishment or the maintenance of a tumor-promoting vascular niche around quiescent cancer cells. Dissecting the molecular pathways regulated by angiogenic EC in tumor cells, such as Notch, will be fundamental for improved understanding of the mechanisms of tumor dormancy.

If inability to activate the angiogenic switch keeps tumors dormant, why shouldn't be possible to induce tumor dormancy by regressing the established tumor vasculature?

The definitive answer to this question is complex and certainly awaits future work. Impact of angiogenesis inhibition on oncogenic signaling pathways and on the behavior of cancer stem cells and metabolic features of tumors are all examples of topics that deserve further investigation.

Finally, a key limitation in the tumor dormancy field is represented by the models. Knowledge on mechanisms of tumor dormancy is generally grounded on observations in tumor xenografts grown in immunodeficient mice, owing to the difficulty of identifying dormant cells seeded in visceral organs such as the lungs or the liver. Hence, the challenge for the future will be to develop new technical approaches which will enable us to dissect the molecular pathways involved in the regulation of tumor dormancy in preclinical models closer to human cancer, in particular tumor-prone genetically engineered immunocompetent mice. A deeper understanding of tumor dormancy will not only lead to future therapeutic targets for cancer, but will identify the minimal number of steps that are required to suppress growth of transformed cells.

References

1. Folkman J, Kalluri R (2004) Cancer without disease. Nature 427(6977):787
2. Uhr JW, Pantel K (2011) Controversies in clinical cancer dormancy. Proc Natl Acad Sci USA 108(30):12396–12400
3. Strauss DC, Thomas JM (2010) Transmission of donor melanoma by organ transplantation. Lancet Oncol 11(8):790–796
4. Pantel K, Brakenhoff RH, Brandt B (2008) Detection, clinical relevance and specific biological properties of disseminating tumour cells. Nat Rev Cancer 8(5):329–340
5. Goldberg SF, Harms JF, Quon K, Welch DR (1999) Metastasis-suppressed C8161 melanoma cells arrest in lung but fail to proliferate. Clin Exp Metastasis 17(7):601–607
6. Goodison S, Kawai K, Hihara J, Jiang P, Yang M, Urquidi V et al (2003) Prolonged dormancy and site-specific growth potential of cancer cells spontaneously disseminated from nonmetastatic breast tumors as revealed by labeling with green fluorescent protein. Clin Cancer Res 9(10 Pt 1):3808–3814
7. Naumov GN, Townson JL, MacDonald IC, Wilson SM, Bramwell VH, Groom AC et al (2003) Ineffectiveness of doxorubicin treatment on solitary dormant mammary carcinoma cells or late-developing metastases. Breast Cancer Res Treat 82(3):199–206
8. Pantel K, Cote RJ, Fodstad O (1999) Detection and clinical importance of micrometastatic disease. J Natl Cancer Inst 91(13):1113–1124
9. Pantel K, Schlimok G, Braun S, Kutter D, Lindemann F, Schaller G et al (1993) Differential expression of proliferation-associated molecules in individual micrometastatic carcinoma cells. J Natl Cancer Inst 85(17):1419–1424
10. Achilles EG, Fernandez A, Allred EN, Kisker O, Udagawa T, Beecken WD et al (2001) Heterogeneity of angiogenic activity in a human liposarcoma: a proposed mechanism for "no take" of human tumors in mice. J Natl Cancer Inst 93(14):1075–1081
11. Almog N, Henke V, Flores L, Hlatky L, Kung AL, Wright RD et al (2006) Prolonged dormancy of human liposarcoma is associated with impaired tumor angiogenesis. FASEB J 20(7):947–949
12. Indraccolo S, Favaro E, Amadori A (2006) Dormant tumors awaken by a short-term angiogenic burst: the spike hypothesis. Cell Cycle 5(16):1751–1755, Georgetown, Tex
13. Naumov GN, Bender E, Zurakowski D, Kang SY, Sampson D, Flynn E et al (2006) A model of human tumor dormancy: an angiogenic switch from the nonangiogenic phenotype. J Natl Cancer Inst 98(5):316–325

14. Udagawa T, Fernandez A, Achilles EG, Folkman J, D'Amato RJ (2002) Persistence of microscopic human cancers in mice: alterations in the angiogenic balance accompanies loss of tumor dormancy. FASEB J 16(11):1361–1370
15. Aguirre-Ghiso JA (2007) Models, mechanisms and clinical evidence for cancer dormancy. Nat Rev Cancer 7(11):834–846
16. Carmeliet P, Jain RK (2000) Angiogenesis in cancer and other diseases. Nature 407(6801): 249–257
17. Relf M, LeJeune S, Scott PA, Fox S, Smith K, Leek R et al (1997) Expression of the angiogenic factors vascular endothelial cell growth factor, acidic and basic fibroblast growth factor, tumor growth factor beta-1, platelet-derived endothelial cell growth factor, placenta growth factor, and pleiotrophin in human primary breast cancer and its relation to angiogenesis. Cancer Res 57(5):963–969
18. O'Reilly MS, Boehm T, Shing Y, Fukai N, Vasios G, Lane WS et al (1997) Endostatin: an endogenous inhibitor of angiogenesis and tumor growth. Cell 88(2):277–285
19. O'Reilly MS, Holmgren L, Shing Y, Chen C, Rosenthal RA, Cao Y et al (1994) Angiostatin: a circulating endothelial cell inhibitor that suppresses angiogenesis and tumor growth. Cold Spring Harb Symp Quant Biol 59:471–482
20. Eyles J, Puaux AL, Wang X, Toh B, Prakash C, Hong M et al (2010) Tumor cells disseminate early, but immunosurveillance limits metastatic outgrowth, in a mouse model of melanoma. J Clin Invest 120(6):2030–2039
21. Koebel CM, Vermi W, Swann JB, Zerafa N, Rodig SJ, Old LJ et al (2007) Adaptive immunity maintains occult cancer in an equilibrium state. Nature 450(7171):903–907
22. de Visser KE, Eichten A, Coussens LM (2006) Paradoxical roles of the immune system during cancer development. Nat Rev Cancer 6(1):24–37
23. Kalluri R, Zeisberg M (2006) Fibroblasts in cancer. Nat Rev Cancer 6(5):392–401
24. Mantovani A, Schioppa T, Porta C, Allavena P, Sica A (2006) Role of tumor-associated macrophages in tumor progression and invasion. Cancer Metastasis Rev 25(3):315–322
25. Orimo A, Weinberg RA (2006) Stromal fibroblasts in cancer: a novel tumor-promoting cell type. Cell Cycle 5(15):1597–1601, Georgetown, Tex
26. Pollard JW (2004) Tumour-educated macrophages promote tumour progression and metastasis. Nat Rev Cancer 4(1):71–78
27. Semenza GL (2003) Targeting HIF-1 for cancer therapy. Nat Rev Cancer 3(10):721–732
28. Murdoch C, Giannoudis A, Lewis CE (2004) Mechanisms regulating the recruitment of macrophages into hypoxic areas of tumors and other ischemic tissues. Blood 104(8):2224–2234
29. Nardo G, Favaro E, Curtarello M, Moserle L, Zulato E, Persano L et al (2011) Glycolytic phenotype and AMP kinase modify the pathologic response of tumor xenografts to VEGF neutralization. Cancer Res 71(12):4214–4225
30. Bergers G, Benjamin LE (2003) Tumorigenesis and the angiogenic switch. Nat Rev Cancer 3(6):401–410
31. Coussens LM, Raymond WW, Bergers G, Laig-Webster M, Behrendtsen O, Werb Z et al (1999) Inflammatory mast cells up-regulate angiogenesis during squamous epithelial carcinogenesis. Genes Dev 13(11):1382–1397
32. Folkman J, Watson K, Ingber D, Hanahan D (1989) Induction of angiogenesis during the transition from hyperplasia to neoplasia. Nature 339(6219):58–61
33. Huss WJ, Hanrahan CF, Barrios RJ, Simons JW, Greenberg NM (2001) Angiogenesis and prostate cancer: identification of a molecular progression switch. Cancer Res 61(6): 2736–2743
34. Lin EY, Li JF, Gnatovskiy L, Deng Y, Zhu L, Grzesik DA et al (2006) Macrophages regulate the angiogenic switch in a mouse model of breast cancer. Cancer Res 66(23):11238–11246
35. Bergers G, Brekken R, McMahon G, Vu TH, Itoh T, Tamaki K et al (2000) Matrix metalloproteinase-9 triggers the angiogenic switch during carcinogenesis. Nat Cell Biol 2(10):737–744
36. Coussens LM, Tinkle CL, Hanahan D, Werb Z (2000) MMP-9 supplied by bone marrow-derived cells contributes to skin carcinogenesis. Cell 103(3):481–490
37. Giraudo E, Inoue M, Hanahan D (2004) An amino-bisphosphonate targets MMP-9-expressing macrophages and angiogenesis to impair cervical carcinogenesis. J Clin Invest 114(5):623–633

38. Nozawa H, Chiu C, Hanahan D (2006) Infiltrating neutrophils mediate the initial angiogenic switch in a mouse model of multistage carcinogenesis. Proc Natl Acad Sci USA 103(33):12493–12498
39. Balkwill F, Charles KA, Mantovani A (2005) Smoldering and polarized inflammation in the initiation and promotion of malignant disease. Cancer Cell 7(3):211–217
40. Lewis C, Murdoch C (2005) Macrophage responses to hypoxia: implications for tumor progression and anti-cancer therapies. Am J Pathol 167(3):627–635
41. Crescenzi M, Persano L, Esposito G, Zulato E, Borsi L, Balza E et al (2011) Vandetanib improves anti-tumor effects of L19mTNFalpha in xenograft models of esophageal cancer. Clin Cancer Res 17(3):447–458
42. Shojaei F, Ferrara N (2008) Refractoriness to antivascular endothelial growth factor treatment: role of myeloid cells. Cancer Res 68(14):5501–5504
43. Indraccolo S, Stievano L, Minuzzo S, Tosello V, Esposito G, Piovan E et al (2006) Interruption of tumor dormancy by a transient angiogenic burst within the tumor microenvironment. Proc Natl Acad Sci USA 103(11):4216–4221
44. Husemann Y, Geigl JB, Schubert F, Musiani P, Meyer M, Burghart E et al (2008) Systemic spread is an early step in breast cancer. Cancer Cell 13(1):58–68
45. Li L, Xie T (2005) Stem cell niche: structure and function. Annu Rev Cell Dev Biol 21:605–631
46. Psaila B, Lyden D (2009) The metastatic niche: adapting the foreign soil. Nat Rev Cancer 9(4):285–293
47. Bao S, Wu Q, Sathornsumetee S, Hao Y, Li Z, Hjelmeland AB et al (2006) Stem cell-like glioma cells promote tumor angiogenesis through vascular endothelial growth factor. Cancer Res 66(16):7843–7848
48. Calabrese C, Poppleton H, Kocak M, Hogg TL, Fuller C, Hamner B et al (2007) A perivascular niche for brain tumor stem cells. Cancer Cell 11(1):69–82
49. Folkins C, Man S, Xu P, Shaked Y, Hicklin DJ, Kerbel RS (2007) Anticancer therapies combining antiangiogenic and tumor cell cytotoxic effects reduce the tumor stem-like cell fraction in glioma xenograft tumors. Cancer Res 67(8):3560–3564
50. Sonoshita M, Aoki M, Fuwa H, Aoki K, Hosogi H, Sakai Y et al (2011) Suppression of colon cancer metastasis by Aes through inhibition of Notch signaling. Cancer Cell 19(1):125–137
51. Kaplan RN, Riba RD, Zacharoulis S, Bramley AH, Vincent L, Costa C et al (2005) VEGFR1-positive haematopoietic bone marrow progenitors initiate the pre-metastatic niche. Nature 438(7069):820–827
52. Kaplan RN, Rafii S, Lyden D (2006) Preparing the "soil": the premetastatic niche. Cancer Res 66(23):11089–11093
53. Oskarsson T, Acharyya S, Zhang XH, Vanharanta S, Tavazoie SF, Morris PG et al (2011) Breast cancer cells produce tenascin C as a metastatic niche component to colonize the lungs. Nat Med 17(7):867–874
54. Patel NS, Li JL, Generali D, Poulsom R, Cranston DW, Harris AL (2005) Up-regulation of delta-like 4 ligand in human tumor vasculature and the role of basal expression in endothelial cell function. Cancer Res 65(19):8690–8697
55. Indraccolo S, Minuzzo S, Masiero M, Pusceddu I, Persano L, Moserle L et al (2009) Cross-talk between tumor and endothelial cells involving the Notch3-Dll4 interaction marks escape from tumor dormancy. Cancer Res 69(4):1314–1323
56. Masiero M, Minuzzo S, Pusceddu I, Moserle L, Persano L, Agnusdei V et al (2011) Notch3-mediated regulation of MKP-1 levels promotes survival of T acute lymphoblastic leukemia cells. Leukemia 25(4):588–598
57. Aguirre Ghiso JA, Kovalski K, Ossowski L (1999) Tumor dormancy induced by downregulation of urokinase receptor in human carcinoma involves integrin and MAPK signaling. J Cell Biol 147(1):89–104
58. Aguirre-Ghiso JA, Estrada Y, Liu D, Ossowski L (2003) ERK(MAPK) activity as a determinant of tumor growth and dormancy; regulation by p38(SAPK). Cancer Res 63(7):1684–1695
59. Aguirre-Ghiso JA, Liu D, Mignatti A, Kovalski K, Ossowski L (2001) Urokinase receptor and fibronectin regulate the ERK(MAPK) to p38(MAPK) activity ratios that determine carcinoma cell proliferation or dormancy in vivo. Mol Biol Cell 12(4):863–879

60. Leong KG, Karsan A (2006) Recent insights into the role of Notch signaling in tumorigenesis. Blood 107(6):2223–2233
61. Sansone P, Storci G, Tavolari S, Guarnieri T, Giovannini C, Taffurelli M et al (2007) IL-6 triggers malignant features in mammospheres from human ductal breast carcinoma and normal mammary gland. J Clin Invest 117(12):3988–4002
62. Serafin V, Persano L, Moserle L, Esposito G, Ghisi M, Curtarello M et al (2011) Notch3 signalling promotes tumour growth in colorectal cancer. J Pathol 224(4):448–460
63. Sivasankaran B, Degen M, Ghaffari A, Hegi ME, Hamou MF, Ionescu MC et al (2009) Tenascin-C is a novel RBPJkappa-induced target gene for Notch signaling in gliomas. Cancer Res 69(2):458–465
64. Lammert E, Cleaver O, Melton D (2001) Induction of pancreatic differentiation by signals from blood vessels. Science 294(5542):564–567, New York, NY
65. Matsumoto K, Yoshitomi H, Rossant J, Zaret KS (2001) Liver organogenesis promoted by endothelial cells prior to vascular function. Science 294(5542):559–563, New York, NY
66. Shen Q, Goderie SK, Jin L, Karanth N, Sun Y, Abramova N et al (2004) Endothelial cells stimulate self-renewal and expand neurogenesis of neural stem cells. Science 304(5675):1338–1340, New York, NY
67. Folkman J (1998) Is tissue mass regulated by vascular endothelial cells? Prostate as the first evidence. Endocrinology 139(2):441–442
68. LeCouter J, Moritz DR, Li B, Phillips GL, Liang XH, Gerber HP et al (2003) Angiogenesis-independent endothelial protection of liver: role of VEGFR-1. Science 299(5608):890–893, New York, NY
69. Zhu AX, Duda DG, Sahani DV, Jain RK (2011) HCC and angiogenesis: possible targets and future directions. Nat Rev Clin Oncol 8(5):292–301
70. Bandyopadhyay S, Zhan R, Chaudhuri A, Watabe M, Pai SK, Hirota S et al (2006) Interaction of KAI1 on tumor cells with DARC on vascular endothelium leads to metastasis suppression. Nat Med 12(8):933–938
71. Faivre S, Demetri G, Sargent W, Raymond E (2007) Molecular basis for sunitinib efficacy and future clinical development. Nat Rev Drug Discov 6(9):734–745
72. Vaupel P (2004) Tumor microenvironmental physiology and its implications for radiation oncology. Semin Radiat Oncol 14(3):198–206
73. Walenta S, Schroeder T, Mueller-Klieser W (2004) Lactate in solid malignant tumors: potential basis of a metabolic classification in clinical oncology. Curr Med Chem 11(16):2195–2204
74. Willett CG, Boucher Y, di Tomaso E, Duda DG, Munn LL, Tong RT et al (2004) Direct evidence that the VEGF-specific antibody bevacizumab has antivascular effects in human rectal cancer. Nat Med 10(2):145–147
75. Keunen O, Johansson M, Oudin A, Sanzey M, Rahim SA, Fack F et al (2011) Anti-VEGF treatment reduces blood supply and increases tumor cell invasion in glioblastoma. Proc Natl Acad Sci USA 108(9):3749–3754
76. Howe FA, Barton SJ, Cudlip SA, Stubbs M, Saunders DE, Murphy M et al (2003) Metabolic profiles of human brain tumors using quantitative in vivo 1H magnetic resonance spectroscopy. Magn Reson Med 49(2):223–232
77. Jain RK (2005) Normalization of tumor vasculature: an emerging concept in antiangiogenic therapy. Science 307(5706):58–62, New York, NY
78. Alexander A, Walker CL (2011) The role of LKB1 and AMPK in cellular responses to stress and damage. FEBS Lett 585(7):952–957
79. Shackelford DB, Shaw RJ (2009) The LKB1-AMPK pathway: metabolism and growth control in tumour suppression. Nat Rev Cancer 9(8):563–575
80. Hawley SA, Pan DA, Mustard KJ, Ross L, Bain J, Edelman AM et al (2005) Calmodulin-dependent protein kinase kinase-beta is an alternative upstream kinase for AMP-activated protein kinase. Cell Metab 2(1):9–19
81. Woods A, Johnstone SR, Dickerson K, Leiper FC, Fryer LG, Neumann D et al (2003) LKB1 is the upstream kinase in the AMP-activated protein kinase cascade. Curr Biol 13(22):2004–2008

82. Zhang BB, Zhou G, Li C (2009) AMPK: an emerging drug target for diabetes and the metabolic syndrome. Cell Metab 9(5):407–416
83. Luo Z, Zang M, Guo W (2010) AMPK as a metabolic tumor suppressor: control of metabolism and cell growth. Future Oncol 6(3):457–470
84. Hardie DG (2007) AMP-activated protein kinase as a drug target. Annu Rev Pharmacol Toxicol 47:185–210
85. Rattan R, Graham RP, Maguire JL, Giri S, Shridhar V (2011) Metformin suppresses ovarian cancer growth and metastasis with enhancement of cisplatin cytotoxicity in vivo. Neoplasia 13(5):483–491
86. Rocha GZ, Dias MM, Ropelle ER, Osorio-Costa F, Rossato FA, Vercesi AE et al (2011) Metformin amplifies chemotherapy-induced AMPK activation and antitumoral growth. Clin Cancer Res 17(12):3993–4005
87. Laderoute KR, Amin K, Calaoagan JM, Knapp M, Le T, Orduna J et al (2006) 5′-AMP-activated protein kinase (AMPK) is induced by low-oxygen and glucose deprivation conditions found in solid-tumor microenvironments. Mol Cell Biol 26(14):5336–5347
88. Tsavachidou-Fenner D, Tannir N, Tamboli P, Liu W, Petillo D, Teh B et al (2010) Gene and protein expression markers of response to combined antiangiogenic and epidermal growth factor targeted therapy in renal cell carcinoma. Ann Oncol 21(8):1599–1606
89. Carretero J, Medina PP, Blanco R, Smit L, Tang M, Roncador G et al (2007) Dysfunctional AMPK activity, signalling through mTOR and survival in response to energetic stress in LKB1-deficient lung cancer. Oncogene 26(11):1616–1625
90. Shaw RJ, Kosmatka M, Bardeesy N, Hurley RL, Witters LA, DePinho RA et al (2004) The tumor suppressor LKB1 kinase directly activates AMP-activated kinase and regulates apoptosis in response to energy stress. Proc Natl Acad Sci USA 101(10):3329–3335
91. Fabian C, Koetz L, Favaro E, Indraccolo S, Mueller-Klieser W, Sattler UG (2012) Protein profiles in human ovarian cancer cell lines correspond to their metabolic activity and to metabolic profiles of respective tumor xenografts. FEBS J 279:882–891

Part II
Stem Cells and Signaling Pathways

Chapter 4
Cancer Stem Cells and Tumor Dormancy

Heiko Enderling

Abstract The cancer stem cell hypothesis postulates that only a subpopulation of cancer cells in a tumor is capable of initiating, sustaining, and reinitiating tumors, while the bulk of the population comprises non-stem cancer cells that lack tumor initiation potential. The interactions of these two phenotypically distinct populations can provoke various nonlinear growth kinetics in the emerging tumor. An environmentally independent, intrinsic dormant state is an inevitable early tumor progression bottleneck within a range of biologically realistic cell kinetic parameters. In certain conditions, cell kinetics can combine to enable escape to tumor progression, yielding morphologically distinct self-metastatic expansion of multiple self-limiting tumor clones.

Keywords Cancer stem cells • Tumor dormancy • Self-metastasis • Agent-based model

Introduction

It has been a long held paradigm that once a cell has transformed into a cancer cell and acquired all required traits [1] it will inevitably proliferate, eventually forming a clinically presenting cancer. Standard treatment options share the philosophy of delivering the maximum tolerable dose to inflict maximum gross tumor reduction. When the tumor shrinks below clinical detection, complete response or complete remission is declared. Often the tumor has been eradicated and treatment indeed was successful; in other cases the tumor will grow back more aggressively than the

H. Enderling, PhD (✉)
Center of Cancer Systems Biology, Steward Research & Specialty Projects Corp.,
St. Elizabeth's Medical Center, Tufts University School of Medicine, 736 Cambridge St.,
Boston, MA 02135, USA.
e-mail: heiko.enderling@tufts.edu

H. Enderling et al. (eds.), *Systems Biology of Tumor Dormancy*,
Advances in Experimental Medicine and Biology 734,
DOI 10.1007/978-1-4614-1445-2_4, © Springer Science+Business Media New York 2013

primary tumor, thus worsening patient prognosis [2, 3]. One potential explanation is that conventional treatment, in general, selects for resistant, aggressive sub-clones in the tumor [4–6], an idea given some support from the tumor-initiating, or "cancer stem cell (CSC)" hypothesis [7–13]. If tumor-initiating cells are the engine of tumor progression [14], then it follows that non-initiating cancer cells may naturally compete with and even hinder tumor-initiating cell dynamics [15]. The extent to which solid tumors remain asymptomatic at small sizes marked by balanced cell proliferation and cell death lends support to this hypothesis [16–19], revealing that cancers can exhibit self-limiting kinetics under generalizable circumstances.

CSCs are often hailed as the "Holy Grail" of cancer research [20], as they are assumed to be the sole initiators and drivers of tumor growth, and thus their eradication may offer targeted tumor treatment without the need to extinguish the entire tumor population. CSCs are not derived from somatic stem cells per se, but nonetheless conceptually exhibit stem-like behavior of symmetric and asymmetric division and possess the unique ability to repopulate a heterogeneous population after deleterious insults. CSCs can be enriched through expression of specific surface markers, for example in breast (CD44+ CD24–/low; [21]), colon (CD133+; [22]), brain (CD133+; [23]), and prostate tumors (CD44+ CD24–; [24]). Serial re-transplantation of populations enriched for CSCs can consistently form tumors using correspondingly fewer numbers of cells per implant. Interestingly, the new tumors mimic the original fraction of cells that express stem cell surface markers [21]. Large numbers of tumor cells not expressing these markers, although apparently transformed, cannot initiate new tumors even in immunocompromised mice [25]. Although contributing in substantial proportion to the bulk tumor, these cells can retard tumor growth—especially in early stages—as they compete with tumor-driving CSCs for space, oxygen, and nutrients [15].

While CSCs have received a lot of attention, the contribution to total tumor growth kinetics of their non-stem counterparts has been less thoroughly investigated, if not ignored outright. If CSCs are the subpopulation in a tumor that drives progression, then how do the non-stem cancer cells contribute to tumor fate? We set out to explore the interactions of CSCs and non-stem cancer cells during early avascular tumor growth. Understanding the interaction of intrinsic cell kinetics that combine to keep tumor populations in a dormant state would clearly be an invaluable augment to current treatment modalities. A worthwhile therapeutic consideration may then be to ask whether a tumor may be controlled through a strategy of chronic containment—i.e., maintaining the tumor in a dormant, nonmalignant state by exploiting existing global tumor kinetics that work to thwart those of the more aggressive subpopulations of the tumor. Clearly, elucidating such a strategy will require a means to identify and track the complicated interactions among the competing subpopulations. To these ends, experimentally validated hybrid mathematical-computational agent-based models have proven to be powerful tools to quickly simulate complex, emerging system dynamics that depend on the behavior of their components and their interactions [26–28]. Before such agent-based models of CSC-initiated tumor growth can be developed, a closer look at the conceptual understanding of CSCs and their non-stem cancer cell counterparts is required.

Cancer Stem Cells and Non-stem Cancer Cells

The CSC hypothesis is a conceptual construct that emerged from a variety of observations. Most assumptions that went into the formulation of the CSC hypothesis are based on individual observations of transient biological information and functional assays, and are yet to be thoroughly validated. As reliable biological markers are still to be identified, purification of CSCs and proper characterization remains elusive. Nevertheless, a conceptual framework about cellular fate has been derived from the following phenomena:

1. A single CSC can initiate and reinitiate a tumor.
2. A single CSC gives rise to a heterogeneous population of CSC and non-stem cancer cells (CC).
3. CC cannot initiate or sustain a tumor.
4. Tumors consisting only of CC will inevitably die out.

The abstract interpretations of these observations are:

(a) CSC are immortal and have infinite proliferation potential.
(b) CSC can divide symmetrically to increase the number of CSCs.
(c) CSC can divide asymmetrically to simultaneously self-renew and produce a CC.
(d) CC only divide symmetrically to produce more CC.
(e) CC have a discrete, finite proliferation potential, and inevitably die.

The different fates and division patterns of CSC and CC are summarized in Fig. 4.1. For a complete understanding of cellular division mechanisms, Fig. 4.1 also includes two additional panels describing symmetric differentiation of CSC (one CSC gives rise to two CC) and dedifferentiation of CC into CSC. Both of these mechanisms are increasingly discussed in the literature in the context of plasticity of cellular fate [29–33]. To rigorously understand the contribution of CSC and CC *phenotypes* to tumor growth dynamics and tumor dormancy, we ignore *cell fate* plasticity and limit our study to the above-described mechanisms.

The division mechanisms and phenotypic differences between CSC and CC introduce two independent parameters. The first parameter is the ratio of symmetric and asymmetric division of CSC. Let p_s denote the probability of symmetric division with $0 \leq p_s \leq 1$, where $p_s = 1$ represents 100% symmetric division. Therefore, the probability of asymmetric division is $1-p_s$. Parameter values of $p_s = 0$ and $p_s = 1$ yield homogeneous populations of CCs and CSCs, respectively. These cases will be discussed separately to set the boundaries of the problem. The second parameter is the proliferation potential of CC or number of mitotic divisions that a cell may undergo before the generational life span is exhausted and subsequent division attempts result in cell death. The progressive reduction in proliferation potential up to cell death offers a quantitative visualization of the "Hayflick limit" [34, 35], a preset lifetime for cells that continuously decrements after mitosis, ostensibly due to the shortening of telomeres, the noncoding replicative protective ends of the DNA

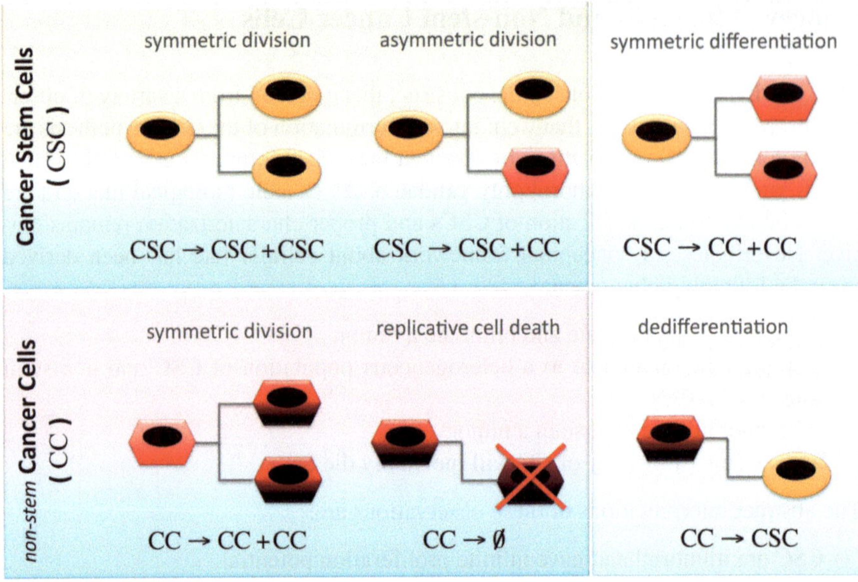

Fig. 4.1 Division fates of cancer stem cells (*yellow ellipse*, *top panel*) and non-stem cancer cells (*red hexagon*, *bottom panel*). Symmetric differentiation of cancer stem cells and dedifferentiation of non-stem cancer cells are shown for completeness only and are omitted in the agent-based model

[36–38]. Let ρ denote the proliferation potential of a CC with $0 \leq \rho \leq \rho_{max}$, where ρ_{max} is the proliferation potential of a daughter CC that arises from an asymmetric CSC division (Fig. 4.2). It is conceivable that once transformed, host cells of different ages and/or from different organs confer a variety of proliferation potentials to their respective progeny [39]. With each division, the proliferation potential of CC decreases and the daughter cells inherit the decremented proliferation potential $\rho-1$. Proliferation attempts at $\rho=0$ result in cell death.

An Agent-Based Model of Cancer Stem Cell-Initiated Tumor Growth

Agent-based modeling (ABM) is a computational methodology to simulate emerging properties of a dynamic system in time and space that has been widely used in a variety of fields such as ecology [40], radiobiology [41], vascular biology [42], or morphogenesis [43, 44]. Such models are well suited to predict the response of complex systems within or out of their equilibrium state. Over the past decade, ABM-based studies have also expanded into the areas of cancer and tumor biology [45–52]. In ABM, the system is represented as a collection of autonomous, decision-making agents, or cells, that have a set of intrinsic state variables and predefined instructions, which determine how they behave and interact with each other and the local environment. Cell properties and rules of interaction are based, when possible,

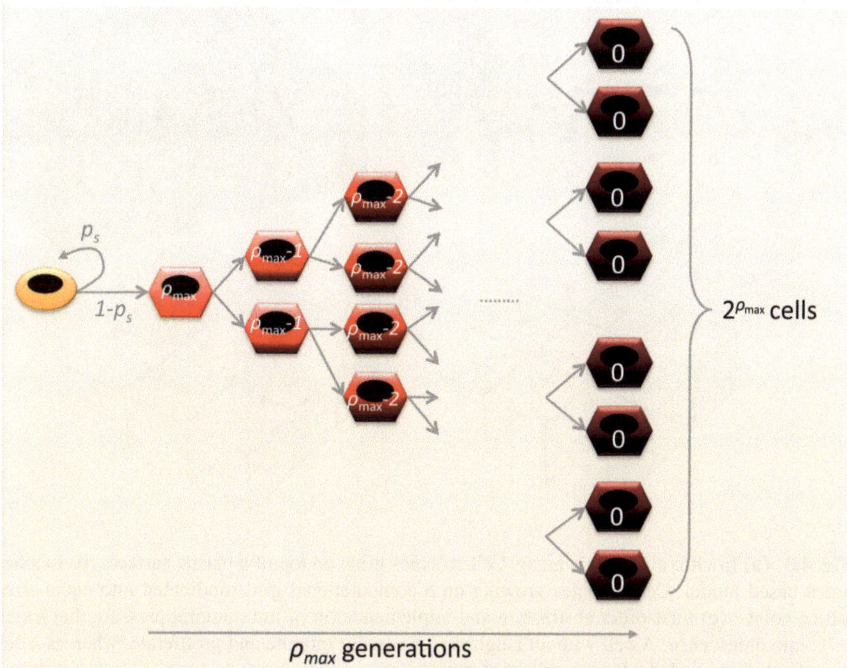

Fig. 4.2 Schematic of the phenotypic hierarchy dependent on proliferative potential ρ_{max}

on empirical observations from in vivo and in vitro biological experiments, with empirically immeasurable phenomena consolidated into hypothesized dynamic parameters. The full details of cells and their interactions as the simulation progresses can be visualized, measured, and perturbed [44]. Simulations of the interaction of multiple cell populations and single cells with one another and with their immediate environment can result in multifaceted population dynamic behaviors such as co-operation or competition [53]. ABM can capture the complex interactive consequences of these dynamics while allowing for the behavioral distinction between the CSC and CC compartments.

In the proposed ABM discussed here, all cells are modeled as individual objects, and the fate of each cell is tracked throughout the simulation. Comparable to many in vitro experiments, the proposed ABM simulates tumor growth in a two-dimensional setup (Fig. 4.3). We introduce a computational lattice with 350×350 grid points, each of which representing (10 μm)^2 to host at the most one cancer cell at any time. Each cell is described by a set of intrinsic properties including proliferation potential ρ, migratory capability μ, and probability of symmetric division p_s. Cells are equipped with a migration rate of $\mu = 150$ μm day^{-1}, traversing one cell width every 96 min. Therefore, simulation time is advanced at small proportional intervals of $\Delta t = 96$ min, reflecting the time it takes for a single cell to migrate to an adjacent lattice point. Cells are assumed to divide on average once per day, and the probability of cell division scaled to the simulation time steps is $1/15 \approx 0.067$ per

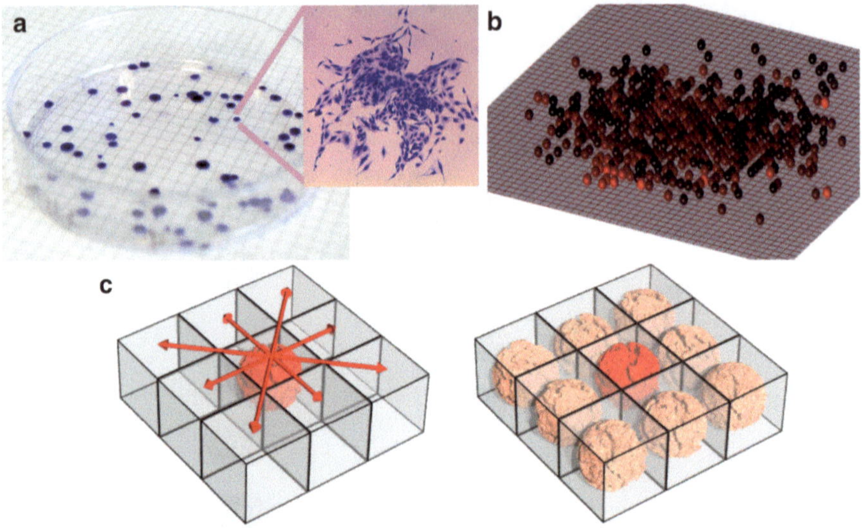

Fig. 4.3 (**a**) In vitro clonogenic assay. Cell colonies grow on top of a plastic surface. (**b**) In silico agent-based model. Cell colonies growing on a computational grid subdivided into equal-sized lattice points. (**c**) First-order abstraction and implementation of intratumoral pressure that forces cells into quiescence. A cell without neighbors can freely migrate and proliferate, whereas a cell completely surrounded adopts a quiescent state

96-min interval. Each cancer cell in the simulation follows these prescribed rules, and emerging tumor population growth dynamics are derived from multiple stochastic simulations. At each time increment, cells are chosen at random and their behavior is updated dependent on their internal state and the local environment. A cell that is completely surrounded by other cells is assumed to be inhibited from migration or proliferation and thus forced into a quiescent state (Fig. 4.3). This is a first-order approximation of the rising physical pressure and subsequent quiescence observed in tumor populations despite sufficient availability of oxygen and nutrients [54, 55]. The simulation framework and decision process for all cells at each time increment is visualized in Fig. 4.4. To understand the kinetics of the individual cell types in the model, simulations are first performed for homogeneous populations of CSC and CC.

A Homogeneous Population of Cancer Stem Cells

We first simulate tumor growth for a homogeneous population of CSCs. At time $t=0$, the simulation is initiated with one CSC in the center of the computational lattice. To ensure a pure CSC population, the probability of symmetric division is set to $p_s=1$. As the simulation progresses, new CSCs are formed and a population develops. Within the growing population, cells occupy lattice points, space becomes limited and cells are forced to become quiescent. Only cells on the outer rim have

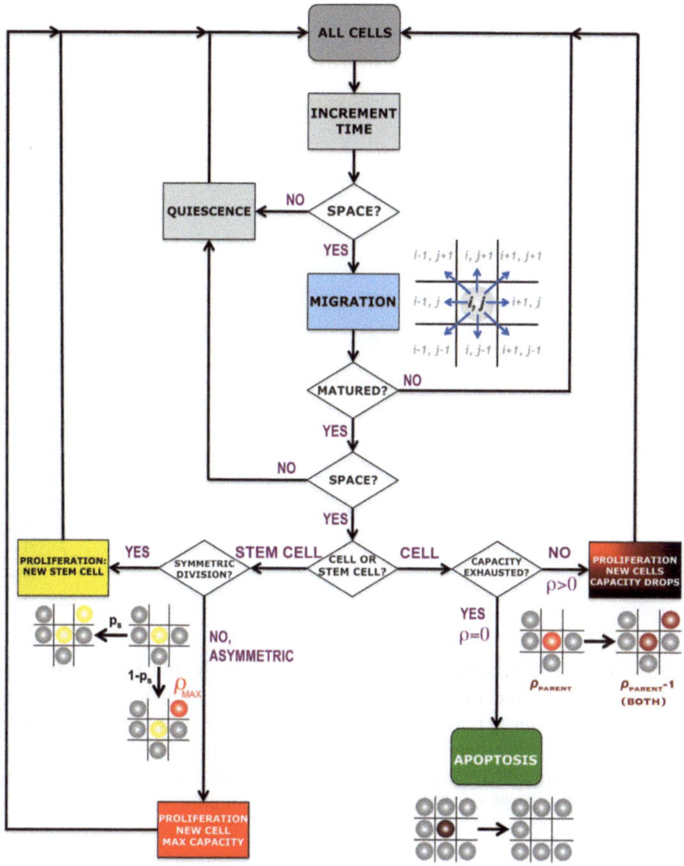

Fig. 4.4 Simulation flowchart for all cell decisions at each time increment (reproduced from [39])

sufficient space to proliferate and/or migrate, and macroscopic expansion is restricted to the outer rim (Fig. 4.5). Under these conditions, the homogeneous population of CSC grows with radial symmetry. As all cells in the population are CSC, any cell can be removed, reseeded, and the observed growth repeats (Fig. 4.5). Due to rising competition for space, intratumoral pressure inhibits proliferation of cells in the interior of the population and thus growth is decremented exponential [56], comparable to early phases of logistic and Gompertz growth [54, 57, 58].

A Homogeneous Population of Non-Stem Cancer Cells and Tumor Dormancy

We now simulate tumor growth for a homogeneous population of CC. Let $\rho_{max} = 20$, arbitrarily chosen. Analogous to homogeneous populations of CSC, the initial tumor expansion is exponential until intratumoral space becomes limited, cells are forced

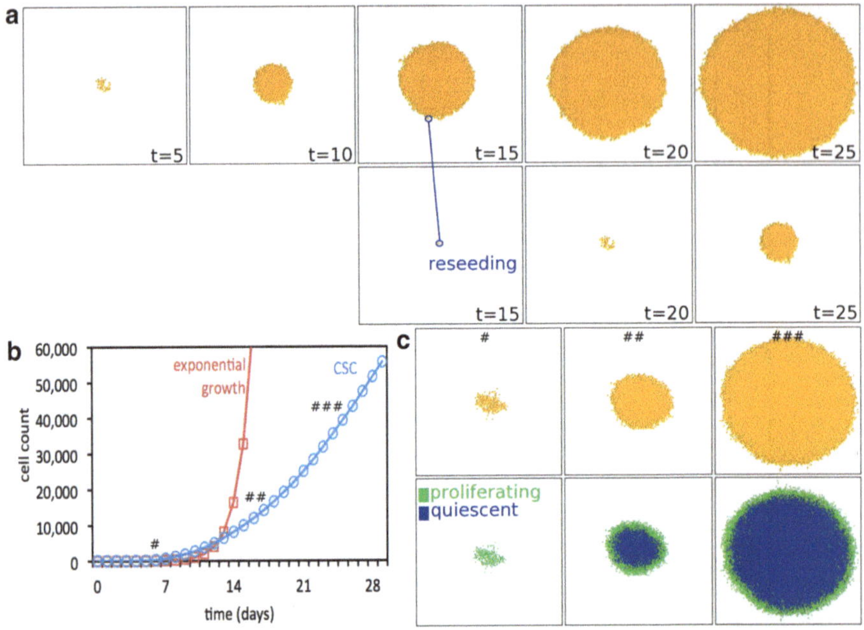

Fig. 4.5 Simulation of colony formation from a single CSC cell with 100% symmetric divisions ($p_s = 1$). (**a**) Representative simulations showing a monotonically growing pure population of CSC. Any cell (shown at $t = 15$) can be removed from the population to seed a new population with identical growth pattern. (**b**) Growth curve of the population shown in (**a**) (*blue circles*) compared to theoretical exponential growth (*red squares*). (**c**) Visualization of the population at three discrete time points marked by the *#, ##, and ###* in (**b**), and differentiation between proliferating (*green*) and quiescent (*blue*) CSC

into quiescence, and proliferation is restricted to cells at the periphery. However, the cells on the periphery decrement their proliferation potential at each division, such that their daughter cells inherit a shortened lifespan. Newly produced cells will be placed towards the population periphery, and thus the cells with the lowest remaining proliferation potential ρ reside at the outer rim. Eventually these cells exhaust their remaining potential, cease proliferation, and die. Net population growth is halted, and a decline in cell number begins (Fig. 4.6). With cell death at the periphery, previously quiescent cells become exposed to space again and reenter the proliferation cycle, until they inevitably exhaust proliferation potential and die. The dynamics of cell death at the periphery and reactivation of interior cells continues, marked by continuously decreasing overall population size due to the progressively declining proliferation potentials of the remaining CC cells. At all times, the tumor can be described as a solid mass of "layers" of cells with decreasing proliferative potentials (visualized by color gradient in Fig. 4.6). The outermost layers are exposed to space and are thus the site of active cell proliferation and death, which accounts for the overall tumor dynamics observed. As layer after layer dies off (due to proliferation

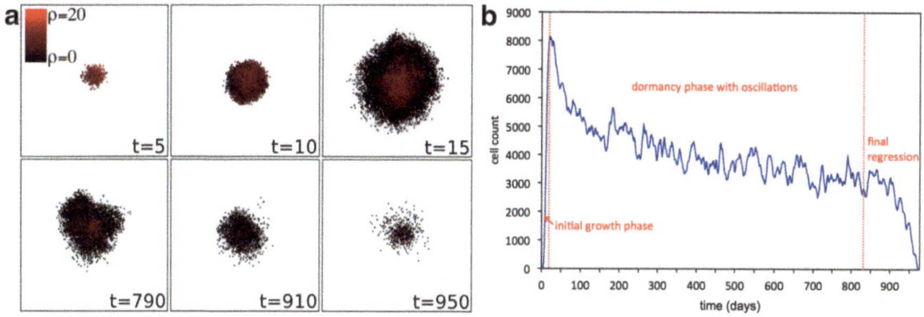

Fig. 4.6 Simulation of colony formation from a single CC cell with $\rho_{max}=20$. (**a**) Simulation snapshots at various time points. (**b**) Temporal evolution of cell counts in a homogeneous population of non-stem cancer cells. An initial growth phase is followed by a decline in number, dormancy phase with oscillations, and final regression

attempts at $\rho=0$), the tumor population, as a whole, exhibits what could be characterized as a protracted period of dormancy, with strong oscillations. Eventually, the last fraction of quiescent cells reenters proliferation, marking the start of the final regression period leading to the complete regression of the tumor population (Fig. 4.6). The maximum size the tumor reaches and the length of the dormancy phase are both dependent on the initial proliferation potential ρ_{max} of the founding cell. Smaller values of ρ_{max} yield smaller tumors and shorter dormancy phases [15], whereas CC with larger ρ_{max} can form tumors that theoretically would grow to sizes above clinical detection thresholds and persist for months or years before a final regression (Fig. 4.7). To grow to a macroscopic size, however, these tumors would need to overcome environmental bottlenecks, such as oxygen and nutrient diffusion limits (angiogenesis-dependent dormancy) and immune surveillance (immune system-dependent dormancy).

This ABM simulation of tumor growth from non-stem cancer cells, albeit naïve and simplistic, may offer an intuitive explanation of two frequently observed phenomena. First, the large frequency of microscopic in situ tumors [18] of the breast, prostate and thyroid revealed at autopsy in non-cancer patients that died of other causes [17, 59] may, in part, be due to transformation events in somatic cells that lack the self-renewal and immortality potential of stem cells. The uncontrolled proliferation of the non-stem cancer cells yields a tumor mass that is self-limiting, can exhibit long phases of dormancy, and will ultimately regress if no further events confer self-renewal and immortality to at least one of the transformed cells.

Second, more recent observations enabled through advances in diagnostic imaging and increased implementation of screening for early detection show that the natural course of some clinically presenting tumors is spontaneous regression [60, 61]. Without CSCs, this simple model suggests that although some tumors have the intrinsic ability to grow beyond screening detection limits, they will ultimately regress. For tumors devoid of CSCs to grow to sizes observable with medical imaging (>1 million cells) they must be initiated with sufficiently large ρ_{max} values; in

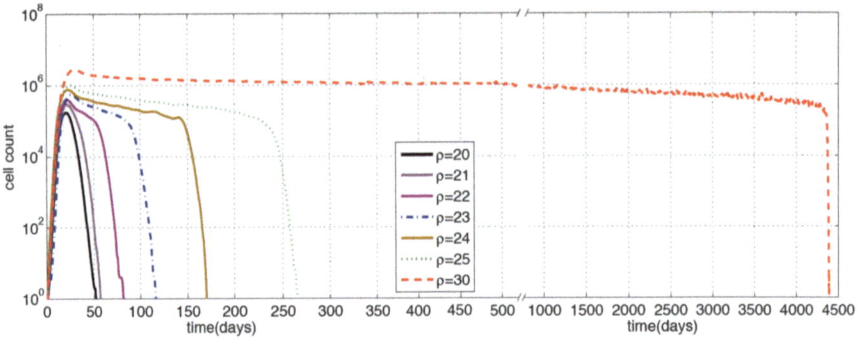

Fig. 4.7 Simulation of tumor growth from a single CC with limited proliferation potential ρ_{max} in a three-dimensional domain of 350×350×350 grid points (≈43 mm^3) to avoid boundary-inflicted growth modulation. With increasing proliferation potential, ρ_{max} tumors grow to larger sizes and exhibit longer dormancy periods until final regression. Oscillations during the dormancy period are not apparent in this plot due to log-scale presentation of cell count

our simulations greater than 25 (Fig. 4.7). The actual proliferative potential of both non-transformed and of cancer cells is not well determined. Seminal studies by Hayflick and Moorhead suggest that human fetal cells divide in culture 40–60 times [34, 35]. Hematopoietic progenitors are estimated to divide between 20 and 30 times [62], whereas progenitors in the colonic crypt may undergo only four to six divisions [63]. These values, however, may further depend on host age [64]. Thorough studies to quantify proliferative potential of cells of different organ, host age, and transformation state are yet to be performed.

Another interpretation of the presented simulation results augments diffusion-limited dormancy and angiogenesis discussions. Although the angiogeneic switch is probably the best-studied bottleneck standing between tumor dormancy and continued tumor progression [16, 57, 65, 66], simulations with our model suggest that not all tumors that acquire angiogenic capability are inevitably fated to progress. Complete regression even after the angiogenic switch is plausible in tumors comprising only non-stem cancer cells that have sufficiently large proliferative potentials to grow beyond diffusion limit.

Tumor Dormancy in Heterogeneous Tumors

Simulations of homogeneous populations of CSCs and CCs revealed different tumor dynamics—monotonic growth for pure CSC populations and growth, dormancy and regression in CC populations, respectively. As the CSC hypothesis proposes tumors are composed of a heterogeneous population of CSC and CC, we now simulate progression of tumors with both subpopulations. A single CSC is initiated at time $t=0$, equipped with a low frequency of symmetric division

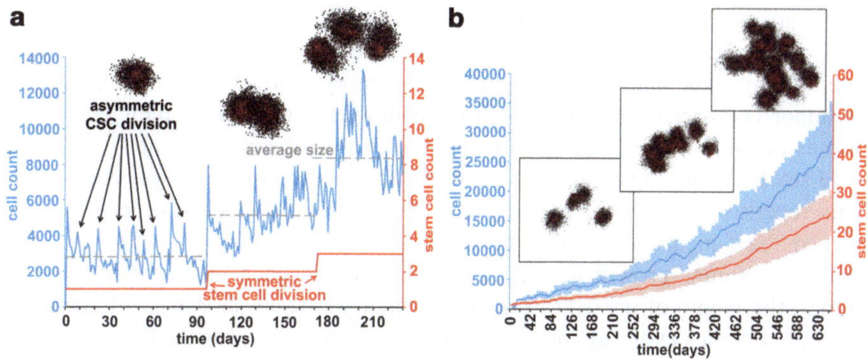

Fig. 4.8 Growth dynamics of a heterogeneous population of cancer (stem) cells. (**a**) Representative simulation result of a slowly progressing tumor with long dormancy periods subject to oscillations. Simulation snapshots (enlarged for visualization purpose) show formation of self-metastases with color code: *yellow*, CSC; *red* to *black* gradient, remaining proliferative potential of CC (*red*=10, *black*=0). (**b**) Simulation of tumor growth to 30,000 cells. Shown are average and standard error for $n=5$ independent simulations, and representative simulation results at $t=200$, 400, and 600. Color code as in (**a**)

$p_s=0.01$ (i.e., 1%; at each CSC division there is a 1% chance of increasing the CSC pool) to reflect the low frequency of CSC reported in the literature [67]. The proliferation potential of all first generation CC, i.e., the CC daughter of an asymmetric CSC division, is designated be $\rho_{max}=10$. The effects of changing the model parameters will be discussed later. With the low probability of symmetric CSC division, the initial offspring of the founding CSC are likely CC, which in turn produce more CC with decremented proliferation potential. As observed in the cases of homogeneous populations, the heterogeneous tumor grows initially exponentially and later decremented exponentially due to the rising competition for space and thus intratumoral quiescence. As the founding CSC is rapidly outnumbered by CC it becomes trapped in the quiescent tumor core. Proliferation is restricted to the outer rim that only consists of CC (Fig. 4.8). With decrementing proliferation potentials in CC, tumor growth is eventually halted analogously to homogeneous CC populations (c.f. Fig. 4.6). Although the tumor contains a CSC with the ability to further tumor growth, the non-stem CC spatially inhibit the CSC and thus negatively modulate tumor progression kinetics. In this CSC/CS simulation, growth of the tumor is seen to be inhibited by its own mass, consistent with empirical and theoretical findings [57, 68] (although in those studies inhibition was proposed to be due to the increasing levels of tumor-inhibitory factors by additional tumor mass rather than cell crowding and spatial inhibition). Again, these initial conditions give rise to a dormancy phase subject to oscillations resulting from cell death at the periphery and the consequential activation of previously quiescent cells. These dynamics repeat until most CC exhaust their proliferation potential and vacate the space they occupied. Eventually, the CSC in the very core of the population can reenter proliferation. If the resulting daughter cell(s) are again CC, tumor growth dynamics repeat. If the CSC divides symmetrically, a second CSC is born. As the simulation progresses, the two CSC can separate via migration and eventually form two nearby spatially distinct populations of tumor cells—each of which is limited in size and entering their respective phases of dormancy. The two subpopulations are subject to the above kinetics until stochastically a new CSC is born and the seeding of a third cluster begins. This repetitive seeding of new clusters in the vicinity of each other has been termed "self-metastatic tumor progression" by us and others [58, 69, 70]. Collective tumor growth accelerates, and although each cluster exhibits dormancy, the length of the overall tumor dormancy periods shorten, with the macroscopic kinetics evolving toward continuous population growth (Fig. 4.8).

Both the emergence from and the length of the tumor dormancy period in heterogeneous populations of CSC and CC are dependent on the interplay of cell-intrinsic parameters. Rigorous computational exploration of the three-dimensional parameter space (cell migration speed μ, probability of symmetric CSC division p_s, and non-stem cancer cell proliferation potential ρ_{max}) yields the following macroscopic conclusions:

1. Population kinetics that inhibit CSC proliferation yield prolonged periods of tumor dormancy and slow tumor progression.
2. Population kinetics that allow for CSC proliferation yield shorter periods of tumor dormancy and faster tumor progression.

On the cellular level, increasing cell migration speed (μ) loosens intratumoral spatial competition, facilitating more frequent CSC division and thus accelerated tumor expansion [69]. Increasing the frequency of symmetric CSC division (p_s) yields shorter dormancy periods and faster self-metastatic expansion [71]. In general, any increase in CC proliferation potential (ρ_{max}) produces larger populations of CC that impede CSC proliferation and thus prolongs tumor dormancy and negatively modulates tumor progression [15]. Reducing ρ_{max}, however, only accelerates tumor progression in certain cases. While a lowered proliferation potential of CC reduces the spatial barrier to symmetric expansion of CSC, a proliferation potential that is too low prevents a significant contribution of CC to overall cell count, and tumor growth becomes increasingly dependent on the comparatively rare symmetric expansion of CSC. An optimal value for ρ_{max} balancing spatial availability with CC persistence can be derived from stochastic simulations and can be shown to be crucially dependent on other model parameters [39].

The complex interplay of CSC and CC yields unexpected emergent growth dynamics of heterogeneous tumors that depends on the interactions of both subpopulations. While CSC are the necessary "engine of tumor progression" [14] tumor growth is surprisingly also modulated by their non-stem CC counterparts. Intrinsic cell kinetics often combine such that dormancy is a predominant, early phase in tumor progression, and escape from dormancy to tumor progression con only be observed in narrow areas of the parameter space (Fig. 4.9) [15].

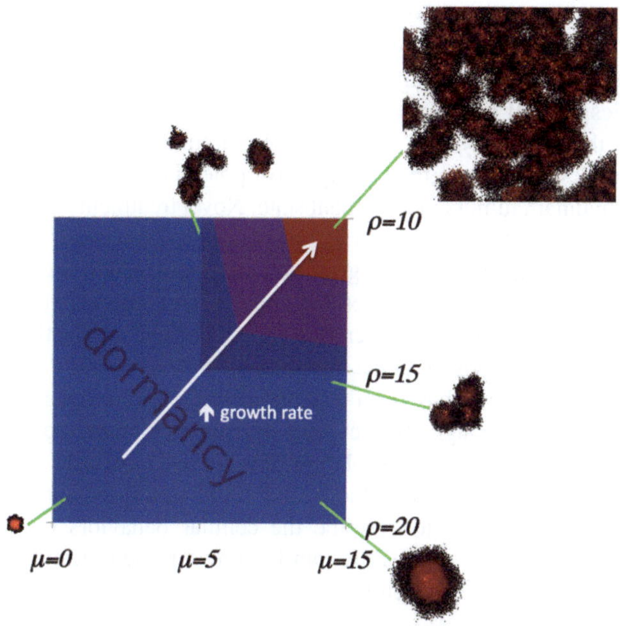

Fig. 4.9 Exploration of parameter space (here migration speed μ and non-stem cancer cell proliferative potential ρ_{max}) reveals large areas of very flat tumor growth rates interpreted as tumor dormancy (increasing growth rate color-coded as heat map from *blue* to *red*). Only when cell kinetics combine in a specific manner escape from dormancy and fast tumor growth can be observed. Representative simulation snapshots are shown

Discussion

The CSC hypothesis is an attractive conceptual construct of tumorigenesis. The interplay of CSC with their non-stem cancer cell (CC) counterparts, however, is not intuitive but can be computationally analyzed utilizing an ABM approach. Simulations of the model revealed novel insights into tumor progression and can describe dormancy during early growth—i.e., before nutrient diffusion limitations and initiation of angiogenesis. These results augment the current understanding of tumor progression, specifically tumor progression within the CSC hypothesis.

The model presented further suggests that populations devoid of CSC can form tumors of sizes equal to or even greater than the oxygen diffusion limit [72, 73], and thus will require angiogenic capacity to further their growth. If successful, however, these tumors will cease expansion at a nontrivial plateau, enter a dormancy phase marked by oscillations, and inevitably regress if no further transformation event confers stem cell properties to the system. Tumors without CSC can grow up to and well beyond sizes detectable by diagnostic imaging. The natural course of their progression, however, is complete regression, which offers an intuitive explanation to observations of this phenomenon in neuroblastoma and breast cancer [60, 61].

The remarkably large number of solid tumors that remain asymptomatic at small sizes [17, 18] and the observation of accelerated repopulation after cytotoxic treatment suggest that CC may naturally compete with and even hinder tumor-initiating cell dynamics [15]. If CC are sensitive to conventional cytotoxic treatments, it is conceivable that they may also be susceptible to other agents. Such agents might alter CC kinetics to improve their competitive potential with CSC to thwart tumor growth and maintain tumors in a dormant state. Novel treatment approaches include migration inhibitors [58, 74] as well as pro-senescence treatment [75] that disable cell proliferation while avoiding the gross cell kill implicated in accelerated repopulation of disease [76, 77] and progression along a more aggressive trajectory [78]. The concept of disease control, rather than comprehensive eradication, is becoming an increasingly popular vision in the literature [79].

The presented study is based on a minimal set of biological assumptions and kinetic parameters. Although abstract, such an approach offers a platform to quantitatively explore the cause–effect relationships between intrinsic cell mechanisms and macroscopic tumor growth. Due to the effort to implement mechanistically nonoverlapping parameters to describe the cellular behaviors of the composite tumor population, robust conclusions can be drawn to help design future experiments and identify novel therapeutic targets and strategies.

Acknowledgments This project was supported by the AACR Centennial Postdoctoral Fellowship in Cancer Research 08-40-02-ENDE and the National Cancer Institute under Award Number U54CA149233 (to L. Hlatky). The content is solely the responsibility of the author and does not necessarily represent the official views of the National Cancer Institute or the National Institutes of Health.

I would like to thank Lynn Hlatky and Philip Hahnfeldt for their mentorship, vision, and contribution to the work discussed and reviewed in this chapter. I would also like to thank Jan Poleszczuk for help with Fig. 4.7, and Charles Morton for critical reading of the manuscript.

References

1. Hanahan D, Weinberg RA (2000) The hallmarks of cancer. Cell 100(1):57–70
2. Withers HR, Taylor JM, Maciejewski B (1988) The hazard of accelerated tumor clonogen repopulation during radiotherapy. Acta Oncol 27(2):131–146
3. Hanin L (2011) Why victory in the war on cancer remains elusive: biomedical hypotheses and mathematical models. Cancers 3(1):340–367
4. Bese NS, Hendry J, Jeremic B (2007) Effects of prolongation of overall treatment time due to unplanned interruptions during radiotherapy of different tumor sites and practical methods for compensation. Int J Radiat Oncol Biol Phys 68(3):654–661
5. Kummermehr JC (2001) Tumour stem cells—the evidence and the ambiguity. Acta Oncol 40(8):981–988
6. Phillips TM, McBride WH, Pajonk F (2006) The response of CD24(–/low)/CD44+ breast cancer-initiating cells to radiation. J Natl Cancer Inst 98(24):1777–1785
7. Allan AL, Vantyghem SA, Tuck AB, Chambers AF (2006) Tumor dormancy and cancer stem cells: implications for the biology and treatment of breast cancer metastasis. Breast Dis 26:87–98
8. Bjerkvig R, Tysnes BB, Aboody KS, Najbauer J, Terzis AJA (2005) Opinion: the origin of the cancer stem cell: current controversies and new insights. Nat Rev Cancer 5(11):899–904

9. Dalerba P, Cho RW, Clarke MF (2007) Cancer stem cells: models and concepts. Annu Rev Med 58:267–284
10. Hill RP (2006) Identifying cancer stem cells in solid tumors: case not proven. Cancer Res 66(4):1891–1896
11. Reya T, Morrison SJ, Clarke MF, Weissman IL (2001) Stem cells, cancer, and cancer stem cells. Nature 414(6859):105–111
12. Takebe N, Ivy SP (2010) Controversies in cancer stem cells: targeting embryonic signaling pathways. Clin Cancer Res 16(12):3106–3112
13. Wicha MS, Liu S, Dontu G (2006) Cancer stem cells: an old idea—a paradigm shift. Cancer Res 66(4):1883–1890, discussion 1895–1896
14. Solé RV, Rodríguez-Caso C, Deisboeck TS, Saldaña J (2008) Cancer stem cells as the engine of unstable tumor progression. J Theor Biol 253(4):629–637
15. Enderling H, Anderson ARA, Chaplain MAJ, Beheshti A, Hlatky L, Hahnfeldt P (2009) Paradoxical dependencies of tumor dormancy and progression on basic cell kinetics. Cancer Res 69(22):8814–8821
16. Almog N, Ma L, Raychowdhury R, Schwager C, Erber R, Short S et al (2009) Transcriptional switch of dormant tumors to fast-growing angiogenic phenotype. Cancer Res 69(3):836–844
17. Black WC, Welch HG (1993) Advances in diagnostic imaging and overestimations of disease prevalence and the benefits of therapy. N Engl J Med 328(17):1237–1243
18. Folkman J, Kalluri R (2004) Cancer without disease. Nature 427(6977):787
19. Weinberg R (2008) The many faces of tumor dormancy. APMIS 116:548–551
20. Misaghian N, Ligresti G, Steelman LS, Bertrand FE, Bäsecke J, Libra M et al (2009) Targeting the leukemic stem cell: the Holy Grail of leukemia therapy. Leukemia 23(1):25–42
21. Al-Hajj M, Wicha MS, Benito-Hernandez A, Morrison SJ, Clarke MF (2003) Prospective identification of tumorigenic breast cancer cells. Proc Natl Acad Sci USA 100(7): 3983–3988
22. Cammareri P, Lombardo Y, Francipane MG, Bonventre S, Todaro M, Stassi G (2008) Isolation and culture of colon cancer stem cells. Methods Cell Biol 86:311–324
23. Singh SK, Clarke ID, Terasaki M, Bonn VE, Hawkins C, Squire J et al (2003) Identification of a cancer stem cell in human brain tumors. Cancer Res 63(18):5821–5828
24. Hurt EM, Kawasaki BT, Klarmann GJ, Thomas SB, Farrar WL (2008) CD44+ CD24(−) prostate cells are early cancer progenitor/stem cells that provide a model for patients with poor prognosis. Br J Cancer 98(4):756–765
25. Shay JW, Wright WE (2010) Telomeres and telomerase in normal and cancer stem cells. FEBS Lett 584(17):3819–3825
26. Anderson ARA, Chaplain MAJ, Rejniak KA (2009) Single-cell-based models in biology and medicine. Birkhauser Basel, Boston, Berlin, p 1–355
27. Bonabeau E (2002) Agent-based modeling: methods and techniques for simulating human systems. Proc Natl Acad Sci USA 99(suppl 3):7280–7287
28. Deutsch A, Dormann S (2005) Cellular automaton modeling of biological pattern formation: characterization, applications, and analysis. Birkhäuser, Boston
29. Alison MR, Lovell MJ, Direkze NC, Wright NA, Poulsom R (2006) Stem cell plasticity and tumour formation. Eur J Cancer 42(9):1247–1256
30. Gupta PB, Fillmore CM, Jiang G, Shapira SD, Tao K, Kuperwasser C et al (2011) Stochastic state transitions give rise to phenotypic equilibrium in populations of cancer cells. Cell 146(4):633–644
31. Leder K, Holland EC, Michor F (2010) The therapeutic implications of plasticity of the cancer stem cell phenotype. PLoS One 5(12):e14366
32. Luo G, Long J, Zhang B, Liu C, Xu J, Yu X et al (2010) Developmental plasticity of stem cells and diseases. Med Hypotheses 75(6):507–510
33. Rapp UR, Ceteci F, Schreck R (2008) Oncogene-induced plasticity and cancer stem cells. Cell Cycle 7(1):45–51
34. Hayflick L, Moorhead PS (1961) The serial cultivation of human diploid cell strains. Exp Cell Res 25:585–621

35. Hayflick L (1965) The limited in vitro lifetime of human diploid cell strains. Exp Cell Res 37:614–636
36. Blackburn EH, Gall JG (1978) A tandemly repeated sequence at the termini of the extrachromosomal ribosomal RNA genes in Tetrahymena. J Mol Biol 120(1):33–53
37. Greider CW, Blackburn EH (1985) Identification of a specific telomere terminal transferase activity in Tetrahymena extracts. Cell 43(2 Pt 1):405–413
38. Olovnikov AM (1973) A theory of marginotomy. The incomplete copying of template margin in enzymic synthesis of polynucleotides and biological significance of the phenomenon. J Theor Biol 41(1):181–190
39. Morton CI, Hlatky L, Hahnfeldt P, Enderling H (2011) Non-stem cancer cell kinetics modulate solid tumor progression. Theor Biol Med Model 8(1):48
40. Robinson EJH, Ratnieks FLW, Holcombe M (2008) An agent-based model to investigate the roles of attractive and repellent pheromones in ant decision making during foraging. J Theor Biol 255(2):250–258
41. Mukhopadhyay R, Costes SV, Bazarov AV, Hines WC, Barcellos-Hoff MH, Yaswen P (2010) Promotion of variant human mammary epithelial cell outgrowth by ionizing radiation: an agent-based model supported by in vitro studies. Breast Cancer Res 12(1):R11
42. Qutub AA, Popel AS (2009) Elongation, proliferation & migration differentiate endothelial cell phenotypes and determine capillary sprouting. BMC Syst Biol 3:13
43. Setty Y, Cohen IR, Dor Y, Harel D (2008) Four-dimensional realistic modeling of pancreatic organogenesis. Proc Natl Acad Sci USA 105(51):20374–20379
44. Tang J, Enderling H, Becker-Weimann S, Pham C, Polyzos A, Chen C-Y et al (2011) Phenotypic transition maps of 3D breast acini obtained by imaging-guided agent-based modeling. Integr Biol (Camb) 3(4):408–421
45. Alarcon T, Byrne HM, Maini PK (2003) A cellular automaton model for tumour growth in inhomogeneous environment. J Theor Biol 225(2):257–274
46. Anderson ARA, Rejniak KA, Gerlee P, Quaranta V (2009) Microenvironment driven invasion: a multiscale multimodel investigation. J Math Biol 58(4–5):579–624
47. Basanta D, Simon M, Hatzikirou H, Deutsch A (2008) Evolutionary game theory elucidates the role of glycolysis in glioma progression and invasion. Cell Prolif 41(6):980–987
48. Enderling H, Alexander NR, Clark ES, Branch KM, Estrada L, Crooke C et al (2008) Dependence of invadopodia function on collagen fiber spacing and cross-linking: computational modeling and experimental evidence. Biophys J 95(5):2203–2218
49. Gerlee P, Anderson ARA (2007) An evolutionary hybrid cellular automaton model of solid tumour growth. J Theor Biol 246(4):583–603
50. Piotrowska MJ, Angus SD (2009) A quantitative cellular automaton model of in vitro multicellular spheroid tumour growth. J Theor Biol 258(2):165–178
51. Rejniak KA, Anderson ARA (2008) A computational study of the development of epithelial acini: I. Sufficient conditions for the formation of a hollow structure. Bull Math Biol 70(3):677–712
52. Wang Z, Zhang L, Sagotsky JA, Deisboeck TS (2007) Simulating non-small cell lung cancer with a multiscale agent-based model. Theor Biol Med Model 4(1):50
53. Power C (2009) A spatial agent-based model of N-Person Prisoner's dilemma cooperation in a socio-geographic community. J Artif Soc Soc 12. http://jasss.soc.surrey.ac.uk/12/1/8/citation.html
54. Brú A, Albertos S, Luis Subiza J, García-Asenjo JL, Brú I (2003) The universal dynamics of tumor growth. Biophys J 85(5):2948–2961
55. Galle J, Hoffmann M, Aust G (2009) From single cells to tissue architecture-a bottom-up approach to modelling the spatio-temporal organisation of complex multi-cellular systems. J Math Biol 58(1–2):261–283
56. Norton L (2008) Cancer stem cells, self-seeding, and decremented exponential growth: theoretical and clinical implications. Breast Dis 29:27–36
57. Hahnfeldt P, Panigrahy D, Folkman J, Hlatky L (1999) Tumor development under angiogenic signaling: a dynamical theory of tumor growth, treatment response, and postvascular dormancy. Cancer Res 59:4770–4775

58. Norton L (2005) Conceptual and practical implications of breast tissue geometry: toward a more effective, less toxic therapy. Oncologist 10(6):370–381
59. Welch HG, Black WC (2010) Overdiagnosis in cancer. J Natl Cancer Inst 102(9):605–613
60. Yamamoto K, Hanada R, Kikuchi A, Ichikawa M (1998) Spontaneous regression of localized neuroblastoma detected by mass screening. J Clin Oncol 16:1265–1269
61. Zahl P-H, Maehlen J, Welch HG (2008) The natural history of invasive breast cancers detected by screening mammography. Arch Intern Med 168(21):2311–2316
62. Bernard S, Bélair J, Mackey M (2003) Oscillations in cyclical neutropenia: new evidence based on mathematical modeling. J Theor Biol 223:283–298
63. Potten CS, Booth C, Hargreaves D (2003) The small intestine as a model for evaluating adult tissue stem cell drug targets. Cell Prolif 36(3):115–129
64. Sharpless NE, DePinho RA (2004) Telomeres, stem cells, senescence, and cancer. J Clin Invest 113(2):160–168
65. Folkman J (1971) Tumor angiogenesis: therapeutic implications. New Engl J Med 285(21): 1182–1186
66. Naumov GN, Bender E, Zurakowski D, Kang S-Y, Sampson D, Flynn E et al (2006) A model of human tumor dormancy: an angiogenic switch from the nonangiogenic phenotype. J Natl Cancer Inst 98(5):316–325
67. Visvader JE, Lindeman GJ (2008) Cancer stem cells in solid tumours: accumulating evidence and unresolved questions. Nat Rev Cancer 8(10):755–768
68. Prehn RT (1991) The inhibition of tumor growth by tumor mass. Cancer Res 51(1):2–4
69. Enderling H, Hlatky L, Hahnfeldt P (2009) Migration rules: tumours are conglomerates of self-metastases. Br J Cancer 100(12):1917–1925
70. Enderling H, Hlatky L, Hahnfeldt P (2010) Tumor morphological evolution: directed migration and gain and loss of the self-metastatic phenotype. Biol Direct 5:23
71. Enderling H, Park D, Hlatky L, Hahnfeldt P (2009) The importance of spatial distribution of stemness and proliferation state in determining tumor radioresponse. Math Model Nat Phenom 4(3):117–133
72. Thomlinson RH, Gray LH (1955) The histological structure of some human lung cancers and the possible implications for radiotherapy. Br J Cancer 9(4):539–549
73. Folkman J, Hochberg M (1973) Self-regulation of growth in three dimensions. J Exp Med 138(4):745–753
74. Basanta D, Hatzikirou H, Deutsch A (2008) Studying the emergence of invasiveness in tumours using game theory. Eur Phys J B 63(3):393–397
75. Nardella C, Clohessy JG, Alimonti A, Pandolfi PP (2011) Pro-senescence therapy for cancer treatment. Nat Rev Cancer 11(7):503–511
76. Roberts SA, Hendry JH (1993) The delay before onset of accelerated tumour cell repopulation during radiotherapy: a direct maximum-likelihood analysis of a collection of worldwide tumour-control data. Radiother Oncol 29(1):69–74
77. Schmidt-Ullrich RK, Contessa JN, Dent P, Mikkelsen RB, Valerie K, Reardon DB et al (1999) Molecular mechanisms of radiation-induced accelerated repopulation. Radiat Oncol Investig 7(6):321–330
78. Enderling H, Chaplain MAJ, Hahnfeldt P (2010) Quantitative modeling of tumor dynamics and radiotherapy. Acta Biotheor 58(4):341–353
79. Gatenby RA (2009) A change of strategy in the war on cancer. Nature 459(7246):508–509

Chapter 5
Regulation of Tumor Cell Dormancy by Tissue Microenvironments and Autophagy

Maria Soledad Sosa, Paloma Bragado, Jayanta Debnath, and Julio A. Aguirre-Ghiso

Abstract The development of metastasis is the major cause of death in cancer patients. In certain instances, this occurs shortly after primary tumor detection and treatment, indicating these lesions were already expanding at the moment of diagnosis or initiated exponential growth shortly after. However, in many types of cancer, patients succumb to metastatic disease years and sometimes decades after being treated for a primary tumor. This has led to the notion that in these patients residual disease may remain in a dormant state. Tumor cell dormancy is a poorly understood phase of cancer progression and only recently have its underlying molecular mechanisms started to be revealed. Important questions that remain to be elucidated include not only which mechanisms prevent residual disease from proliferating but also which mechanisms critically maintain the long-term survival of these disseminated residual cells. Herein, we review recent evidence in support of genetic and epigenetic mechanisms driving dormancy. We also explore how therapy may cause the onset of dormancy in the surviving fraction of cells after treatment and how autophagy may be a mechanism that maintains the residual cells that are viable for prolonged periods.

Keywords Quiescence • Minimal residual disease • Cellular stress • p38 • MAPK • Metastasis

M.S. Sosa, PhD • P. Bragado, PhD • J.A. Aguirre-Ghiso, PhD (✉)
Departments of Medicine and Otolaryngology, Tisch Cancer Institute, Black Family Stem Cell Institute, Mount Sinai School of Medicine, New York, NY, USA
e-mail: julio.aguirre-ghiso@mssm.edu

J. Debnath, MD
Department of Pathology, Helen Diller Family Comprehensive Cancer Center, University of California San Francisco, San Francisco, CA, USA

H. Enderling et al. (eds.), *Systems Biology of Tumor Dormancy*,
Advances in Experimental Medicine and Biology 734,
DOI 10.1007/978-1-4614-1445-2_5, © Springer Science+Business Media New York 2013

Introduction

Metastasis is responsible for the majority of cancer-related deaths. However, our understanding of this complex process is incomplete, which limits our opportunities to prevent metastatic development. There are several fundamental questions that remain mostly unanswered in this field: How does early dissemination contribute to a dormant cell population and what are the underlying mechanisms? How does the tumor microenvironment aid this process? Are primary tumor niches responsible for programming disseminated tumor cells (DTCs) to grow or enter quiescence in target organs? What role does the microenvironment of the target organ play in determining the timing or extent of DTC dormancy?

The "seed and soil" theory of metastasis proposes that a natural match exists between the DTCs (the seeds) and the target organ (the soil) in which they can grow into overt lesions [1]. This theory is derived from the relatively predictable pattern of target organ metastasis that depends on the tissue origin of the primary tumor. However, it remains difficult to predict the timing of metastasis because, even in those sites propitious for growth, it can take years to decades for metastases to develop [1]. Indeed, certain cancers, such as breast carcinoma and melanoma, are well known for their propensity to relapse after a long disease-free period, often decades after initial diagnosis and treatment of the primary tumor. Moreover, it has been proposed that these long periods of asymptomatic disease are due to minimal residual disease (MRD), because DTCs enter a nonproductive or dormant state [1, 2].

In cancer patients, DTCs can be found in sites where they typically form secondary lesions, as well as in sites where they rarely do [1]. Thus, despite being able to disseminate, these DTCs are presumably "growth-suppressed" by the microenvironments of certain organs. Insights into these mechanisms should lead to the identification of novel biomarkers that indicate whether patients harbor dormant disease, and should uncover new signaling pathways that can be modulated to either maintain the dormancy of DTCs or eliminate them entirely by blocking critical survival pathways.

To date, several mechanisms have been proposed to explain clinical dormancy (i.e., asymptomatic disease) in cancers. The lack of proliferation markers in surviving DTCs obtained from patients and findings from experimental studies suggest that solitary DTC dormancy may be controlled by mechanisms of quiescence [1], a reversible growth arrest that can be induced by different signals [3]. Angiogenic dormancy or immune system-mediated tumor mass dormancy may also be responsible for maintaining the dormancy of residual disease [4, 5] (see Almog and Quesnel chapters in this book).

The primary tumor and target organ microenvironments are intimately interconnected by the biology of DTCs (Fig. 5.1). Three potential scenarios that relate to this concept may explain DTC dormancy. First, DTCs from invasive cancers activate stress signals in response to the dissemination process and/or due to a growth-suppressive microenvironment of the target organ (see "The Target Organ Microenvironment and DTC Dormancy" section for examples of such microenvironments and their

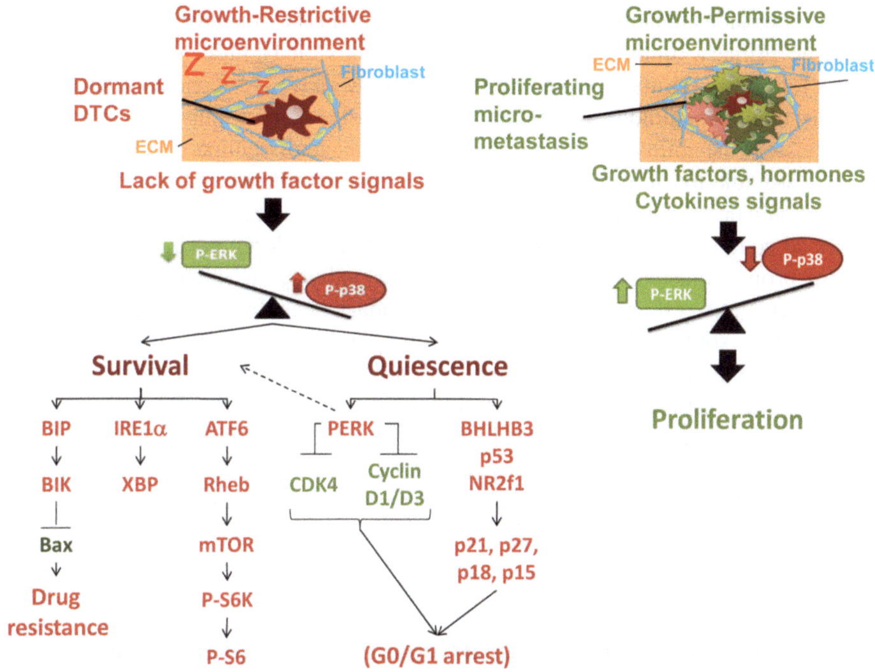

Fig. 5.1 Upon arrival at secondary sites, the crosstalk between the DTCs and the new microenvironment will determine the fate of the DTCs: In a permissive microenvironment (*right*), such as the lungs, interactions with the extracellular matrix (ECM), and stromal cells of the favorable microenvironment will allow DTCs to adapt and integrate growth-promoting signals, such as those derived from fibronectin, which will result in activation of mitogenic signaling (high ERK/low p38 ratio), thereby promoting DTC proliferation and the formation of micrometastasis. On the contrary, in restrictive microenvironments (*left*) such as bone marrow or liver for some cancers, either the loss of surface receptors or the interaction with non-growth-permissive ligands will result in activation of stress signaling (low ERK/high p38 ratio) that will induce both quiescence and survival signals, which will in turn lead to a prolonged phase of dormancy. Activation of p38 induces a G_0–G_1 arrest that is partly mediated by transcriptional activation of BHLHB3, NR2F1, and p53, which control the expression of different regulators of the cell cycle, such as p21, p27, p15, and p18, which mediate tumor cell growth arrest. Furthermore, active p38α induces an ER-stress response that coordinates growth arrest and survival through the activation of PERK, IRE-1, and ATF6. PERK contributes to both quiescence and survival of DTCs. Upon activation, PERK induces phosphorylation of EIF2α and attenuation of translation initiation, which leads to downregulation of cyclin D1/D3 and CDK4 and to the induction of quiescence. On the other hand, the other arms of the ER-stress pathways, ATF6α and IRE1α, contribute to DTC dormancy by promoting survival. IRE1α activation leads to the induction of XBP and the activation of the transcription of survival genes, whereas activation of ATF6α induces survival through the upregulation of Rheb and activation of mTOR signaling, allowing DTCs to adapt to the in vivo microenvironment. In addition to this, as part of the ER-stress response, the chaperone BiP/Grp78 is also activated, and this leads to inhibition of Bax activation to prevent apoptosis and thus, promote survival and drug resistance

components), ultimately leading to induction of dormancy [1]. Second, therapy and/or microenvironmental stress conditions (e.g., hypoxia, reactive oxygen species) acting on tumor cells in the primary lesion endow these tumor cells with specific gene expression signatures that prime newly formed DTCs to enter dormancy. Here, specific primary tumor "stress microenvironments" may influence the DTCs to enter long-term dormancy when the cells initially arrive at secondary sites. Third, lesions that are pathologically defined as noninvasive carry a subpopulation of cells that possess the ability to undergo micro-invasion and disseminate. Although these DTCs are able to intravasate into and extravasate out of the systemic circulation, they remain unfit for expansion in secondary sites. Nonetheless, they can survive in an arrested state over an extended period and perhaps undergo occasional cell divisions, progressing via epigenetic and genetic pathways to eventually become a fully metastatic cell able to grow at the secondary sites. In this chapter, we focus both on how solitary DTC fate is influenced by tumor–host interactions occurring in primary tumors and target organs, and on how autophagy may serve as a cell-autonomous survival function in residual disease (Fig. 5.2). We propose that DTCs undergo dormancy to survive specific stressful microenvironments (see section "The Target Organ Microenvironment and DTC Dormancy") and, therefore, that blockade of the survival signals in dormant cells will ultimately lead to their eradication.

Early Dissemination as a Contributing Factor to Dormancy and MRD

The present paradigm proposes that metastases arise from rare clones that evolve in the primary tumor and acquire characteristics that allow them to disseminate and grow in secondary sites [6, 7]. This somewhat linear model motivates the prediction that tumor cells will emerge with metastatic capacity only if they are derived from evolutionarily "late-progressed" tumors (i.e., those with multiple malignancy-associated genetic alterations). It also suggests that tumor cells endowed with metastatic capacity should be absent or infrequent in patients carrying premalignant/invasive lesions (with fewer genetic alterations, see below) [6, 7]. However, a major challenge to this theory was posed by a series of studies in breast cancer from the Klein lab, which suggested that dissemination had already occurred in lesions that were considered to develop "early" in tumor progression and were pathologically defined as noninvasive, such as atypical ductal hyperplasia (ADH) and ductal carcinoma in situ (DCIS) [6, 8–11]. These studies also indicated that the pause observed in the progression of early DTCs may be due to a "lead time" [6, 8–11]. This refers to the time when DTCs with a limited number of genetic alterations are able to survive, but are unable to efficiently proliferate to gain additional mutations that would favor growth ectopically. This is supported by the fact that genetic alterations in DTCs detected in patients with ADH or DCIS are very heterogeneous [12]. In contrast, genetic anomalies in DTCs from patients carrying diagnosed metastatic disease are significantly more homogeneous, suggesting that certain genetic traits are selected for active expansion in the secondary site and that the original heterogeneity in DTCs is reduced [6, 8–12].

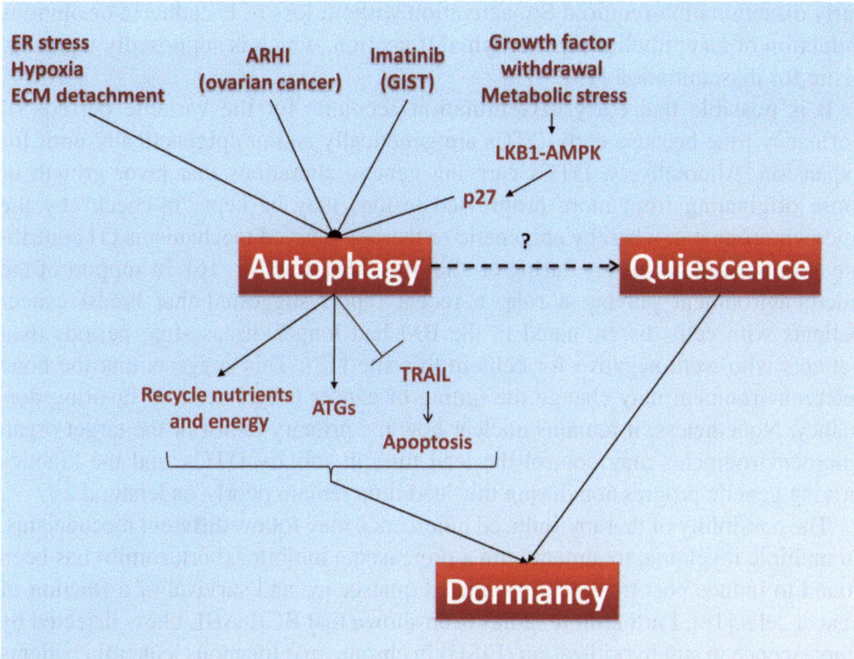

Fig. 5.2 Activation of autophagy in response to different forms of stress can promote survival during growth arrest, making autophagy a component of dormant tumor cell survival. Autophagy is a stress-response mechanism that can be activated in response to various microenvironmental stresses such as hypoxia, extracellular matrix (ECM) detachment, endoplasmic reticulum (ER) stress, growth factor withdrawal, metabolic stress, activation of tumor suppressor genes (aplasia Ras homolog member I [*ARHI*]), or therapy-induced stress (Gleevec). Once activated, autophagy can mediate cell survival through different mechanisms that usually involve the activation of *ATG* genes, although it can also inhibit TRAIL-mediated apoptosis, for example. Some evidence exists that autophagy might contribute to tumor dormancy through the induction of tumor cell survival; for example, in ovarian carcinoma, ARHI-induced autophagy was shown to contribute to cell survival and tumor dormancy through restoration of PI3K signaling. Furthermore, in gastrointestinal stromal tumors (GIST), autophagy is induced in response to Gleevec, which leads to the induction of a dormant state in which these tumor cells can survive for extremely prolonged periods. Because autophagy can protect cells from different microenvironmental induced stresses, one can speculate that autophagy might be one of the mechanisms activated to promote the survival of dormant disseminated tumor cells in restrictive microenvironments

Modeling of tumor cell dissemination during early stages of cancer progression in MMTV-Neu (Neu) mice showed that premalignant lesions contained micro-invasive cells and that dissemination to lungs and bone marrow (BM) was readily detected [9]. In uveal melanoma, a cancer with 50% incidence of late liver recurrence (>10 years) in humans [13], analysis of tumor doubling times led to the conclusion that dissemination had occurred at least half a decade before diagnosis. In an uveal melanoma mouse model [14], it was shown that dissemination occurred early and dormant (i.e., growth-arrested) DTCs were commonplace. In a *Drosophila melanogaster* model using CSK (C-terminal Src kinase)-null flies, early dissemination required Src activation without loss of E-cadherin or obvious induction of an epithelial-mesenchymal transition, which is supposedly a prerequisite for dissemination [15].

It is possible that early dissemination accounts for the variable periods of dormancy time because early DTCs are genetically and/or epigenetically unfit for expansion. Alternatively, DTCs carrying genetic alterations that favor growth or those originating from more progressed lesions may be kept "in-check" by the microenvironment, whereby epigenetic or therapy-derived mechanisms [1] contribute to tumor cell dormancy during or after the "lead time" [1, 16]. In support of the microenvironment playing a role, a recent report suggested that breast cancer patients with cells disseminated to the BM had longer disease-free periods than patients who were negative for cells in this site [17]. This suggests that the bone microenvironment may change the timing of cancer progression by favoring dormancy. Nonetheless, it remains unclear how the primary tumor or the target organ microenvironments may control the lead time in solitary DTCs, and the kinetics driving genetic progression during this lead time remain poorly understood.

The possibility of therapy-induced quiescence may follow different mechanisms. In multiple myeloma, treatment with a proteasome inhibitor (bortezomib) has been found to induce post treatment protracted quiescence and survival of a fraction of cancer cells [18]. Furthermore, it has been shown that BCR-ABL blasts detected by fluorescence in situ hybridization (FISH) in chronic myelogenous leukemia patients who had responded to interferon-γ treatment 5–10 years earlier had no detectable mRNA for the oncogene [19, 20]. This suggests that epigenetic or post-transcriptional mechanisms may be dominant and suppress gene expression, including even those genes that are mutated or amplified. This potentially explains why, despite the presence of genetic alterations, these cells remain at a residual level. This dormancy may be explained by mechanisms similar to those controlling hematopoietic stem cell dormancy, whereby inactive STAT1 and Akt1 as well as low Sca-1 levels apparently maintain dormancy of these cells. In fact, it has been proposed that treatment with interferon-α may break the dormancy of leukemic stem cells by activating (activity and expression) the above-mentioned molecules, and that these cells are now prone to being targeted by BCR-ABL inhibitors [21]. This also suggests that, while chemotherapeutic drugs or other treatments kill a large fraction of cells, they can also cause induction of a residual dormant cell population that may subsequently be poised for recurrence (see below).

The Target Organ Microenvironment and DTC Dormancy

Solitary DTCs in target organs can establish interactions with the extracellular matrix (ECM), immune cells, and vasculature [22]. Studies using breast cancer cell lines selected for vigorous growth in target organs identified gene expression profiles that favored organ-specific colonization [23]. On the contrary, some genes including the metastasis suppressor gene (MSG) *MKK4*, via *p38*, can suppress metastases

[24], and this seems to depend on stress signals from the microenvironment (see Fig. 5.1) [25]. *MKK4* belongs to a family of genes that selectively blocks metastatic growth, and includes *KISS1*, *MKK6*, *BHLHLB3/Sharp-1* (another *p38*-induced gene [26]), and *Nm23-H1*, among others [25, 27]. Because these genes suppress the growth and expansion of DTCs at target organs, yet fail to impede primary tumor growth, this further supports that the target organs with specific tissue microenvironments are required for these molecules to exert their growth-suppressing functions.

In squamous carcinoma cells (HEp3), reduced expression of urokinase (uPA) receptor (uPAR) deactivates α5β1 integrins, which makes these cells incapable of binding efficiently to fibronectin [28]. This results in reduced focal adhesion kinase (FAK) and epidermal growth factor receptor signaling, as well as in p38 activation. Thus, a failure by tumor cells to establish appropriate interactions with the ECM may induce growth-restrictive signals that fuel a quiescence state [1]. Furthermore, the loss of β1 integrin or FAK signaling in breast cancer models can induce dormancy, and activation of the Src-MLKC pathway can prevent dormancy [1, 29]. In addition, a fibrous collagen-I-enriched microenvironment in lungs can trigger intravenously injected mouse breast cancer cells to exit dormancy [29]. In contrast, microenvironments rich in fibrillar collagen-I induce melanoma quiescence by activating the discoidin domain receptor 2 and p15INK4b induction [13]. Collectively, these studies demonstrate that the loss of growth pathways induced by either therapies or a restrictive (i.e., fibrotic or non-fibrotic target tissues depending on the tumor type) tissue microenvironment is accompanied by the activation of stress pathways; this immediately motivates the hypothesis that the integration of these two types of signals within a DTC is responsible for both entry into and exit from a dormant state (see Fig. 5.1).

In HEp3 squamous carcinoma cells, while the activation of p38α/β inhibits ERK1/2 signaling, it also activates a stress-adaptive response known as the unfolded protein response (UPR) [26, 30, 31]. These signals lead to an epigenetic reprogramming and induction of survival and quiescence of dormant HEp3 (D-HEp3) cells [32]. D-HEp3 cells inoculated in vivo enter a deep G_0–G_1 arrest characterized by induction of *p21*, *p27*, *p18*, and *p15* [26]. At least three transcription factors (TFs), p53, BHLHB3/41/Sharp1 and NR2F1, are regulated by p38α/β and required for dormancy of tumor cells in vivo [26]. This program is activated in dormant DTCs recovered from the bone marrow (BM) but is reversed when tumor cells exit dormancy or grow persistently in lungs (our unpublished results) (see Fig. 5.1). BM-derived dormant HEp3 cells display a low ERK/p38 signaling ratio and induction of BHLHB3/41/Sharp-1, NR2F1, and p53. Interestingly, MSGs, such as *MKK4* and *MKK6*, are upstream activators of *p38* [25], whereas BHLHB3 is a target of *p38* required for quiescence induction [26]. Thus, it seems that different mechanisms converge in the regulation of the ERK/p38 signaling ratio and result in induction of either proliferation or dormancy.

An important question is whether the target organ microenvironment, where DTCs reside, induces dormancy programs, and if so, how? In tumors like those in head and neck squamous cell carcinoma and breast cancer, bone metastasis occurs at a frequency of 10–30% [7, 33, 34]. However, the detection of BM DTCs is much higher (>50% of patients) [6, 35]. This suggests that not all DTCs ultimately form overt metastasis and/or that a delay takes place. In mouse models of cancer (xenografts or transgenic), BM metastases are rarely observed. For example, in MMTV-Neu transgenic mice, BM DTCs are readily detected but mice never develop bone metastasis [9]. However, if the BM microenvironment is modified via irradiation [9] or if p38α/β is systemically inhibited, then DTCs expand ([9] and our unpublished data). Thus, in certain organs, restrictive signals mediated at least by p38α/β signaling can prevent occult DTCs from expanding.

In the search for signaling mediators that play a role in dormancy of DTCs in the BM, transforming growth factor-beta (TGFβ), a cytokine rich in the BM microenvironment [36–39], has emerged as a potential factor. Although tumors have been shown to depend on TGFβ to metastasize [40, 41], this ligand, depending on the degree of progression of tumors, can also be a potent inhibitor of epithelial tumor cell proliferation [42, 43]. TGFβ is also required to maintain the quiescence of stem cells and progenitors in the BM [36–39]. Thus, some tumors may remain sensitive to TGFβ growth inhibition in microenvironments where this factor is present (i.e., BM) [44]. In early-stage melanoma, TGFβ is anti-proliferative, thus functioning as a tumor-suppressor, but in advanced melanoma it is pro-invasive [45–47]. How these two opposing scenarios develop is not entirely clear [45, 48]. Furthermore, there is clinical evidence of early spread of uveal melanoma and, in a smaller proportion of patients, cutaneous melanoma thinner than 0.76 mm in depth [49–51]. It is possible that, similar to early dissemination in breast cancer [10], melanoma may spread before the conversion from TGFβ-inhibitory phenotype to pro-invasive behavior is activated, and when single cells arrive at distant sites, such as the liver or BM [9], they may remain in cell cycle arrest for prolonged periods due to high levels of and/or high responsiveness to TGFβ.

ER-Stress Signaling Pathways Contribute to Growth Arrest and Survival Programs During Tumor Cell Dormancy

While exploring the mechanisms that drive quiescence and survival of dormant HEp3 cells, the Aguirre-Ghiso lab discovered that HEp3 cells display a high ERK1/2 to p38α/β signaling ratio that favors proliferation in vivo [52–54]. The reprogramming of cells into dormancy (D-HEp3 cells) results in a reversion of this ratio, and now p38 signaling predominates over ERK. In addition, p38 appears to activate a negative feedback loop [28, 55, 56]. Using proteomics and microarray studies, the same group revealed that D-HEp3 cells develop an UPR characterized by enhanced endoplasmic reticulum (ER) signaling (see Fig. 5.1). In fact, all three arms of the UPR—ATF6α, IRE1α, and PERK—are activated in these cells [30, 31, 57, 58]. These studies led to the discovery that, in addition to inducing growth arrest, dormant cells utilized these signals to robustly withstand stress insults and survive in vivo for months. Among the three ER transmembrane signaling molecules, only

PERK was found to contribute to the quiescence of D-HEp3 cells [30, 31, 57, 58]. It did so by attenuating translation initiation, which resulted in the downregulation of cyclin D1/D3 and CDK4 in these cells [30]. In fact, inducible activation of PERK signaling using a dimerizable Fv2E-PERK fusion protein and the divalent ligand AP20187 was sufficient to fully abrogate tumorigenicity and induce growth arrest, in some cases irreversibly [30]. PERK also contributes survival signals for D-HEp3 cells. In fact, inhibition of PERK made these cells susceptible to both glucose deprivation and chemotherapeutic drug-induced killing (see Fig. 5.1) [30, 31].

The other arms of the ER-stress pathways, ATF6α and IRE1α, were also found to regulate tumor cell dormancy by promoting survival and adaptation to the in vivo microenvironment [57]. RNA interference (RNAi)-mediated targeting of ATF6α caused a decrease in the number of viable D-HEp3 cells in vivo without interrupting their dormancy [57]. RNAi targeting of XBP-1, a transcription factor (TF) that is exclusively activated by IRE1α through noncanonical splicing, also induced dormant D-HEp3 cell killing [57] (and unpublished results). Neither RNAi to ATF6α or XBP-1 affected the tumorigenicity of T-HEp3 cells. Thus, the survival capacity of these genes seems to operate primarily in the cells that enter quiescence and not in the proliferative counterpart (see Fig. 5.1).

The mechanism of survival for ATF6α has also been explored in more detail. It has been shown the basal survival capacity of D-HEp3 cells to adapt and enter dormancy in vivo is not mediated by classical target genes regulated by ATF6α during the UPR, including genes for the chaperone BiP/Grp78, secretogranin II, and a glucose transporter [57]. We found that ATF6α induced Rheb, a small GTPase of the Ras family that directly activates the survival protein mTOR. Indeed, analysis of the mechanisms revealed that p38-dependent activation of ATF6α results in Rheb induction and stronger activation of mTOR→P-S6K→P-S6 signaling [57]. This pathway confers only dormant cells with resistance to rapamycin, as RNAi targeting of Rheb or ATF6 restored sensitivity to the mTOR inhibitor. Most importantly, dormant D-HEp3 cells can no longer adapt to the in vivo microenvironment and die at least in part through a caspase-3–dependent apoptotic pathway (see Fig. 5.1) [57].

Moreover, p38 also induced the expression of the chaperone BiP/Grp78 (see Fig. 5.1). This chaperone is induced during ER-stress and is an essential survival factor as it is a primary regulator of protein folding in the ER lumen. Numerous studies have shown that BiP serves as a survival factor not only in response to ER-stress but also to other damaging agents, such as chemotherapeutic drugs [31]. The upregulation of BiP in dormant HEp3 cells and its induction by p38 suggests that p38 signaling, like ATF6 activation, could tap into BiP function to provide survival signals [31]. However, as mentioned above, BiP did not provide a survival advantage for basal in vivo adaptation [57], which raises the possibility that BiP may only protect dormant cells under extreme damaging conditions such as those encountered during chemotherapy [31]. In fact, it has been demonstrated that dormant D-HEp3 cells were inherently resistant to chemotherapy compared with their tumorigenic counterpart, and that this was not due to enhanced expression of ATP-binding cassette transporters [31]. Furthermore, RNAi targeting of BiP greatly sensitized dormant D-HEp3 cells to etoposide and doxorubicin treatment. In contrast, the lower levels of BiP in the T-HEp3 cells, when further decreased by RNAi, had no effect on the sensitivity of these cells to chemotherapy. Analysis of the mechanism revealed that BiP inhibited the activation of the pro-apoptotic factor Bax [31]. Recent follow-up on our studies by other investigators revealed that in fact BiP inhibits Bax by regulating its inhibitor Bik (see Fig. 5.1) [59].

These studies highlight a mostly overlooked aspect of dormancy: cells, from either early primary lesions or more advanced tumors, must survive for prolonged periods before resuming growth. Our results suggest that there may be mechanisms that selectively protect quiescent cells from a hostile microenvironment or from stress imposed by the therapies used to treat different cancers. This may be an evolutionary conserved response to stress. For example, organisms like *Caenorhabditis elegans* are able to pause development and enter a dormant dauer stage in response to nutritional stress or oxidative stress derived from the environment [1, 60–62]. Numerous studies in yeast also suggest that stress signaling and induction of quiescent growth are coupled with the induction of survival pathways that protect the organism from stress conditions during growth arrest [63]. This prompts the question of whether these mechanisms are active in DTCs in patients and whether they can be exploited therapeutically.

Autophagy and Survival of Residual Disease

With increasing scrutiny on how fundamental cellular stress-response pathways impact survival and expansion of dormant tumor cells, autophagy has emerged as an attractive target against dormant tumor cells (see Fig. 5.2). Importantly, multiple routes of autophagic degradation exist within cells, including: (1) macroautophagy, in which cytoplasmic contents are sequestered in double membrane autophagosomes and subsequently delivered to the lysosome; (2) microautophagy, where cytoplasm is directly engulfed by the lysosomal membrane; and (3) chaperone-mediated autophagy, where proteins with a specific signal sequence are transported to the lysosomal lumen by a receptor-mediated process [64]. Of these routes, macroautophagy (hereafter called autophagy) has been most extensively studied for its potential functions in cancer. Macroautophagy is tightly regulated by a limited number of highly conserved genes called *ATG*s (*AuTophaGy*-related genes), which were originally identified in yeast [65]. These landmark studies have led to numerous recent breakthroughs in mammals, demonstrating a critical role for autophagy in both physiological and pathological processes, including cancer initiation and progression [64].

The bulk degradation of cellular material through autophagy allows cells to recycle both nutrients and energy during starvation and stress; in this regard, autophagy is proposed to function as a fitness mechanism that allows tumor cells to survive provided the offending stressor is removed in a timely manner [66, 67]. This indispensable contribution of autophagy as a stress-response mechanism is poignantly illustrated by studies in mice, in which the genetic deletion of critical *ATG*s results in neonatal lethality within a day of birth [68, 69]. A potential role for autophagy in dormancy was originally broached in *C. elegans* during dauer diapause, a stress-induced, dormancy-like state that occurs when larvae are exposed to hostile environments [70]. Notably, in this model, defective autophagy (achieved via RNAi against multiple *ATG*s) potently compromised survival during dauer, implying a conserved mechanism by which autophagy promotes survival during quiescent states [70]. Since autophagy is activated in response to various microenvironmental stresses implicated in tumor dormancy, including the UPR (ER-stress), hypoxia, and ECM detachment, an important outstanding issue is how autophagy impacts the survival, as well as the maintenance of the quiescent state, in dormant tumor cells.

Studies in breast cancer models suggest that decreased mitogenic signaling resulting from impaired integrin and growth factor signaling facilitates tumor dormancy [71, 72]. Specifically, suppression of β1-integrin signaling induces dormancy in the MMTV-PyMT model of breast cancer and squamous carcinoma [28, 72]. Thus, it is possible that, because DTCs cannot efficiently engage a foreign ECM, impaired integrin signaling may stimulate autophagy for survival and maintenance of the dormant state. Consistent with this hypothesis, β1-integrin signaling blockade is a potent inducer of autophagy in ECM-detached cells, and autophagy protects cells from detachment-induced apoptosis (anoikis) (see Fig. 5.2) [73]. Moreover, autophagy may contribute to the ability of solitary dormant cells to resist extrinsic apoptotic stimuli. In breast cancer metastases to bone, where DTCs remain dormant in the BM for extended periods of time, the tissue necrosis factor (TNF) ligand TRAIL is abundantly expressed in the BM microenvironment and can kill tumor cells; nonetheless, mechanisms involving Src-mediated TRAIL resistance promote the survival of indolent cells in the BM [74]. Because autophagy can protect cells from TRAIL-induced apoptosis, one can speculate that autophagy may similarly promote the survival of dormant cells in the BM [75, 76]. Interestingly, we found that D-HEp3 cells have constitutively higher levels of autophagy, as measured by green fluorescence protein-tagged LC3 and endogenous LC3 incorporation into autophagosomes, as well as elevated expression of specific autophagy-regulating genes including *ATG6*, *ATG7*, and *ATG8* (unpublished results). Our ongoing studies also reveal that ATF6, but not PERK, is responsible for LC3 processing into autophagosomes.

Recently, autophagy has been shown to be crucial for the survival of dormant cells in models of ovarian cancer and gastrointestinal stromal tumor (GIST) [77, 78]. The tumor suppressor aplasia Ras homolog member I (*ARHI*) is downregulated in over 60% of ovarian cancers and the re-expression of *ARHI* in a variety of human ovarian cancer cell lines induces autophagy (see Fig. 5.2). In xenograft ovarian tumors, *ARHI* overexpression promotes the formation of dormant tumors, which correlates with an increased level of autophagosome formation; accordingly, when *ARHI* expression is subsequently reduced, the tumor regains proliferative potential and rapidly re-grows. However, upon treatment of *ARHI*-induced dormant tumors with the lysosomal inhibitor chloroquine, this regrowth is dramatically reduced, suggesting that autophagy contributes to survival during *ARHI*-induced dormancy [77].

Another demonstration of autophagy as a survival pathway in quiescent cells comes from studies of GIST, the first solid tumor to be treated successfully with the small-molecule tyrosine kinase inhibitor imatinib mesylate (Gleevec) (see Fig. 5.2) [78]. However, less than 5% of GISTs regress significantly upon Gleevec treatment; rather, in the vast majority of patients, tumor cells indefinitely remain in a dormant, quiescent state in the presence of imatinib. Recent work indicates that this dormant state, termed stable disease, is closely associated with the induction of autophagy in response to imatinib. Upon inhibiting autophagy using RNAi-mediated *ATG* depletion or antimalarials, such as hydroxychloroquine and quinacrine, GIST cells undergo high levels of apoptosis both in vitro and in vivo. Thus, autophagy appears critical for the establishment of a dormant state in which GIST cells can survive indefinitely [78]. Moreover, these results in GIST broach the exciting idea that autophagy can be more widely exploited to kill or prevent the expansion of quiescent or dormant cancer cells, which are notorious for their resistance to both conventional and targeted therapies [79].

Tumor dormancy is also postulated to be a stress-management mechanism adopted by DTCs to cope with an unfavorable microenvironment by completely withdrawing from the cell cycle [1]. $p27^{Kip1}$, the cyclin-dependent kinase inhibitor involved in G_0/G_1 cell cycle arrest, was identified as a downstream target of the energy-sensing LKB1-AMPK pathway, as well as shown to induce autophagy and facilitate cell survival in response to growth factor withdrawal and metabolic stress (see Fig. 5.2) [80]. Thus, DTCs may depend on p27-mediated autophagy to survive in an inhospitable microenvironment and to resist chemotherapy. HEp3 cells in which p38 signaling induces dormancy also have induced strong expression of p27 during their prolonged dormancy, further supporting this notion [26]. However, the exact biological role for autophagy during quiescence remains largely unknown; if autophagy promotes growth suppression in quiescent cells, one can alternatively hypothesize that it may limit the outgrowth of dormant cells into frank macrometastases. These questions are important, and it will be critical to determine whether autophagy plays a quiescence or survival-inducing role (or both) in quiescent DTCs. If autophagy induces a pro-survival state then strategies to block it could eradicate DTCs. In the case that it contributes to both quiescence and survival, then more detailed mechanistic analysis of these pathways will be required to reveal ways to block only the survival signals without interrupting quiescence.

Overall, these results motivate future work, especially those using in vivo preclinical models, to assess how autophagy influences the quiescence and/or survival and biological behavior of dormant breast cancer cells, and specifically whether autophagy inhibition can be exploited to prevent the development of macrometastases in cancer patients.

Concluding Remarks

Our knowledge on how the biology and genetics of DTCs influence dormancy and progression of metastasis remains limited. Many open questions still exist, which will likely become central themes in the future. For example, how is DTC fate affected by the primary tumor microenvironment, how do therapies applied to patients affect DTCs, and how do the target organs condition these responses? If DTCs are indeed the "seeds" of metastases, it will be imperative to directly investigate these questions by analyzing DTCs from patients. Importantly, an analysis of DTCs that survive therapy of the primary tumor will inform us on how these treatments, as well as target organs, impact adaptation and/or selection of subsequent recurrent metastatic disease. For example, the demonstration that DTCs undergo autophagy or tap into UPR survival signals to survive and persist for prolonged periods will be a promising finding that will motivate clinical trials targeting specific components of the autophagy or UPR machinery to eradicate these cells (i.e., maintenance therapy). Studies on dormancy may also yield information on how to maintain signals that propel quiescence, such as a combination of MEK inhibitors and agonists that mimic p38α/β activation. A deeper understanding of the signals that maintain dormancy may lead to the identification of drugs that should be avoided in patients because of their potential to break this state, and thus, enhance disease progression. Although the study of DTCs and dormant disease is difficult, unraveling the inherent complexity of this poorly understood step of metastasis biology should profoundly impact cancer patients.

Acknowledgements This work is supported by grants from the Samuel Waxman Cancer Research Foundation Tumor Dormancy Program, NIH/National Cancer Institute (CA109182, CA163131), NIEHS (ES017146), and NYSTEM to J.A.A-G, DoD Breast Cancer Postdoctoral Fellowship to M.S.S. and NIH RO1 CA126792, CA126792-S1 (ARRA), and a DOD BCRP Era of Hope Scholar Award (W81XWH-11-1-0310) to J.D.

References

1. Aguirre-Ghiso JA (2007) Models, mechanisms and clinical evidence for cancer dormancy. Nat Rev Cancer 7:834–846
2. Klein CA (2011) Framework models of tumor dormancy from patient-derived observations. Curr Opin Genet Dev 21(1):42–49
3. Sang L, Coller HA, Roberts JM (2008) Control of the reversibility of cellular quiescence by the transcriptional repressor HES1. Science 321:1095–1100
4. Almog N, Ma L, Raychowdhury R, Schwager C, Erber R, Short S et al (2009) Transcriptional switch of dormant tumors to fast-growing angiogenic phenotype. Cancer Res 69:836–844
5. Mahnke YD, Schwendemann J, Beckhove P, Schirrmacher V (2005) Maintenance of long-term tumour-specific T-cell memory by residual dormant tumour cells. Immunology 115:325–336
6. Klein CA (2008) The direct molecular analysis of metastatic precursor cells in breast cancer: a chance for a better understanding of metastasis and for personalised medicine. Eur J Cancer 44(18):2721–2725
7. Pantel K, Brakenhoff RH (2004) Dissecting the metastatic cascade. Nat Rev Cancer 4: 448–456
8. Stoecklein NH, Hosch SB, Bezler M, Stern F, Hartmann CH, Vay C et al (2008) Direct genetic analysis of single disseminated cancer cells for prediction of outcome and therapy selection in esophageal cancer. Cancer Cell 13:441–453
9. Husemann Y, Geigl JB, Schubert F, Musiani P, Meyer M, Burghart E et al (2008) Systemic spread is an early step in breast cancer. Cancer Cell 13:58–68
10. Schardt JA, Meyer M, Hartmann CH, Schubert F, Schmidt-Kittler O, Fuhrmann C et al (2005) Genomic analysis of single cytokeratin-positive cells from bone marrow reveals early mutational events in breast cancer. Cancer Cell 8:227–239

11. Schmidt-Kittler O, Ragg T, Daskalakis A, Granzow M, Ahr A, Blankenstein TJ et al (2003) From latent disseminated cells to overt metastasis: genetic analysis of systemic breast cancer progression. Proc Natl Acad Sci USA 100:7737–7742
12. Klein CA (2009) Parallel progression of primary tumours and metastases. Nat Rev Cancer 9:302–312
13. Ossowski L, Aguirre-Ghiso JA (2010) Dormancy of metastatic melanoma. Pigment Cell Melanoma Res 23:41–56
14. Eyles J, Puaux AL, Wang X, Toh B, Prakash C, Hong M et al (2010) Tumor cells disseminate early, but immunosurveillance limits metastatic outgrowth, in a mouse model of melanoma. J Clin Invest 120:2030–2039
15. Vidal M, Larson DE, Cagan RL (2006) Csk-deficient boundary cells are eliminated from normal Drosophila epithelia by exclusion, migration, and apoptosis. Dev Cell 10:33–44
16. Klein CA, Hölzel D (2006) Systemic cancer progression and tumor dormancy: mathematical models meet single cell genomics. Cell Cycle 5(16):1788–1798
17. Bidard FC, Vincent-Salomon A, Sigal-Zafrani B, Rodrigues M, Dieras V, Mignot L et al (2008) Time to metastatic relapse and breast cancer cells dissemination in bone marrow at metastatic relapse. Clin Exp Metastasis 25:871–875
18. Schewe DM, Aguirre-Ghiso JA (2009) Inhibition of eIF2alpha dephosphorylation maximizes bortezomib efficiency and eliminates quiescent multiple myeloma cells surviving proteasome inhibitor therapy. Cancer Res 69:1545–1552
19. Kitzis A, Brizard F, Dascalescu C, Chomel JC, Guilhot F, Brizard A (2001) Persistence of transcriptionally silent BCR-ABL rearrangements in chronic myeloid leukemia patients in sustained complete cytogenetic remission. Leuk Lymphoma 42:933–944
20. Talpaz M, Estrov Z, Kantarjian H, Ku S, Foteh A, Kurzroc R (1994) Persistence of dormant leukemic progenitors during interferon-induced remission in chronic myelogenous leukemia. Analysis by polymerase chain reaction of individual colonies. J Clin Invest 94:1383–1389
21. Essers MA, Offner S, Blanco-Bose WE, Waibler Z, Kalinke U, Duchosal MA et al (2009) IFNalpha activates dormant haematopoietic stem cells in vivo. Nature 458:904–908
22. Joyce JA, Pollard JW (2009) Microenvironmental regulation of metastasis. Nat Rev Cancer 9:239–252
23. Nguyen DX, Bos PD, Massague J (2009) Metastasis: from dissemination to organ-specific colonization. Nat Rev Cancer 9:274–284
24. Hickson JA, Huo D, Vander Griend DJ, Lin A, Rinker-Schaeffer CW, Yamada SD (2006) The p38 kinases MKK4 and MKK6 suppress metastatic colonization in human ovarian carcinoma. Cancer Res 66:2264–2270
25. Taylor J, Hickson J, Lotan T, Yamada DS, Rinker-Schaeffer C (2008) Using metastasis suppressor proteins to dissect interactions among cancer cells and their microenvironment. Cancer Metastasis Rev 27:67–73
26. Adam AP, George A, Schewe D, Bragado P, Iglesias BV, Ranganathan AC et al (2009) Computational identification of a p38SAPK-regulated transcription factor network required for tumor cell quiescence. Cancer Res 69:5664–5672
27. Adorno M, Cordenonsi M, Montagner M, Dupont S, Wong C, Hann B et al (2009) A mutant-p53/Smad complex opposes p63 to empower TGFbeta-induced metastasis. Cell 137:87–98
28. Aguirre Ghiso JA, Kovalski K, Ossowski L (1999) Tumor dormancy induced by downregulation of urokinase receptor in human carcinoma involves integrin and MAPK signaling. J Cell Biol 147:89–104
29. Barkan D, El Touny LH, Michalowski AM, Smith JA, Chu I, Davis AS et al (2010) Metastatic growth from dormant cells induced by a col-I-enriched fibrotic environment. Cancer Res 70:5706–5716
30. Ranganathan AC, Ojha S, Kourtidis A, Conklin DS, Aguirre-Ghiso JA (2008) Dual function of pancreatic endoplasmic reticulum kinase in tumor cell growth arrest and survival. Cancer Res 68:3260–3268
31. Ranganathan AC, Zhang L, Adam AP, Aguirre-Ghiso JA (2006) Functional coupling of p38-induced up-regulation of BiP and activation of RNA-dependent protein kinase-like

endoplasmic reticulum kinase to drug resistance of dormant carcinoma cells. Cancer Res 66: 1702–1711
32. Ranganathan AC, Adam AP, Aguirre-Ghiso JA (2006) Opposing roles of mitogenic and stress signaling pathways in the induction of cancer dormancy. Cell Cycle 5(16):1799–1807
33. Harrison LB, Sessions RB, Ki-Hong W (2003) Head and neck cancer. A multidisciplinary approach, 3rd edn. Lippincott Williams & Wilkins, Philadelphia
34. Gath HJ, Brakenhoff RH (1999) Minimal residual disease in head and neck cancer. Cancer Metastasis Rev 18:109–126
35. Wikman H, Vessella R, Pantel K (2008) Cancer micrometastasis and tumour dormancy. APMIS 116:754–770
36. Fan X, Valdimarsdottir G, Larsson J, Brun A, Magnusson M, Jacobsen SE et al (2002) Transient disruption of autocrine TGF-beta signaling leads to enhanced survival and proliferation potential in single primitive human hemopoietic progenitor cells. J Immunol 168:755–762
37. Fortunel N, Hatzfeld J, Kisselev S, Monier MN, Ducos K, Cardoso A et al (2000) Release from quiescence of primitive human hematopoietic stem/progenitor cells by blocking their cell-surface TGF-beta type II receptor in a short-term in vitro assay. Stem Cells 18:102–111
38. Scandura JM, Boccuni P, Massague J, Nimer SD (2004) Transforming growth factor beta-induced cell cycle arrest of human hematopoietic cells requires p57KIP2 up-regulation. Proc Natl Acad Sci USA 101:15231–15236
39. Yamazaki S, Iwama A, Takayanagi S, Eto K, Ema H, Nakauchi H (2009) TGF-beta as a candidate bone marrow niche signal to induce hematopoietic stem cell hibernation. Blood 113:1250–1256
40. Muraoka-Cook RS, Kurokawa H, Koh Y, Forbes JT, Roebuck LR, Barcellos-Hoff MH et al (2004) Conditional overexpression of active transforming growth factor beta1 in vivo accelerates metastases of transgenic mammary tumors. Cancer Res 64:9002–9011
41. Kang Y, Siegel PM, Shu W, Drobnjak M, Kakonen SM, Cordon-Cardo C et al (2003) A multigenic program mediating breast cancer metastasis to bone. Cancer Cell 3:537–549
42. Dumont N, Arteaga CL (2003) Targeting the TGF beta signaling network in human neoplasia. Cancer Cell 3:531–536
43. Siegel PM, Massague J (2003) Cytostatic and apoptotic actions of TGF-beta in homeostasis and cancer. Nat Rev Cancer 3:807–821
44. Hideshima T, Podar K, Chauhan D, Anderson KC (2005) Cytokines and signal transduction. Best Pract Res Clin Haematol 18:509–524
45. Javelaud D, Alexaki VI, Mauviel A (2008) Transforming growth factor-beta in cutaneous melanoma. Pigment Cell Melanoma Res 21:123–132
46. Hussein MR (2005) Transforming growth factor-beta and malignant melanoma: molecular mechanisms. J Cutan Pathol 32:389–395
47. Hsu MY, Rovinsky S, Penmatcha S, Herlyn M, Muirhead D (2005) Bone morphogenetic proteins in melanoma: angel or devil? Cancer Metastasis Rev 24:251–263
48. Reed JA, Bales E, Xu W, Okan NA, Bandyopadhyay D, Medrano EE (2001) Cytoplasmic localization of the oncogenic protein Ski in human cutaneous melanomas in vivo: functional implications for transforming growth factor beta signaling. Cancer Res 61:8074–8078
49. Zapas JL, Coley HC, Beam SL, Brown SD, Jablonski KA, Elias EG (2003) The risk of regional lymph node metastases in patients with melanoma less than 1.0 mm thick: recommendations for sentinel lymph node biopsy. J Am Coll Surg 197:403–407
50. Gamel JW, George SL, Edwards MJ, Seigler HF (2002) The long-term clinical course of patients with cutaneous melanoma. Cancer 95:1286–1293
51. Eskelin S, Pyrhonen S, Summanen P, Hahka-Kemppinen M, Kivela T (2000) Tumor doubling times in metastatic malignant melanoma of the uvea: tumor progression before and after treatment. Ophthalmology 107:1443–1449
52. Aguirre-Ghiso JA, Ossowski L, Rosenbaum SK (2004) Green fluorescent protein tagging of extracellular signal-regulated kinase and p38 pathways reveals novel dynamics of pathway activation during primary and metastatic growth. Cancer Res 64:7336–7345
53. Aguirre-Ghiso JA, Estrada Y, Liu D, Ossowski L (2003) ERK(MAPK) activity as a determinant of tumor growth and dormancy; regulation by p38(SAPK). Cancer Res 63:1684–1695

54. Aguirre-Ghiso JA, Liu D, Mignatti A, Kovalski K, Ossowski L (2001) Urokinase receptor and fibronectin regulate the ERK(MAPK) to p38(MAPK) activity ratios that determine carcinoma cell proliferation or dormancy in vivo. Mol Biol Cell 12:863–879
55. Liu D, Aguirre Ghiso J, Estrada Y, Ossowski L (2002) EGFR is a transducer of the urokinase receptor initiated signal that is required for in vivo growth of a human carcinoma. Cancer Cell 1:445–457
56. Aguirre Ghiso JA (2002) Inhibition of FAK signaling activated by urokinase receptor induces dormancy in human carcinoma cells in vivo. Oncogene 21:2513–2524
57. Schewe DM, Aguirre-Ghiso JA (2008) ATF6alpha-Rheb-mTOR signaling promotes survival of dormant tumor cells in vivo. Proc Natl Acad Sci USA 105:10519–10524
58. Ranganathan AC, Adam AP, Zhang L, Aguirre-Ghiso JA (2006) Tumor cell dormancy induced by p38(SAPK) and ER-stress signaling: an adaptive advantage for metastatic cells? Cancer Biol Ther 5:729–735
59. Fu Y, Li J, Lee AS (2007) GRP78/BiP inhibits endoplasmic reticulum BIK and protects human breast cancer cells against estrogen starvation-induced apoptosis. Cancer Res 67:3734–3740
60. Fukuyama M, Rougvie AE, Rothman JH (2006) C. elegans DAF-18/PTEN mediates nutrient-dependent arrest of cell cycle and growth in the germline. Curr Biol 16:773–779
61. Wang J, Kim SK (2003) Global analysis of dauer gene expression in Caenorhabditis elegans. Development 130:1621–1634
62. Long X, Spycher C, Han ZS, Rose AM, Muller F, Avruch J (2002) TOR deficiency in C. elegans causes developmental arrest and intestinal atrophy by inhibition of mRNA translation. Curr Biol 12:1448–1461
63. Lopez-Maury L, Marguerat S, Bahler J (2008) Tuning gene expression to changing environments: from rapid responses to evolutionary adaptation. Nat Rev Genet 9:583–593
64. Mizushima N, Levine B, Cuervo AM, Klionsky DJ (2008) Autophagy fights disease through cellular self-digestion. Nature 451:1069–1075
65. Klionsky DJ, Cregg JM, Dunn WA Jr, Emr SD, Sakai Y, Sandoval IV et al (2003) A unified nomenclature for yeast autophagy-related genes. Dev Cell 5:539–545
66. Lum JJ, Bauer DE, Kong M, Harris MH, Li C, Lindsten T et al (2005) Growth factor regulation of autophagy and cell survival in the absence of apoptosis. Cell 120:237–248
67. Roy S, Debnath J (2010) Autophagy and tumorigenesis. Semin Immunopathol 32:383–396
68. Komatsu M, Waguri S, Ueno T, Iwata J, Murata S, Tanida I et al (2005) Impairment of starvation-induced and constitutive autophagy in Atg7-deficient mice. J Cell Biol 169:425–434
69. Kuma A, Hatano M, Matsui M, Yamamoto A, Nakaya H, Yoshimori T et al (2004) The role of autophagy during the early neonatal starvation period. Nature 432:1032–1036
70. Melendez A, Talloczy Z, Seaman M, Eskelinen E-L, Hall DH, Levine B (2003) Autophagy genes are essential for dauer development and life-span extension in C. elegans. Science 301:1387–1391
71. Korah R, Boots M, Wieder R (2004) Integrin alpha5beta1 promotes survival of growth-arrested breast cancer cells: an in vitro paradigm for breast cancer dormancy in bone marrow. Cancer Res 64:4514–4522
72. White DE, Kurpios NA, Zuo D, Hassell JA, Blaess S, Mueller U et al (2004) Targeted disruption of beta1-integrin in a transgenic mouse model of human breast cancer reveals an essential role in mammary tumor induction. Cancer Cell 6:159–170
73. Fung C, Lock R, Gao S, Salas E, Debnath J (2008) Induction of autophagy during extracellular matrix detachment promotes cell survival. Mol Biol Cell 19:797–806
74. Zhang XH, Wang Q, Gerald W, Hudis CA, Norton L, Smid M et al (2009) Latent bone metastasis in breast cancer tied to Src-dependent survival signals. Cancer Cell 16:67–78
75. Han J, Hou W, Goldstein LA, Lu C, Stolz DB, Yin XM et al (2008) Involvement of protective autophagy in TRAIL resistance of apoptosis-defective tumor cells. J Biol Chem 283:19665–19677
76. Herrero-Martin G, Hoyer-Hansen M, Garcia-Garcia C, Fumarola C, Farkas T, Lopez-Rivas A et al (2009) TAK1 activates AMPK-dependent cytoprotective autophagy in TRAIL-treated epithelial cells. EMBO J 28:677–685

77. Lu Z, Luo RZ, Lu Y, Zhang X, Yu Q, Khare S et al (2008) The tumor suppressor gene ARHI regulates autophagy and tumor dormancy in human ovarian cancer cells. J Clin Invest 118: 3917–3929
78. Gupta A, Roy S, Lazar AJ, Wang WL, McAuliffe JC, Reynoso D et al (2010) Autophagy inhibition and antimalarials promote cell death in gastrointestinal stromal tumor (GIST). Proc Natl Acad Sci USA 107:14333–14338
79. Rubin BP, Debnath J (2010) Therapeutic implications of autophagy-mediated cell survival in gastrointestinal stromal tumor after treatment with imatinib mesylate. Autophagy 6: 1190–1191
80. Liang J, Shao SH, Xu ZX, Hennessy B, Ding Z, Larrea M et al (2007) The energy sensing LKB1-AMPK pathway regulates p27(kip1) phosphorylation mediating the decision to enter autophagy or apoptosis. Nat Cell Biol 9:218–224

Chapter 6
Tumor Dormancy, Oncogene Addiction, Cellular Senescence, and Self-Renewal Programs

David I. Bellovin, Bikul Das, and Dean W. Felsher

Abstract Cancers are frequently addicted to initiating oncogenes that elicit aberrant cellular proliferation, self-renewal, and apoptosis. Restoration of oncogenes to normal physiologic regulation can elicit dramatic reversal of the neoplastic phenotype, including reduced proliferation and increased apoptosis of tumor cells (Science 297(5578):63–64, 2002). In some cases, oncogene inactivation is associated with compete elimination of a tumor. However, in other cases, oncogene inactivation induces a conversion of tumor cells to a dormant state that is associated with cellular differentiation and/or loss of the ability to self-replicate. Importantly, this dormant state is reversible, with tumor cells regaining the ability to self-renew upon oncogene reactivation. Thus, understanding the mechanism of oncogene inactivation-induced dormancy may be crucial for predicting therapeutic outcome of targeted therapy. One important mechanistic insight into tumor dormancy is that oncogene addiction might involve regulation of a decision between self-renewal and cellular senescence. Recent evidence suggests that this decision is regulated by multiple mechanisms that include tumor cell-intrinsic, cell-autonomous mechanisms and host-dependent, tumor cell-non-autonomous programs (Mol Cell 4(2):199–207, 1999; Science 297(5578):102–104, 2002; Nature 431(7012):1112–1117, 2004; Proc Natl Acad Sci U S A 104(32):13028–13033, 2007). In particular, the tumor microenvironment, which is known to be critical during tumor initiation (Cancer Cell 7(5):411–423, 2005; J Clin Invest 121(6):2436–2446, 2011), prevention (Nature 410(6832):1107–1111, 2001), and progression (Cytokine Growth Factor Rev 21(1):3–10, 2010), also appears to dictate when oncogene inactivation elicits the permanent loss of self-renewal through induction of cellular senescence (Nat Rev Clin Oncol 8(3):151–160, 2011; Science 313(5795):1960–1964, 2006; N Engl

---

D.I. Bellovin, PhD • B. Das, MBBS, PhD • D.W. Felsher, MD, PhD (✉)
Division of Oncology, Departments of Medicine and Pathology, Stanford University School of Medicine, 269 Campus Drive, CCSR 1105, Stanford, CA 94305-5151, USA
e-mail: dfelsher@stanford.edu

H. Enderling et al. (eds.), *Systems Biology of Tumor Dormancy*,
Advances in Experimental Medicine and Biology 734,
DOI 10.1007/978-1-4614-1445-2_6, © Springer Science+Business Media New York 2013

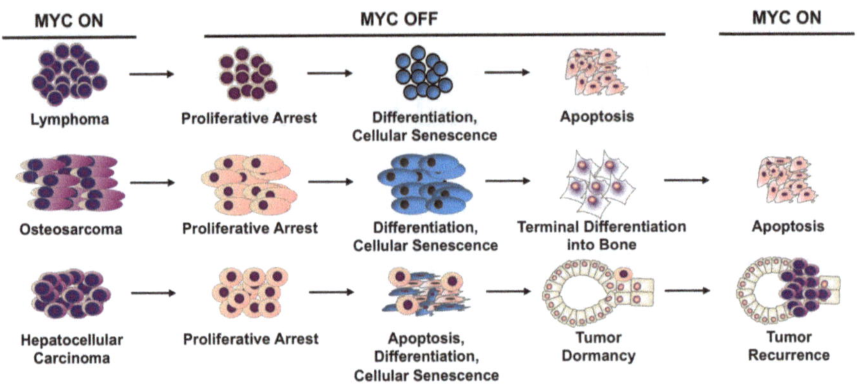

Fig. 6.1 Oncogene addiction elicits tissue-specific effects. *MYC* inactivation has been shown to have different outcomes depending upon the tissue origin of tumors, including proliferative arrest, differentiation, apoptosis, and/or cellular senescence. The specific consequences of oncogene addiction have a dramatic impact on whether targeted therapies will result in tumor dormancy or tumor elimination. In particular, although oncogene inactivation in *MYC*-induced lymphoma, osteosarcoma elicits complete tumor elimination due to a combination of proliferative arrest, differentiation, cellular senescence, and apoptosis. In contrast, upon *MYC* inactivation in hepatocellular carcinoma, a small population of cells possessing self-renewal capacity persists that therefore gives rise to a new tumor upon oncogene reactivation

J Med 351(21):2159–21569, 2004). Thus, oncogene addiction may be best modeled as a consequence of the interplay amongst cell-autonomous and host-dependent programs that define when a therapy will result in tumor dormancy.

Oncogene Addiction: A Mechanism of Tumor Regression

Tumors are induced either by the alteration of normal cellular genes into mutant, active forms, termed oncogenes, or by the loss of expression of tumor-suppressor genes, that elicit neoplastic features including increased cellular proliferation and resistance to programmed cell death (apoptosis). Oncogene addiction is the phenomenon by which the resulting tumor cells, as a consequence of a multitude of genetic and epigenetic changes, remain exquisitely dependent upon a single genetic mutation in oncogenes for the persistence of their neoplastic phenotype (Figs. 6.1 and 6.2) [13, 14]. The first indication that tumor cells could be addicted to oncogenes came from in vitro observations that tumor-derived cell lines sometimes exhibited proliferative arrest and/or apoptosis upon the suppression of an oncogene or the restoration of expression of a tumor suppressor [15]. These observations hinted that therapeutic agents targeting the repair or suppression of these mutant gene products could be generally effective for the treatment of cancer.

Later, the development of transgenic mice that can conditionally express oncogenes enabled the direct in situ interrogation of the role of specific oncogenes in the initiation and maintenance of tumorigenesis. Many mouse models have been

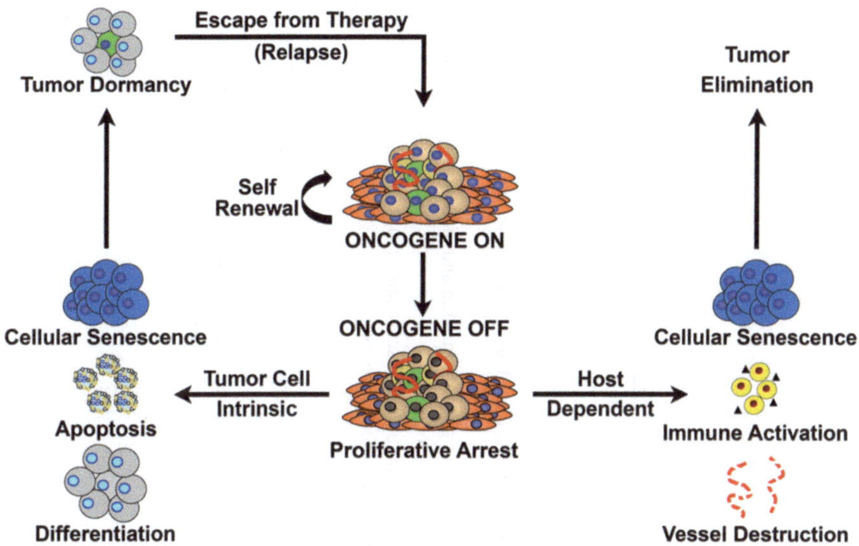

Fig. 6.2 Oncogene addiction comprises both cancer cell-autonomous and non-cell-autonomous mechanisms of tumor regression. Oncogene inactivation reverses many of the hallmarks of cancer to lead to tumor regression. Cell-intrinsic programs such as proliferative arrest, followed by apoptosis, differentiation, and self-renewal vs. cellular senescence are induced upon oncogene inactivation (*left*). Host-dependent phenomena, including immune system activation and destruction of the tumor vasculature, are also crucial for oncogene inactivation-induced tumor regression (*right*). Notably, cellular senescence is both a tumor cell-intrinsic and a host-regulated consequence of oncogene addiction; however, recent evidence indicates that only in an immune-intact host will targeted therapy result in complete tumor elimination

generated to explore the tumor-specific consequences of the suppression of oncogenes, including *MYC*, *RAS*, *BRAF*, and *BCR-ABL* (Table 6.1) [2–4, 16–19]. In these models, the consequences of oncogene inactivation include proliferative arrest, apoptosis [2], differentiation [3, 4], and senescence [5], as well as the inhibition of angiogenesis [20, 21]. These observations implied that many cancers are addicted to a single oncogene.

Further studies established that the specific consequences of oncogene inactivation are highly dependent upon the tissue or origin from which the cancer was initiated. Even upon brief inactivation of an oncogene, the diversity of these outcomes is evidenced by the induction of a permanent loss of the neoplastic phenotype in osteosarcoma and lymphoma (Fig. 6.1) [2, 3]. In marked contrast, oncogene suppression in hepatocellular and breast carcinoma [4, 19] induced regression of tumors, whereas restoration of oncogene activity restored their neoplastic features, suggesting a state of tumor dormancy (Fig. 6.1). In yet other cases, the inactivation of the oncogene failed to cause significant tumor regression, such as in a murine model of *MYC*-induced lung adenocarcinoma [22]. Thus, the instances in which inactivation of a specific oncogene that initiated tumorigenesis is sufficient to reverse tumorigenesis depend upon both cellular and genetic context.

Table 6.1 Examples of immune system-mediated oncogene addiction

	Oncogene	Tumor type	Immune compartment	Immune-mediated mechanism	References
Adaptive immunity	*MYC*	T cell acute lymphoblastic lymphoma	$CD4^+$ T cells	Induction of senescence and suppression of angiogenesis	[76]
	BCR-ABL	Pro-B cell acute lymphocytic leukemia	$CD4^+$ T cells	Induction of senescence and suppression of angiogenesis	[76]
	PML	Acute promyelocytic leukemia; prostate carcinoma	$CD8^+$ T cells	Influencing MHC class I antigen presentation	[97, 98]
	BRAF	Melanoma	Dendritic cells, CTLs	Inhibition of antigen presentation and induction of IL-10, IL-6	[99, 100]
Innate immunity	*MYC*	B cell lymphoma; pancreatic islet cell tumor	Macrophages, mast cells	Macrophage induction of senescence; mast cell promotion of angiogenesis	[73, 107]
	P53	Hepatocellular carcinoma	Neutrophils, macrophages, NK cells	Tumor clearance	[77]
	RAS	Cervical cancer; renal cell carcinoma	Neutrophils	Promotion of angiogenesis and tumor growth due to IL-6 and IL-8	[102, 103]

Murine models of *MYC* and *BCR-ABL* inactivation, as well as *P53* restoration, directly implicate immune involvement in implementation of the consequences of oncogene addiction. Oncogenes whose tumorigenicity relies heavily on immune evasion and/or pro-tumor inflammation highlight the potentially broad generalizability of the concept of immune system-mediated oncogene addiction

Importantly, the clinical relevance of oncogene addiction has been established through the development of effective therapeutics [23, 24]. The identification of potent agents, such as imatinib for chronic myelogenous leukemia and gastrointestinal stromal tumors [25], trastuzumab for breast cancer [26], and vemurafenib for melanoma [27], amongst other drugs [28], supports the paradigm of exploiting oncogene addiction for cancer therapy. Moreover, these successes underscore how elucidating the underlying principles of oncogene addiction may be generally exploited as a strategy for treating a broad spectrum of cancers.

Oncogene addiction had been presumed to be largely a consequence of cell-autonomous mechanisms that occur through processes intrinsic and exclusively dependent upon biological programs that are governed within a tumor cell, including proliferation and apoptosis (Fig. 6.2). Several mechanisms have been proposed for oncogene addiction, including the notion of abnormal tumor cell genetic circuitry [1], reversibility of tumorigenesis [29], oncogenic shock [30], and synthetic lethality [31]. More recently, the host microenvironment has been shown to play a critical role in how oncogenes initiate and maintain tumorigenesis (Fig. 6.2) [32–35].

Oncogene Addiction, Tumor Dormancy, Senescence, and Self-renewal Programs

Even brief inactivation of an oncogene may result in tumor regression [2]. However, in some instances, although oncogene inactivation appears to result in the complete loss of the neoplastic properties of a tumor, the reactivation of the oncogene results in rapid restoration of neoplastic properties [4]. Tumors that are not fully eliminated may also recur due to resistance to the targeted therapy conferred by mutation(s) in either the targeted gene or a downstream pathway [36–38]. This tumor dormancy, or the persistence of a state of minimal residual disease (MRD), therefore represents an immense hurdle to tumor elimination and ultimately patient survival [39].

One convergent feature of oncogene addiction appears to be the loss of self-renewal capacity (Figs. 6.1 and 6.2) [40–43]. Self-renewal is defined as a cell's ability to replicate limitlessly. The importance of self-renewal as an essential feature of cancer cells has been appreciated for decades [44]. Self-renewing cancer cells give rise to other self-renewing cells, as well as transient-amplifying cells, that are characterized by their rapid, yet limited, proliferation. This latter population makes up the bulk of a tumor. Recently, it was shown that a subpopulation of tumor cells retains a limitless lifespan potential. These cells are deemed cancer stem cells (CSCs) [45–50]. CSCs are maintained by multiple signaling pathways, regulated by transcription factors, including *MYC* [50–53]. The regression of tumors upon oncogene inactivation is likely a consequence of an inhibition of the self-renewal capacity of CSCs [45, 48, 50].

At least in some cases, the loss of self-renewal in cancer cells has been associated with molecular and morphological features that have been described as cellular senescence [41]. Senescence is a cellular program that was first described as a barrier to limitless proliferation of normal cells grown in vitro [54, 55], and subsequently was shown to be a conserved response to many types of cellular stress, including telomere shortening [56, 57], DNA damage [58, 59], chemotherapy treatment [60–62], and oncogene activation [63–66]. Cellular senescence is associated with permanent changes in gene expression, chromatin condensation, and induction of cell cycle arrest programs that involve p15(INK4b), p16(INK4a), and/or p53, and it is correlated with an increase in acidic beta-galactosidase enzymatic activity [67–71].

Oncogene addiction may elicit cellular senescence through at least four different mechanisms: first, through induction of expression of cell cycle arrest proteins including p15(INK4b), p16(INK4a), and p21(CIP1) [5]; second, through restoration of autocrine programs that induce cellular senescence, such as TGF-beta (TGF-β) signaling [72, 73]; third, through unopposed MAPK signaling [74, 75]; and fourth, via immune mechanisms that are apparently mediated through secreted cytokines, such as thrombospondin-1 (TSP-1) [76, 77]. These mechanisms might also be involved in inducing senescence in self-renewing CSCs, as well as the other tumor cells.

Thus, oncogene addiction could be modeled as a consequence of the balance between self-renewal and cellular senescence programs (Fig. 6.2). Cellular senescence is defined by its irreversibility, and it is the converse of self-renewal. Furthermore, cellular senescence appears to be a likely mechanism that would dictate tumor dormancy, and it would necessarily eliminate self-renewing CSCs. The absence of senescence could promote the persistence of dormant tumor cells with the latent ability to give rise to a new tumor upon cessation of a therapy. Hence, whether oncogene inactivation induces tumor elimination or tumor dormancy also appears to depend upon the balance between self-renewal and cellular senescence programs.

Oncogene Addiction, Tumor Dormancy, and the Tumor Microenvironment

Tumor cells evolve in a host with an intact immune system [78]. Coevolution of incipient tumors cells with host cells is integral to each step of tumorigenesis, including tumor initiation [6, 7], prevention [8], and progression [9]. Tumors appear to undergo immune editing, which is important for both their generation and therapeutic destruction [79, 80]. Thus, tumorigenesis is a consequence of interactions between neoplastic cells and host cells [32] that interact to regulate tumorigenesis [81].

Specific immune effectors and secreted factors have been implicated in the initiation of tumorigenesis [6, 7], as well as tumor growth, survival, and metastasis [81]. Immune effectors, including macrophages, T-cells and B-cells, have been shown to have a role in either promoting [82–84] or inhibiting [73, 85–87] tumorigenesis. For example, NK (natural killer) cells [88] can inhibit metastasis, whereas $CD4^+$ T-cells [89] and macrophages [90] have been shown to promote metastasis. Similarly, in human patients, autoimmune stimulation or inflammation can be associated with increased tumorigenesis [78, 91–93]. Immunocompromised hosts exhibit an increased incidence of certain tumors [79]. Consequently, the presence or absence of immune effectors, such as CD4+ T-cells, is associated with a favorable [94] or non-favorable prognosis [95] depending on tissue type, thereby indicating the complexity of the interaction between the host immune system and the evolving tumor. Indeed, immune cells and cytokines are important to the pathogenesis of tumorigenesis.

Oncogene activation can directly influence the immune response [96–100]. The *RET* oncogene in normal human thymocytes induces an inflammatory response that leads to tumor tissue remodeling, angiogenesis, and metastasis [101]. *RAS* upregulates expression of the cytokines IL-6 [102] and IL-8 [103], which contribute to tumorigenesis. Furthermore, *MYC* can suppress CD4+ T-cells to maintain the angiogenic tumor microenvironment in multiple tumor models [76, 104]. However, *MYC*-dependent activation of macrophages is also associated with tumor suppression [73]. Hence, oncogene activation and inactivation can have dramatic consequences on both the tumor cells and tumor microenvironment (see Table 6.1).

The host immune system is also important for the efficacy of therapeutics [10–12]. Patients with impaired immunity have decreased overall and progression-free survival in a variety of solid and hematologic malignancies [105, 106]. In colorectal carcinomas, the type, density, and intratumoral location of the T-cell infiltrate has proven a more robust predictor of patient outcome than the TNM or Duke's classification [11]. More generally, the host immune status influences the efficacy of conventional chemotherapy and radiation therapies [106].

In mouse models, the immune system can be directly interrogated mechanistically to define its role in therapeutic response [11]. For example, in mouse models of hepatocellular carcinoma, pancreatic cancer and B-cell lymphoma, innate immune components such as mast cells [107] and macrophages [73] have been implicated as barriers to tumor growth and facilitators of tumor regression. In models of colon and breast adenocarcinomas, chemotherapeutic agents and radiation therapies have been shown to elicit immunogenic apoptosis of cancer cells [108].

Multiple mechanisms of the immune system contribution to therapeutic response have been suggested, including both innate and adaptive immune effector cells, as well as specific cytokines [10–12]. Recently, it was proposed that restoration of tumor cells' "find me" and "eat me" immune stimulatory signals could potentially be used to treat cancer [108, 109]. Hence, the promotion of both the adaptive and innate arms of host immunity might be highly useful toward the complete elimination of tumor cells [108, 109].

Immune Effectors and Tumor Dormancy Versus Elimination

Specific cellular and cytokine-mediated immune effectors might define the consequences of oncogene inactivation. In experimental mouse models of *MYC- or BCR-ABL*-induced hematopoietic tumorigenesis, CD4+ T-cells appear to be essential in

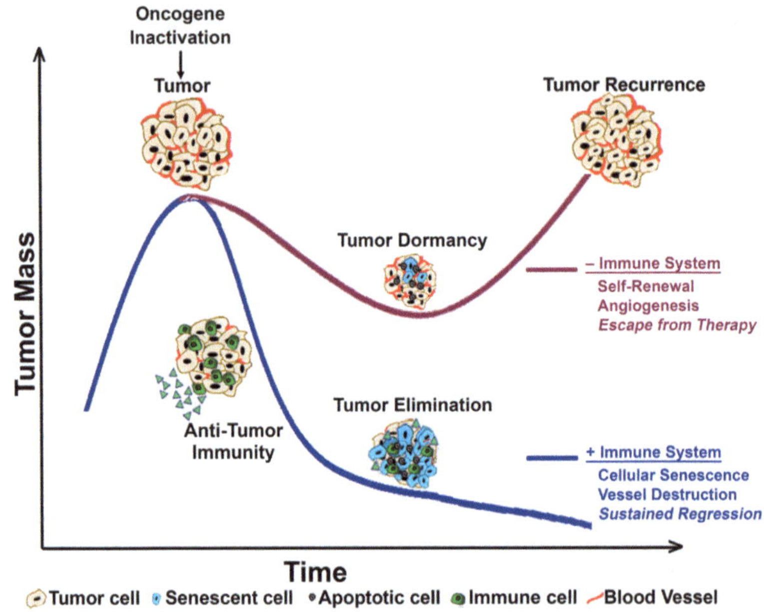

Fig. 6.3 Tumor dormancy vs. tumor elimination is regulated by an intact host immune system. Oncogene inactivation results in tumor regression regardless of host immune status due to proliferative arrest, apoptosis, and/or differentiation. In an intact host, the immune system is subsequently activated to facilitate elimination of the residual tumor cells via induction of cellular senescence and destruction of the tumor vasculature. However, in the absence of an immune system, tumors establish a state of dormancy, due to lack of inhibition of self-renewal and angiogenesis, followed by eventual escape from therapy and tumor recurrence

the mechanism of tumor regression upon oncogene inactivation (Fig. 6.3, Table 6.1) [76]. In a previous study, oncogene inactivation in *MYC*-induced tumors in CD4$^+$ T-cell immunodeficient mice resulted in significantly delayed kinetics of tumor regression and failed to completely eradicate tumor cells, leaving up to 1,000-fold more MRD than in wild-type hosts [76]. Other effectors are also recruited to the tumor site, suggesting their possible contribution [110].

CD4$^+$ T-cells contribute to sustained tumor regression by at least two mechanisms: enforcing both the induction of cellular senescence and the suppression of angiogenesis (Fig. 6.3, Table 6.1) [76]. Of importance, both of these processes have previously been characterized as hallmarks of oncogene addiction (Figs. 6.2 and 6.3). CD4$^+$ T-cells may mediate their influence on the tumor and its microenvironment directly or indirectly through the expression of many cytokines [111–114].

TSP-1 was found to be a critical mediator of CD4$^+$ T-cell-induced, sustained tumor regression upon *MYC* inactivation (Fig. 6.3). Furthermore, TSP-1 might contribute to remodeling of the tumor microenvironment upon oncogene inactivation [76, 115], and it is a potent anti-angiogenic and immune-modulatory cytokine that can induce apoptosis of endothelial cells and regulate T-cell chemotaxis [116].

TSP-1 may also mediate its effects through the regulation of TGF-β [117]. TGF-β can play a tumor-suppressive role in the tumor microenvironment [118, 119]. In particular, TGF-β can contribute to both the restraint of tumor onset and oncogene addiction through the regulation of cellular senescence upon *MYC* activation and inactivation [72, 73].

Additional cytokines and effectors may be involved in $CD4^+$ T-cell-mediated oncogene addiction. Cytokines that appear to play a role include eotaxin-1, IL-5, interferon-gamma (IFN-γ), and tissue necrosis factor-alpha (TNF-α); the down-regulation of "pro-tumor" cytokines, such as vascular epidermal growth factor (VEGF), IL-1β, and MCP-1 upon *MYC* inactivation may also play a role [76]. Whether any of these cytokines contribute more generally to the phenomenon of oncogene addiction remains to be seen.

CD4+ T-cells coordinate multiple components of both the innate and adaptive immune system [120], suggesting a likely contribution of other immune effectors to oncogene addiction. Indeed, in oncogene-induced hepatocellular carcinoma, pancreatic cancer and B-cell lymphoma, innate immune cell types such as mast cells [107] and macrophages [73] have been implicated as barriers to tumor growth and facilitators of tumor regression.

Notably, the restoration of the p53 tumor suppressor has been shown previously to induce tumor senescence, elicit chemokine expression, and induce the activation and recruitment of innate immune cells that contribute to tumor clearance [77]. Thus, the restoration of normal cellular function of a single tumor suppressor or oncogene can elicit oncogene addiction through changes in the tumor microenvironment, dependent upon various host immune effectors.

Both cellular and cytokine-associated immune mechanisms are essential components of oncogene addiction. They define the kinetics, extent, and durability of tumor elimination (Fig. 6.3). In the absence of an immune system, and upon oncogene inactivation, tumor cells persist, in a dormant state. However, in the presence of a fully intact immune system, there is complete elimination of tumor cells.

Therapeutic Implications: Tumor Dormancy Versus Elimination

For maximal clinical efficacy, a therapeutic for cancer would ideally either completely eliminate a tumor or induce a permanent state of dormancy. Since both tumor cell-intrinsic and host-dependent programs appear to be required to elicit oncogene addiction, it seems that it would be critical to consider both the tumor and the host in order to design a therapeutic that is most efficacious. Therapies that target programs in cancer cells but suppress the immune system, or those that stimulate the immune system but have no effect on the biology of a tumor cell, may not be as effective as therapies that modulate both processes in concert. In particular, therapies that target the tumor but suppress the immune system could blunt their overall efficacy. Many existing anticancer therapies cause immunosuppression and lymphodepletion that may undermine their efficacy [10].

To best identify anticancer therapies, it is critical to perform preclinical evaluation in host model systems that have an intact immune system and recapitulate a tumor microenvironment. In vitro or animal models in which a host is immunocompromised would not correctly identify the best therapeutic agents precisely because the kinetics of tumor cell elimination, the degree of tumor elimination, the ablation of MRD, and the duration of a clinical response could all be dictated by mechanisms related to the host.

The ability to identify whether a therapy will induce dormancy vs. elimination seems critical to evaluating potential therapeutics. The regulation of self-renewal vs. cellular senescence appears to be the key determinant of the fate of a tumor. The ability to interrogate self-renewal may be intrinsic to evaluating and predicting therapeutic activity. Therefore, the direct targeting of self-renewal/cellular senescence programs through the inactivation of particular oncogenes or other gene products may be a particularly effective strategy for treating cancer. This critical decision in cell fate appears to be tightly coupled to interactions between tumor cells, host cells, and cytokines, and it defines whether a tumor expands, regresses, or becomes dormant (Fig. 6.3) [39]. Hence, therapeutics that target self-renewal and/or activate cellular senescence could be very effective; for example, therapeutics that induce p53 or modulate genes that regulate the cell cycle machinery would likely be effective [41]. Therapeutic strategies that modulate the tumor microenvironment may also be useful adjuncts, including drugs that target angiogenesis [121]. A combination of approaches would likely be most effective for tumor elimination.

Finally, the appreciation that immune mechanisms can dictate the balance between self-renewal and senescence suggests that therapeutic manipulation of a host's immune system and secreted cytokines may be an important treatment strategy. Specific host immune effectors and chemokines profoundly influence the consequences of therapeutic oncogene inactivation, radiation therapy, and chemotherapy [11]. The integration of targeted and immune therapies may be the most efficacious strategy in treating cancer [122].

Modeling and Predicting Tumor Dormancy Versus Elimination

The mechanistic understanding of oncogene addiction should make it possible to predict therapeutic efficacy. Oncogene addiction involves both tumor cell-intrinsic and host-dependent programs that regulate self-renewal, proliferation, apoptosis, and cellular senescence. Hence, the incorporation of these cellular programs into a model may yield a model that can predict oncogene addiction [30, 123]. One such possible approach would be to presume that cancer cells behave stochastically and can exist in three different states: proliferating, apoptotic, or quiescent/dormant. Then, the acquisition of even very simple measurements of proliferation and apoptosis combined with assessments of tumor size could be used to mathematically predict oncogene addiction [123].

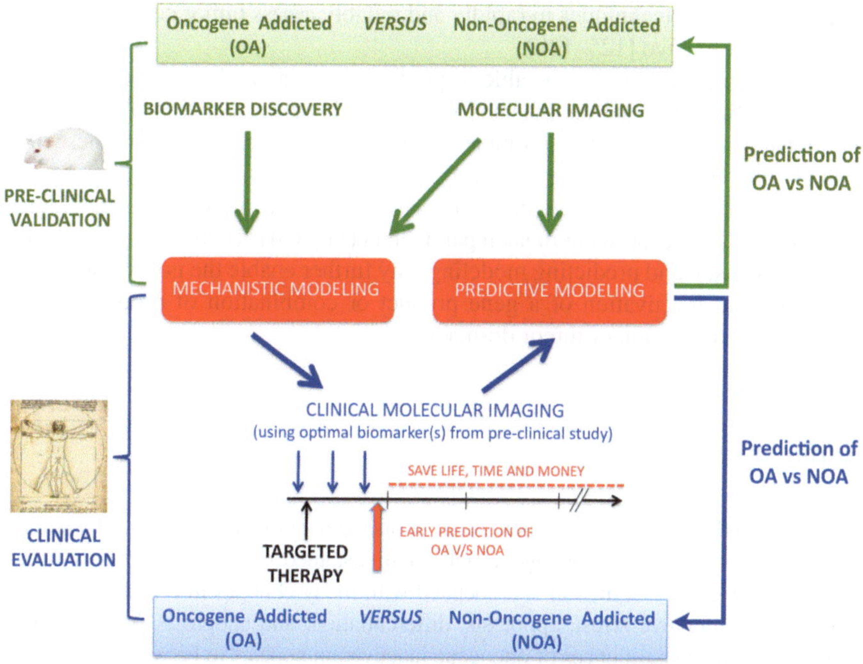

Fig. 6.4 Utilizing modeling to predict when oncogene inactivation will result in tumor dormancy vs. tumor elimination. A combinatorial, iterative approach can be used to model the consequences of oncogene addiction, thereby allowing for prediction of response to targeted therapy in patients. Continuing progress in molecular imaging and biomarker discovery will be crucial for the validation of mathematical models of oncogene addiction in primary animal models of cancer. In parallel, the advent of new molecular imaging tools in the clinic will allow for incorporation of discoveries in these preclinical models into assessment of human cancer patient response to therapy

Such modeling has revealed some possible insights into the mechanism of oncogene addiction and tumor regression following oncogene inactivation [123]. A simple differential decay between pro-survival and pro-death signals is sufficient to explain the majority of what occurs upon oncogene inactivation. A decay of both pro-survival and pro-death signals follows targeted oncogene inactivation. Although, the final level of the pro-death signal is comparable to the pro-survival signal, it is precisely because the death signals induced by the oncogene are extinguished more slowly after oncogene inactivation than the survival signals that tumors regress. These results support the oncogenic shock hypothesis, first suggested by both Settleman and Kaelin [30, 31].

Mathematical modeling of the response to targeted therapy indicates that simple measurements of tumors before and after initiation of a therapeutic may be useful toward predicting therapeutic outcome [124]. A variety of different computational approaches could be used, and this could potentially be useful in enabling the more rapid identification of therapeutics as well as the more rapid discontinuation of therapies that are not effective (Fig. 6.4). This approach would exploit both existing and emerging imaging techniques to rapidly and reliably assess tumor cell proliferation and apoptosis ex vivo [125–129].

Even simple models may be able to predict oncogene addiction with measurement of proliferation and apoptosis alone [123]. However, the inclusion of certain additional parameters, such as immune cell infiltration, onset of cellular senescence, loss of self-renewal, and suppression of angiogenesis, would likely improve the modeling. New molecular imaging approaches, as well as proteomic technologies, may enable the measurement of such parameters (Fig. 6.4). Then, the application of both mechanistic and predicting modeling may further enable the goal of predicting when targeted inactivation of a gene product or combination of products would elicit tumor elimination or tumor dormancy.

Summary

An understanding of the mechanistic basis of oncogene addiction will reveal when the inactivation of a particular oncogene will elicit sustained tumor regression vs. tumor dormancy. Oncogene addiction is associated with many changes in cellular programs, including proliferative arrest, apoptosis, differentiation, and apoptosis. Yet, central to determining the consequences of oncogene inactivation appears to be whether this elicits a permanent loss of self-renewal through the program of cellular senescence. Importantly, both cell-autonomous tumor-intrinsic mechanisms and host-dependent mechanisms contribute to the consequences of oncogene inactivation and of self-renewal. Hence, modeling approaches that incorporate both tumor and host factors appear to likely be able to predict the consequences of targeted oncogene inactivation, their influence on self-renewal programs, and thus the clinical efficacy of therapeutics.

Acknowledgements The authors would like to acknowledge current members of the Felsher laboratory for critical discussion and previous members who have contributed to characterizing various models of oncogene addiction.

References

1. Weinstein IB (2002) Cancer. Addiction to oncogenes—the Achilles heal of cancer. Science 297(5578):63–64
2. Felsher DW, Bishop JM (1999) Reversible tumorigenesis by MYC in hematopoietic lineages. Mol Cell 4(2):199–207
3. Jain M et al (2002) Sustained loss of a neoplastic phenotype by brief inactivation of MYC. Science 297(5578):102–104
4. Shachaf CM et al (2004) MYC inactivation uncovers pluripotent differentiation and tumour dormancy in hepatocellular cancer. Nature 431(7012):1112–1117
5. Wu CH et al (2007) Cellular senescence is an important mechanism of tumor regression upon c-Myc inactivation. Proc Natl Acad Sci U S A 104(32):13028–13033
6. de Visser KE, Korets LV, Coussens LM (2005) De novo carcinogenesis promoted by chronic inflammation is B lymphocyte dependent. Cancer Cell 7(5):411–423

7. Dougan M et al (2011) A dual role for the immune response in a mouse model of inflammation-associated lung cancer. J Clin Invest 121(6):2436–2446
8. Shankaran V et al (2001) IFNgamma and lymphocytes prevent primary tumour development and shape tumour immunogenicity. Nature 410(6832):1107–1111
9. Ruffell B et al (2010) Lymphocytes in cancer development: polarization towards pro-tumor immunity. Cytokine Growth Factor Rev 21(1):3–10
10. Zitvogel L, Kepp O, Kroemer G (2011) Immune parameters affecting the efficacy of chemotherapeutic regimens. Nat Rev Clin Oncol 8(3):151–160
11. Galon J et al (2006) Type, density, and location of immune cells within human colorectal tumors predict clinical outcome. Science 313(5795):1960–1964
12. Dave SS et al (2004) Prediction of survival in follicular lymphoma based on molecular features of tumor-infiltrating immune cells. N Engl J Med 351(21):2159–2169
13. Felsher DW (2003) Cancer revoked: oncogenes as therapeutic targets. Nat Rev Cancer 3(5):375–380
14. Weinstein IB, Joe A (2008) Oncogene addiction. Cancer Res 68(9):3077–3080; discussion 3080
15. Sharma SV, Settleman J (2007) Oncogene addiction: setting the stage for molecularly targeted cancer therapy. Genes Dev 21(24):3214–3231
16. Huettner CS et al (2000) Reversibility of acute B-cell leukaemia induced by BCR-ABL1. Nat Genet 24(1):57–60
17. Chin L et al (1999) Essential role for oncogenic Ras in tumour maintenance. Nature 400(6743):468–472
18. Hoeflich KP et al (2006) Oncogenic BRAF is required for tumor growth and maintenance in melanoma models. Cancer Res 66(2):999–1006
19. Boxer RB et al (2004) Lack of sustained regression of c-MYC-induced mammary adenocarcinomas following brief or prolonged MYC inactivation. Cancer Cell 6(6):577–586
20. Giuriato S et al (2006) Sustained regression of tumors upon MYC inactivation requires p53 or thrombospondin-1 to reverse the angiogenic switch. Proc Natl Acad Sci USA 103(44): 16266–16271
21. Shchors K et al (2006) The Myc-dependent angiogenic switch in tumors is mediated by interleukin 1beta. Genes Dev 20(18):2527–2538
22. Tran PT et al (2008) Combined inactivation of MYC and K-Ras oncogenes reverses tumorigenesis in lung adenocarcinomas and lymphomas. PLoS One 3(5):e2125
23. Hait WN, Hambley TW (2009) Targeted cancer therapeutics. Cancer Res 69(4):1263–1267; discussion 1267
24. Sawyers C (2004) Targeted cancer therapy. Nature 432(7015):294–297
25. Druker BJ et al (2001) Efficacy and safety of a specific inhibitor of the BCR-ABL tyrosine kinase in chronic myeloid leukemia. N Engl J Med 344(14):1031–1037
26. Hudis CA (2007) Trastuzumab—mechanism of action and use in clinical practice. N Engl J Med 357(1):39–51
27. Chapman PB et al (2011) Improved survival with vemurafenib in melanoma with BRAF V600E mutation. N Engl J Med 364(26):2507–2516
28. Cataldo VD et al (2011) Treatment of non-small-cell lung cancer with erlotinib or gefitinib. N Engl J Med 364(10):947–955
29. Felsher DW (2004) Reversibility of oncogene-induced cancer. Curr Opin Genet Dev 14(1):37–42
30. Sharma SV, Settleman J (2006) Oncogenic shock: turning an activated kinase against the tumor cell. Cell Cycle 5(24):2878–2880
31. Kaelin WG Jr (2005) The concept of synthetic lethality in the context of anticancer therapy. Nat Rev Cancer 5(9):689–698
32. Albini A, Sporn MB (2007) The tumour microenvironment as a target for chemoprevention. Nat Rev Cancer 7(2):139–147
33. Hanahan D, Weinberg RA (2011) Hallmarks of cancer: the next generation. Cell 144(5):646–674
34. Coussens LM et al (2000) MMP-9 supplied by bone marrow-derived cells contributes to skin carcinogenesis. Cell 103(3):481–490

35. Bissell MJ, Radisky D (2001) Putting tumours in context. Nat Rev Cancer 1(1):46–54
36. Choi PS et al (2011) Lymphomas that recur after MYC suppression continue to exhibit oncogene addiction. Proc Natl Acad Sci U S A 108(42):17432–17437
37. Gorre ME et al (2001) Clinical resistance to STI-571 cancer therapy caused by BCR-ABL gene mutation or amplification. Science 293(5531):876–880
38. Pao W et al (2005) KRAS mutations and primary resistance of lung adenocarcinomas to gefitinib or erlotinib. PLoS Med 2(1):e17
39. Aguirre-Ghiso JA (2007) Models, mechanisms and clinical evidence for cancer dormancy. Nat Rev Cancer 7(11):834–846
40. Felsher DW (2010) MYC inactivation elicits oncogene addiction through both tumor cell-intrinsic and host-dependent mechanisms. Genes Cancer 1(6):597–604
41. Nardella C et al (2011) Pro-senescence therapy for cancer treatment. Nat Rev Cancer 11(7):503–511
42. Felsher DW (2008) Reversing cancer from inside and out: oncogene addiction, cellular senescence, and the angiogenic switch. Lymphat Res Biol 6(3–4):149–154
43. Felsher DW (2008) Tumor dormancy and oncogene addiction. APMIS 116(7–8):629–637
44. Bishop JM (1991) Molecular themes in oncogenesis. Cell 64(2):235–248
45. Reya T et al (2001) Stem cells, cancer, and cancer stem cells. Nature 414(6859):105–111
46. Dick JE (2008) Stem cell concepts renew cancer research. Blood 112(13):4793–4807
47. Wang JC, Dick JE (2005) Cancer stem cells: lessons from leukemia. Trends Cell Biol 15(9):494–501
48. Sell S (2004) Stem cell origin of cancer and differentiation therapy. Crit Rev Oncol Hematol 51(1):1–28
49. Nguyen LV et al (2012) Cancer stem cells: an evolving concept. Nat Rev Cancer 12(2): 133–143
50. Das B et al (2009) The idea and evidence for the tumor stemness switch. In: Rajasekhar V, Vemuri M (eds) Regulatory networks in stem cells. Humana Press, New York, pp 473–487
51. Zheng H et al (2008) Pten and p53 converge on c-Myc to control differentiation, self-renewal, and transformation of normal and neoplastic stem cells in glioblastoma. Cold Spring Harb Symp Quant Biol 73:427–437
52. Wang J et al (2008) c-Myc is required for maintenance of glioma cancer stem cells. PLoS One 3(11):e3769
53. Marquardt JU et al (2011) Human hepatic cancer stem cells are characterized by common stemness traits and diverse oncogenic pathways. Hepatology 54(3):1031–1042
54. Hayflick L, Moorhead PS (1961) The serial cultivation of human diploid cell strains. Exp Cell Res 25:585–621
55. Hayflick L (1965) The limited in vitro lifetime of human diploid cell strains. Exp Cell Res 37:614–636
56. Lundblad V, Szostak JW (1989) A mutant with a defect in telomere elongation leads to senescence in yeast. Cell 57(4):633–643
57. Yu GL et al (1990) In vivo alteration of telomere sequences and senescence caused by mutated tetrahymena telomerase RNAs. Nature 344(6262):126–132
58. Chen Q, Ames BN (1994) Senescence-like growth arrest induced by hydrogen peroxide in human diploid fibroblast F65 cells. Proc Natl Acad Sci U S A 91(10):4130–4134
59. Di Leonardo A et al (1994) DNA damage triggers a prolonged p53-dependent G1 arrest and long-term induction of Cip1 in normal human fibroblasts. Genes Dev 8(21):2540–2551
60. Schmitt CA et al (2002) A senescence program controlled by p53 and p16INK4a contributes to the outcome of cancer therapy. Cell 109(3):335–346
61. Chang BD et al (1999) A senescence-like phenotype distinguishes tumor cells that undergo terminal proliferation arrest after exposure to anticancer agents. Cancer Res 59(15): 3761–3767
62. Michishita E et al (1999) 5-Bromodeoxyuridine induces senescence-like phenomena in mammalian cells regardless of cell type or species. J Biochem 126(6):1052–1059

63. O'Brien W, Stenman G, Sager R (1986) Suppression of tumor growth by senescence in virally transformed human fibroblasts. Proc Natl Acad Sci U S A 83(22):8659–8663
64. Serrano M et al (1997) Oncogenic ras provokes premature cell senescence associated with accumulation of p53 and p16INK4a. Cell 88(5):593–602
65. Zhu J et al (1998) Senescence of human fibroblasts induced by oncogenic Raf. Genes Dev 12(19):2997–3007
66. Braig M et al (2005) Oncogene-induced senescence as an initial barrier in lymphoma development. Nature 436(7051):660–665
67. Dimri GP et al (1995) A biomarker that identifies senescent human cells in culture and in aging skin in vivo. Proc Natl Acad Sci U S A 92(20):9363–9367
68. Narita M et al (2003) Rb-mediated heterochromatin formation and silencing of E2F target genes during cellular senescence. Cell 113(6):703–716
69. Collado M et al (2005) Tumour biology: senescence in premalignant tumours. Nature 436(7051):642
70. Chan HM et al (2005) The p400 E1A-associated protein is a novel component of the p53 ->p21 senescence pathway. Genes Dev 19(2):196–201
71. Chen Z et al (2005) Crucial role of p53-dependent cellular senescence in suppression of Pten-deficient tumorigenesis. Nature 436(7051):725–730
72. van Riggelen J et al (2010) The interaction between Myc and Miz1 is required to antagonize TGFbeta-dependent autocrine signaling during lymphoma formation and maintenance. Genes Dev 24(12):1281–1294
73. Reimann M et al (2010) Tumor stroma-derived TGF-beta limits Myc-driven lymphomagenesis via Suv39h1-dependent senescence. Cancer Cell 17(3):262–272
74. Zhuang D et al (2008) C-MYC overexpression is required for continuous suppression of oncogene-induced senescence in melanoma cells. Oncogene 27(52):6623–6634
75. Lin AW et al (1998) Premature senescence involving p53 and p16 is activated in response to constitutive MEK/MAPK mitogenic signaling. Genes Dev 12(19):3008–3019
76. Rakhra K et al (2010) CD4(+) T cells contribute to the remodeling of the microenvironment required for sustained tumor regression upon oncogene inactivation. Cancer Cell 18(5): 485–498
77. Xue W et al (2007) Senescence and tumour clearance is triggered by p53 restoration in murine liver carcinomas. Nature 445(7128):656–660
78. Balkwill F, Mantovani A (2001) Inflammation and cancer: back to Virchow? Lancet 357(9255):539–545
79. Dunn GP et al (2002) Cancer immunoediting: from immunosurveillance to tumor escape. Nat Immunol 3(11):991–998
80. Schreiber RD, Old LJ, Smyth MJ (2011) Cancer immunoediting: integrating immunity's roles in cancer suppression and promotion. Science 331(6024):1565–1570
81. de Visser KE, Eichten A, Coussens LM (2006) Paradoxical roles of the immune system during cancer development. Nat Rev Cancer 6(1):24–37
82. Andreu P et al (2010) FcRgamma activation regulates inflammation-associated squamous carcinogenesis. Cancer Cell 17(2):121–134
83. Girardi M et al (2004) Characterizing the protective component of the alphabeta T cell response to transplantable squamous cell carcinoma. J Invest Dermatol 122(3):699–706
84. Lin EY et al (2001) Colony-stimulating factor 1 promotes progression of mammary tumors to malignancy. J Exp Med 193(6):727–740
85. Hung K et al (1998) The central role of CD4(+) T cells in the antitumor immune response. J Exp Med 188(12):2357–2368
86. Martin-Manso G et al (2008) Thrombospondin 1 promotes tumor macrophage recruitment and enhances tumor cell cytotoxicity of differentiated U937 cells. Cancer Res 68(17): 7090–7099
87. Zhou P et al (2010) Mature B cells are critical to T-cell-mediated tumor immunity induced by an agonist anti-GITR monoclonal antibody. J Immunother 33(8):789–797

88. Wu J, Lanier LL (2003) Natural killer cells and cancer. Adv Cancer Res 90:127–156
89. DeNardo DG et al (2009) CD4(+) T cells regulate pulmonary metastasis of mammary carcinomas by enhancing protumor properties of macrophages. Cancer Cell 16(2):91–102
90. Condeelis J, Pollard JW (2006) Macrophages: obligate partners for tumor cell migration, invasion, and metastasis. Cell 124(2):263–266
91. Coussens LM, Werb Z (2002) Inflammation and cancer. Nature 420(6917):860–867
92. Greten FR et al (2004) IKKbeta links inflammation and tumorigenesis in a mouse model of colitis-associated cancer. Cell 118(3):285–296
93. Ekbom A et al (1990) Ulcerative colitis and colorectal cancer. A population-based study. N Engl J Med 323(18):1228–1233
94. Wakabayashi O et al (2003) CD4+ T cells in cancer stroma, not CD8+ T cells in cancer cell nests, are associated with favorable prognosis in human non-small cell lung cancers. Cancer Sci 94(11):1003–1009
95. Zhang JP et al (2009) Increased intratumoral IL-17-producing cells correlate with poor survival in hepatocellular carcinoma patients. J Hepatol 50(5):980–989
96. Mantovani A et al (2008) Cancer-related inflammation. Nature 454(7203):436–444
97. Zhang H et al (2003) Concordant down-regulation of proto-oncogene PML and major histocompatibility antigen HLA class I expression in high-grade prostate cancer. Cancer Immun 3:2
98. Zheng P et al (1998) Proto-oncogene PML controls genes devoted to MHC class I antigen presentation. Nature 396(6709):373–376
99. Sumimoto H et al (2006) The BRAF-MAPK signaling pathway is essential for cancer-immune evasion in human melanoma cells. J Exp Med 203(7):1651–1656
100. Boni A et al (2010) Selective BRAFV600E inhibition enhances T-cell recognition of melanoma without affecting lymphocyte function. Cancer Res 70(13):5213–5219
101. Borrello MG et al (2005) Induction of a proinflammatory program in normal human thyrocytes by the RET/PTC1 oncogene. Proc Natl Acad Sci U S A 102(41):14825–14830
102. Ancrile B, Lim KH, Counter CM (2007) Oncogenic Ras-induced secretion of IL6 is required for tumorigenesis. Genes Dev 21(14):1714–1719
103. Sparmann A, Bar-Sagi D (2004) Ras-induced interleukin-8 expression plays a critical role in tumor growth and angiogenesis. Cancer Cell 6(5):447–458
104. Sodir NM et al (2011) Endogenous Myc maintains the tumor microenvironment. Genes Dev 25(9):907–916
105. Boshoff C, Weiss R (2002) AIDS-related malignancies. Nat Rev Cancer 2(5):373–382
106. Ray-Coquard I et al (2009) Lymphopenia as a prognostic factor for overall survival in advanced carcinomas, sarcomas, and lymphomas. Cancer Res 69(13):5383–5391
107. Soucek L et al (2007) Mast cells are required for angiogenesis and macroscopic expansion of Myc-induced pancreatic islet tumors. Nat Med 13(10):1211–1218
108. Shiao SL, Coussens LM (2010) The tumor-immune microenvironment and response to radiation therapy. J Mammary Gland Biol Neoplasia 15(4):411–421
109. Obeid M et al (2007) Calreticulin exposure dictates the immunogenicity of cancer cell death. Nat Med 13(1):54–61
110. Restifo NP (2010) Can antitumor immunity help to explain "oncogene addiction"? Cancer Cell 18(5):403–405
111. Acosta JC et al (2008) Chemokine signaling via the CXCR2 receptor reinforces senescence. Cell 133(6):1006–1018
112. Beatty G, Paterson Y (2001) IFN-gamma-dependent inhibition of tumor angiogenesis by tumor-infiltrating CD4+ T cells requires tumor responsiveness to IFN-gamma. J Immunol 166(4):2276–2282
113. Kuilman T et al (2008) Oncogene-induced senescence relayed by an interleukin-dependent inflammatory network. Cell 133(6):1019–1031
114. Muller-Hermelink N et al (2008) TNFR1 signaling and IFN-gamma signaling determine whether T cells induce tumor dormancy or promote multistage carcinogenesis. Cancer Cell 13(6):507–518
115. Li SS et al (2002) T lymphocyte expression of thrombospondin-1 and adhesion to extracellular matrix components. Eur J Immunol 32(4):1069–1079

116. Li SS et al (2006) Endogenous thrombospondin-1 is a cell-surface ligand for regulation of integrin-dependent T-lymphocyte adhesion. Blood 108(9):3112–3120
117. Young GD, Murphy-Ullrich JE (2004) The tryptophan-rich motifs of the thrombospondin type 1 repeats bind VLAL motifs in the latent transforming growth factor-beta complex. J Biol Chem 279(46):47633–47642
118. Becker C et al (2004) TGF-beta suppresses tumor progression in colon cancer by inhibition of IL-6 trans-signaling. Immunity 21(4):491–501
119. Tang B et al (2007) Transforming growth factor-beta can suppress tumorigenesis through effects on the putative cancer stem or early progenitor cell and committed progeny in a breast cancer xenograft model. Cancer Res 67(18):8643–8652
120. Muranski P, Restifo NP (2009) Adoptive immunotherapy of cancer using CD4(+) T cells. Curr Opin Immunol 21(2):200–208
121. Gonzalez-Angulo AM, Hortobagyi GN, Ellis LM (2011) Targeted therapies: peaking beneath the surface of recent bevacizumab trials. Nat Rev Clin Oncol 8(6):319–320
122. Wrzesinski C et al (2010) Increased intensity lymphodepletion enhances tumor treatment efficacy of adoptively transferred tumor-specific T cells. J Immunother 33(1):1–7
123. Tran PT et al (2011) Survival and death signals can predict tumor response to therapy after oncogene inactivation. Sci Transl Med 3(103):103ra99
124. Califano A (2011) Striking a balance between feasible and realistic biological models. Sci Transl Med 3(103):103ps39
125. Willmann JK et al (2008) Molecular imaging in drug development. Nat Rev Drug Discov 7(7):591–607
126. Pysz MA, Gambhir SS, Willmann JK (2010) Molecular imaging: current status and emerging strategies. Clin Radiol 65(7):500–516
127. Massoud TF, Gambhir SS (2003) Molecular imaging in living subjects: seeing fundamental biological processes in a new light. Genes Dev 17(5):545–580
128. Nguyen QD, Aboagye EO (2010) Imaging the life and death of tumors in living subjects: preclinical PET imaging of proliferation and apoptosis. Integr Biol 2(10):483–495
129. Michalski MH, Chen X (2011) Molecular imaging in cancer treatment. Eur J Nucl Med Mol Imaging 38(2):358–377

Part III
Immune System

Chapter 7
Multifaceted Kinetics of Immuno-Evasion from Tumor Dormancy

Alberto d'Onofrio

Abstract Tumor progression is subject to modulation by the immune system. The immune system can eliminate tumors or keep them at a dormant equilibrium size, while some tumors escape immunomodulation and advance to malignancy. Herein, we discuss some aspects of immune evasion of dormant tumors from a theoretical biophysics point of view that can be modeled mathematically. We go on to analyze the mathematical system on multiple timescales. First, we consider a long timescale where tumor evasion is likely due to adaptive (and somewhat deterministic) immuno-editing. Then, we consider the temporal mesoscale and hypothesize that extrinsic noise could be a major factor in induction of immuno-evasion. Implications of immuno-evasive mechanisms for the outcome of immunotherapies are also discussed. In addition, we discuss the ideas that population level tumor dormancy may not be a quiescence phenomenon and that dormant tumors can, at least if modulated by the immune system, live a very active and noisy life!

Keywords Tumor dormancy • Immune system • Immuno-evasion • Immuno-editing • Systems biomedicine

Introduction

Tumor cells express specific antigens, for example through under- or over-expressed proteins, mutated proteins, or many other factors, that can trigger a response from both the innate and adaptive immune system (IS) [29, 61, 80]. This finding led to the hypothesis of immune surveillance formulated in 1908 by the Nobel laureate Paul Ehrlich [48]. According to this hypothesis, the IS can act to potently inhibit

A. d'Onofrio (✉)
Department of Experimental Oncology, European Institute of Oncology,
Via Ripamonti 435, Milano 20141, Italy
e-mail: alberto.donofrio@ifom-ieo-campus.it

H. Enderling et al. (eds.), *Systems Biology of Tumor Dormancy*, Advances in Experimental Medicine and Biology 734, DOI 10.1007/978-1-4614-1445-2_7,
© Springer Science+Business Media New York 2013

and even eliminate tumors [15, 48]. The interest in tumor immune surveillance had a resurgence after the late 1950s, thanks to studies by Burnet [15–17] and Thomas [90], who must be considered the other two fathers of immuno-oncology in addition to Ehrlich. Immune surveillance remained a very debated hypothesis up to the late 1990s due to inconsistent or inadequate animal models [46, 47]. Only in the past decade or so has a large amount of experimental and epidemiological data been accumulated which evidently demonstrated that the IS can suppress tumors [47]. Tumor suppression, however, is only one phase of tumor–immune interactions. Termed as the three *"E"*s of immunomodulation, the IS can *E*liminate tumors, keep them in a state of *E*quilibrium, or allow them to *E*vade and progress to malignancy [47].

The competitive interaction between tumor cells and the IS involves a large number of players and events, and is as such extremely complex. Furthermore, the interactions between tumor cells and the IS are strongly nonlinear and *time-varying* phenomena at two different scales, slow and fast, which offer possible causes of tumor evasion from immune control [47]. The *fast scale* corresponds to early phases of growth when cells of the IS learn to recognize and target tumor cells [52, 86], and neoplasms may *E*scape and progress before an effective immune response can be launched [52, 86]. Additionally, tumor cells might be identified as "self" by the IS and hence easily escape [29, 57]. On the other hand, a series of adaptive phenomena shape the onco-immune interplay at the *slow scale*, enabling a dynamic *E*quilibrium such that the tumor survives in a dormant state undetectable by diagnostic equipment [47]. The presence of an IS-induced dormant tumor state was indirectly inferred from clinical data for many decades and predicted by mathematical models [28, 71, 73, 88]. Only recently Koebel and coworkers [64] were able to show experimentally in a mouse model such an IS-induced *E*quilibrium state. The IS-induced dormant state can remain at a constant size or be subject to "short term-small amplitude" oscillations [53, 59, 75, 93], making microscopically oscillating dormant states conceivable.

Externally induced dormant equilibria and oscillating micro-attractors can be disrupted. Disease-related impairments of innate and adaptive ISs or immunosuppressive treatments preceding organ transplantations have been shown epidemiologically and in mouse studies to allow tumors to escape from dormancy and continue growth [47, 89].

However, there is a major class of causes of disruption of the equilibrium that are not related to immunosuppression. Indeed, dormant tumors may themselves develop strategies to escape IS control [47, 64, 80]. From an ecological point of view, it is conceivable that tumors adapt to survive in a hostile, antitumor immune-mediated environment [47]. For example, tumors may develop mechanisms to spread by reducing their immunogenicity over a long period of time [47, 80]. The escaping tumors are called to be "edited"—or, more suggestively said, "sculpted"—by the host IS [47]. A possibility, which we only partially share, is that the IS involuntarily promotes tumor progression by selecting for immunoneutral tumor cell subtypes that unavoidably arise due to the intrinsic genetic instability of tumor cells [47]. A fourth *"E"* in the immunomodulation process is therefore argued to be immuno-*E*diting [47].

For the above-outlined reasons, the current theory including the whole possible time-course of the interaction between a tumor and the IS has been defined as "immuno-editing" [47]. In the immuno-editing framework, no permanent dormancy can exist: the equilibrium is a transitory state whose fate unavoidably ends with evasions in the long run. Of course, due to the slowness of the adaptation process, the length of the equilibrium phase can in many plausible conditions yield tumor dormancy lasting years or decades, thus offering an explanation for the observations that many adults bear asymptomatic dormant tumors [47]. Finally, it is important to remark that in the long temporal range not only the above-mentioned slow evolutionary processes but also IS degradation due to natural senescence [70] may theoretically contribute to long-term evasions of dormant tumors.

A large body of biomedical research on immuno-evasive strategies has accumulated, and a recent, highly recommendable monograph [52] was devoted to some of these strategies and their correlation with the effectiveness of immunotherapies. Adaptive evasion phenomena do not stop in the presence of therapies, so tumor adaptation can be seen as a process competing with treatment [52].

The concept underlying immunotherapy is modification of the natural interplay between tumor cells and both the innate and specific immune system in favor of the latter [31]. This simple and promising idea, however, has only yielded globally controversial medical investigations [1, 10, 58], partly due to huge inter-subject variability, and partly due to variations in specific kinds of immunotherapy.

Immunotherapies are divided into two classes: passive and active therapies [30, 31]. Probably the most important passive therapy is *Adoptive Cellular Immunotherapy*, where cultured, activated immune effector cells are injected into the diseased host [67]. Active immunotherapies aim at boosting the immune response, for example by expanding the proliferation of cytotoxic T lymphocytes (CTLs). Among these therapies, a prominent role is played by *interleukin-based therapies* [20, 76].

The complexity and nonlinearity of the interplay between a tumor and the immune system lend itself to quantitative modeling [23,81,83]. Differential equation models have been developed with constant [28,32,35,36,45,62,63,65,71,73,74,88] or stochastically varying [41, 72] parameters, and hybrid stochastic-deterministic models [21]. The basic idea of most of these models is simple: tumor cells and the effector cells of the IS are competing populations. Tumor cells are the prey of the immune effectors, whose proliferation and local recruitment are stimulated, in turn, by the presence of the tumor. Tumor cells, however, also induce a loss of effector cells; and the influx of effectors may depend on the size and/or growth rate of the tumor population [35, 38, 71, 88]. Tumor dynamics can be studied in greater detail when cellular interactions are represented by generalized kinetic models of nonlinear statistical mechanics [7–9] or agent-based approaches [79, 82]. Owen and Sherrat developed a spatiotemporal model focusing on the role of macrophage dynamics in tumors [77, 78]. These investigators showed that macrophage chemo-attraction toward tumor cells yields both traveling wave solutions and a heterogeneous spatial distribution of tumor cells. Matzavinos, Chaplain, and Kuznetsov proposed a spatiotemporal model of the interactions between tumor cells and CTLs that includes spatial motility of both cell types and CTL chemotaxis [24, 74]. The model is used to study IS-modulated dormant tumor states and reveals through a series of simulations in which tumor cells are spatially distributed in an irregular pattern during dormancy and their number oscillates in a non-periodic fashion. We extended an earlier general framework model [35] with the empirical inclusion of the immuno-editing phenomenon simply by introducing slowly time-varying generic parameters in the metamodels (timescales significantly longer than those typical of the tumor–IS interaction) [38]. The behavioral strategies interrelated with phenotypic changes were described similar to the Lotka–Volterra models with adaptively changing interaction strength [26, 66] with slowly varying parameters [26] and non-monotonically varying parameters [22]. Bellomo recently proposed a generic kinetic approach for immune effector *learning* and tumor cells *hiding* based on the concept of mutual change of activity levels [7].

As a case study of mathematical modeling of immuno-evasion and its implications for therapies, we herein review and extend a recent mathematical model [43] that introduces a dynamic description of tumor evasion, based on the concept that a tumor cell that survived an encounter with a CTL earns a decreased immunosensitivity. This "information gain" can be communicated to other tumor cells, thereby reducing the overall tumor sensitivity to immune-induced cell killing.

This novel hypothesis is in line with the general idea of tumor escape from immune response. Indeed, Stewart and Abrams stressed that tumor cells might escape from immune control through two pathways: (a) mechanisms that involve the secretion of soluble factors; and (b) mechanisms that are dependent on the contact between the tumor cells and the effectors and that are aimed at reducing antigen recognition/adhesion and apoptotic resistance [89]. Such factors are currently believed to induce the evolution of immunosuppressive networks [60]. The model hypothesizes that soluble factors also contribute to intercellular communication. This could, for example, be related to experimental findings by Kurnick et al. [68], who showed that melanoma cells produce soluble factors that diminish Melanoma-A/MART-1 Ag expression with the concomitant loss of recognition by the specific CTLs. *Immunolearning* and *oncohiding* are important and fascinating mechanisms that describe evasion of dormant tumors at a temporal mesoscale, and might be responsible for immune surveillance failure in general.

Another important factor that has been extensively investigated is stochastic fluctuations of tumor proliferation rates [2, 6, 14, 55, 56, 72, 94], as those fluctuations can trigger escape as well as elimination of the neoplasm. Given the complexity and multistability of the tumor–IS interplay, it is conceivable that statistical fluctuations of immune levels trigger noise-induced transitions from dormancy.

We recently showed that the classical method of noise-induced transition must be adapted for immune surveillance models [41]. Classical approaches based on Gaussian perturbations in biology have a limited range of application ([37, 39, 44] and references therein), and suitable bounded noises should be used. The classical theory of noise-induced transitions [56] refers to the study of the qualitative changes in stationary probability densities: $P_{\mathrm{st}}(x) = \lim_{t\to\infty} P(x,t)$ and is therefore asymptotic [41]. Humans and all other living beings, however, have a finite lifespan, and thus the lifespan of the host organisms must set a natural limit, which makes the velocity of convergence to $P_{st}(x)$ an essential parameter. If this velocity is low and the attractor is practically reached in times that are greater than the average lifespan of the organisms being studied, one must investigate the possible qualitative changes of $P(x,t)$ during its transitory period, namely at some given realistic times.

The influence of IS fluctuations has been studied with the result that an increase in the variance of the noise yields (partial) rejection of a tumor [72]. The effect of the IS, however, has been assumed to be independent of tumor size, and is therefore applicable primarily to largely immunogenic tumors. We extended a phenomenological model [14,94] to account for the decreased cytotoxic effectiveness of immune effector cells [41], and focused on the response of this biological system to stochastic bounded perturbation in immune levels. As we will discuss later, immuno-evasion can be induced by noise. This escape depends on the probability density of the "input" noise, and not only on its variance as in classical models.

Case Study 1: Mathematical Modeling of Immuno-Editing: Tumor Evasion at Large Timescales

The Kuznetsov Model of Tumor–Immune System Interplay

In this section, we briefly introduce and discuss the main properties of the well-known model by Vladimir A. Kuznetsov and colleagues [71, 74], describing the growth of an immunogenic tumor and the interplay with CTLs. Let $x(t)$ denote the size of a tumor population at time t, $y(t)$ the size of the CTL compartment at time t, and $C(t)$ the number of tumor cell–CTL complexes. Tumor growth under immunosurveillance can be described by:

$$
\begin{aligned}
x' &= rx\left(1-\frac{x}{K}\right) - kxy + k_{-1}C + k_2(1-p)C \\
y' &= \frac{fC}{a+x} - \mu_0 y - kxy + k_{-1}C + k_2 pC + \sigma, \\
C' &= kxy - \delta C - k_{-1}C - k_2 C,
\end{aligned}
\tag{7.1}
$$

where

- r represents the baseline tumor growth rate (for small x, $x' \approx rx$)
- K represents the host carrying capacity. For low immunogenic tumors, $x(t) \to K$. The carrying capacity is a theoretical endpoint to tumor growth, as the host often dies before this carrying capacity can be reached. In the absence of an immune reaction, the tumor follows a logistic law: $x' = rx\left(1-\frac{x}{K}\right)$, which is one of many possible phenomenological laws describing the dynamics of tumor growth [35,40]

- kxy represents the binding rate between tumor cells and immune effector cells
- Tumor cell–CTL complexes are cleared at a constant rate δ and have a total rate of unbinding $k_{-1}+k_2$
- k_{-1} is the rate at which neither the tumor cell nor the CTL are damaged
- k_2 is the rate at which either the tumor cell [29, 61] (with probability p) or the CTL [27] (with rate $1-p$) are lethally damaged. The counterattack of tumor cells leading to the death of the CTL was theoretically predicted by Kuznetsov in 1979 [69]. p is typically very large. A value fit from data of lymphoma in chimeric mice is $p \approx 0.9997$ [25, 74]. We shall refer to p as the "probability of dying"
- The tumor-stimulated recruitment rate of CTL is modeled by: $fC/(a+x)$
- The baseline death rate of CTLs is μ_0
- σ is the external inflow of CTLs in the region where the tumor is localized [71]

This model was validated with data from experiments on the dynamics of growth of a BCL_1 lymphoma in the spleen of chimeric mice [24, 25, 71, 74]. Since the lifespan of the complexes is very short [49], it can be assumed that complexes are at quasi-equilibrium [71], such that:

$$(\delta + k_{-1} + k_2)C \approx kxy,$$

thus reducing system (7.1) to the following bidimensional system:

$$\begin{aligned} x' &= rx\left(1-\frac{x}{K}\right) - k\frac{k_2}{\delta+k_{-1}+k_2}pxy \\ y' &= \frac{\beta x}{a+x}y - \left(\mu_0 + k\frac{k_2}{\delta+k_{-1}+k_2}(1-p)x\right)y + \sigma. \end{aligned} \tag{7.2}$$

The dynamics of this and variations of this model are characterized by a vast repertoire of nonlinear behaviors that mimic the complex interactions between tumors and CTLs [35, 71, 74].

Modeling Immuno-Escape

In this section, we propose and analyze a novel mechanism that potentially underlies tumor escape from immune surveillance. The proposed mechanism may augment and is not mutually exclusive with more classical mechanisms, such as spontaneous emergence of low immunogenic clones due to, for example, genetic instability of tumor cells [47].

A shortcoming of the above-introduced model of tumor–IS interplay is the time-independent, constant number of tumor cells being killed (k_2pC in the model (7.1)) and number of tumor-stimulated effectors being born and/or recruited per time unit ($fC/(a+x)$ in the model (7.1)) [38]. Initially focusing on the interpretation of the biological natural history of tumors, a rough "kinematic" approach was employed by qualitatively introducing some time-varying, explicit parameters. For immuno-editing models, time-decreasing parameters were introduced in the family of bidimensional models [35], and both simulations and a bifurcation analysis were performed [38]. These models showed that the aggressive growth of a tumor can be understood biophysically as a catastrophic transition from a locally stable dormant state to a globally attractive macroscopic steady state near carrying capacity in the absence of immune reactions.

Here we are interested in offering an explicit and biophysically grounded model of long-term evasion from immune control. We hypothesize that tumor cells that encounter and survive a CTL interaction "learn" how to evade immune surveillance and are consequently equipped with a decreased probability p of being killed by CTLs in the future:

$$p(t+\mathrm{d}t) = p(t) - \eta(p) \times (\mathrm{d}t\,(k_{-1} + k_2(1-p))C) \times p,$$

where $\eta(p) \geq 0$, such that:

$$p' = -\eta(p)\,(k_{-1} + k_2(1-p))Cp. \tag{7.3}$$

We implicitly assume that the information acquired by surviving tumor cells after successful detachment from CTL is transmitted to other cells via rapid intercellular communication, which is a rapid process. The same mechanism might also act upon encounter with effectors, for example, of the innate IS.

Note that, as $p(t)$ is a probability, (7.3) must be such that if $p(0) \in [0,1]$, then $p(t) \in [0,1]$ for all times $-\infty < t < +\infty$. This is trivially verified for $t \geq 0$. In order for that condition to also be true for $t < 0$, η has to be set as:

$$\eta(1) = 0. \tag{7.4}$$

Thus we set:

$$\eta(p) = \eta_0(p)(1-p), \tag{7.5}$$

where $\eta_0(p) \geq 0$ is bounded, for example: $\eta_0(p) = const.$
Thus (7.3) reads:

$$p' = -\eta_0(p)\,(k_{-1} + k_2(1-p))C(1-p)p. \tag{7.6}$$

Assuming that the dynamics of the complexes C is very fast $(k_{-1} + k_2) \gg 1$, they may considered at quasi-equilibrium, i.e., $C \approx kxy/(\delta + k_{\ 1} + k_2)$. Then (7.6) becomes:

$$p' = -\eta_0(p)\,(\delta + k_{-1} + k_2(1-p))\frac{k}{\delta k_{-1} + k_2}xy(1-p)p. \tag{7.7}$$

Note that in a microscopic tumor without appreciable immuno-evasion $p(0) \approx 1$. Furthermore, since the adaptive rate of tumor cells at each encounter is arguably small, it follows that for a long time-interval $p' \approx 0$, which is in line with the assumption that immune evasion occurs on long time scales.

The total number of complexes that do not lead to the death of the involved tumor cell, i.e., the number of nonlethal encounters for tumor cells, is given by:

$$N'(t) = (k_{-1} + k_2(1 - p(t)))C(t). \tag{7.8}$$

By combining (7.6) and (7.8), one straightforwardly obtains:

$$\frac{\mathrm{d}}{\mathrm{d}N} p = -\eta_0(p)p(1-p). \tag{7.9}$$

In the case of constant η_0, (7.9) yields:

$$p(N) = \frac{p(0)}{p(0) + (1 - p(0))\mathrm{Exp}(\eta_0 N)}, \tag{7.10}$$

which is the mathematical realization of the intuitive fact that the probability q is a decreasing function of the total number of nonlethal complex-forming encounters N.

Remark. Equation (7.6) is formally the model of an evolutionary imitation game [50], which in this case is asymmetric since the positive payoff is 0 and the negative payoff is proportional to the encounter rate kxy. ♢

The tumor–CTL binding rate k encodes two distinct phenomena: the baseline rate k^0, at which a tumor cell encounters an immune cell, as well as the probability z that an immune cell recognizes the tumor cell, such that k should be modeled as a time-varying function as follows:

$$k(t) = k^0 z(t).$$

Additionally, the probability z may also be subject to evolutionary changes. Thus, similar to (7.3), we may write:

$$z' = -\gamma_0(z)\,(k_{-1} + k_2(1-p))\,C(1-z)z. \tag{7.11}$$

In contrast to previous studies where f was assumed to be constant [43], here we introduce $f(t)$ to be time varying. This is motivated by tumor-stimulated recruitment of CTLs, which is modeled by means of the function $fC/(a+x)$, being dependent on the antigenic exposition of tumor cells. Thus, we shall adopt the simplest possible model:

$$f(t) = f^0 z(t).$$

Quite interestingly, the use of a quasi-equilibrium approximation reveals further insights into the dynamics of z:

$$z' = -\gamma_0(z)\,(\delta + k_{-1} + k_2(1-p))\,\frac{k^0}{\delta k_{-1} + k_2} xy(1-z)z^2. \tag{7.12}$$

Finally, we briefly discuss the release of immunosuppressive factors by tumor cells [52] and the link to our hypothesis on the onset of immuno-evasion. We assume that surviving tumor cells acquire and transmit information, allowing an increase in the production of an immunosuppressive factor that may induce apoptosis in CTLs. This requires new terms in the equation for $y(t)$ of the static model [74]. We assume that surviving tumor cells produce a factor W, which is taken up by and induces toxicity in CTLs. Let $\beta(t)$ be the production rate of this factor:

$$W' = \beta(t)x - qyW - \mathrm{d}W,$$

where q is the uptake rate by CTLs and d is the factor's natural decay. Assuming that W rapidly degrades, we can set $W \approx (\beta(t)/d)x/(1+\varepsilon y)$, where $\varepsilon = q/d$. By further assuming that the CTL death rate induced by the factor W is proportional to its uptake rate $\gamma^* qyW$, the dynamics of CTLs are governed by:

$$y' = \frac{f(t)C}{a+x} - \mu_0 y - k(t)xy + k_{-1}C + k_2 pC + \sigma - b(t)b_{\mathrm{M}}\frac{x}{1+\varepsilon y}y, \tag{7.13}$$

where $b_{\mathrm{M}} = (\gamma^*\varepsilon)\beta_{\mathrm{M}}$, with β_{M} being the maximum production rate of W, and $0 \leq b(t) \leq 1$.
Thus, proceeding as in the previous section yields the following equation:

$$b' = \xi_0(b)\,(k_{-1} + k_2(1-p))C(1-b)b, \tag{7.14}$$

where $(1-b)$ follows from the saturation in the production of the immunotoxic factor.

Simulations

The asymptotic behavior of the above-discussed models is trivial, as $\lim_{t\to+\infty} p(t) = 0^+$ and $\lim_{t\to+\infty} z(t) = 0$. Of interest to tumor dormancy escape is to assess the typical transitory behaviors during simulated realistic lifespan of the host organism. We first simulated models (7.1)–(7.6) using parameters that were estimated in [74] for the model (7.1) with constant p. The values and meanings of the parameters, and of the state variables, are reported in Tables 7.1 and 7.2. For notational convenience, we rescaled the system with a unit of 10^6 cells and nondimensionalized the variables x, y, and c.

Table 7.1 State variable and parameters, mainly related to the model (7.1)

Variable	Values	Source	Meaning
x	$0 < x < K$	[43, 71, 74]	Nondimensional size of tumor cells compartment
y	$y > 0$	[43, 71, 74]	Nondimensional size of CTLs compartment
c	$c > 0$	[43, 71, 74]	Nondimensional size of TC–CTL complexes compartment
r	$0.18\,\text{day}^{-1}$	Fit from BCL_1 mouse data [25, 71]	Baseline exponential growth rate of the tumor
K	500	Fit from BCL_1 mouse data [25, 71, 74]	Carrying capacity
k^0	$k \in (0.1, 0.4)$	Fit from BCL_1 mouse data [25, 71, 74]	Binding rate at $t = 0$ between tumor cells and CTLs
f^0	$29.88\,\text{day}^{-1}$	Fit from BCL_1 mouse data [25, 71, 74]	Parameter related to CTLs recruitment
a	20.19	Fit from BCL_1 mouse data [25, 71, 74]	Parameter related to CTLs recruitment
μ_0	$0.0412\,\text{day}^{-1}$	Fit from BCL_1 mouse data [25, 71, 74]	Baseline death rate of CTLs
σ	$0.0136\,\text{day}^{-1}$	Fit from BCL_1 mouse data [25, 71, 74]	Local influx of CTLs
k_{-1}	$24.0\,\text{day}^{-1}$	Fit from BCL_1 mouse data [25, 71, 74]	Unbinding rate without cellular damages
k_2	$7.2\,\text{day}^{-1}$	Fit from BCL_1 mouse data [25, 71, 74]	Unbinding rate with cellular damages
δ	$0\,\text{day}^{-1}$	Fit from BCL_1 mouse data [25, 71]	Loss rate of complexes
p^0	0.9997	Fit from BCL_1 mouse data [25, 71, 74]	Probability at $t = 0$ that a tumor cells dies after meeting a CTL

Table 7.2 State variable and parameters, mainly related to our time-varying adaptive extension of model (7.1)

Variable	Values	Source	Meaning
$p(t)$	$0 < p(t) < p^0$	Defined here	Time varying probability that a tumor cells dies after meeting a CTL
η_0	$0 < \eta_0 \ll 1$	Defined here	Baseline decrease rate of $p(t)$
$z(t)$	$0 < z(t) < 1$	Fit from BCL_1 mouse data [25, 71, 74]	Probability that a tumor cells is recognized by a CTL
γ_0	$0 < \gamma_0 \ll 1$	Defined here	Baseline decrease rate of $z(t)$
b_M	$b_M = 0.4\,\text{day}^{-1}$	Assumed	Cytotoxic chemicals-related parameter (see the text)
ε	$\varepsilon = 0.01$	Assumed	Cytotoxic chemicals-related parameter (see the text)
$b(t)$	$0 < b(t) < 1$	Defined here	Proportional to the time varying production rate of the cytotoxic chemical (see the text)
ξ_0	$0 < \xi_0 \ll 1$	Defined here	Baseline decrease rate of $b(t)$

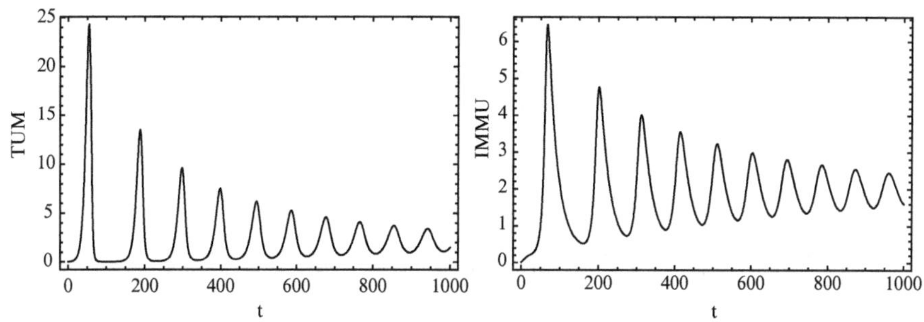

Fig. 7.1 *Immunomodulation of a tumor without adaptive escape learning for a strong binding rate* $k = 0.4\,day^{-1}$. **(a)** Tumor size $x(t)$ in the absence of dynamical changes in $p(t)$ and $z(t)$. **(b)** Corresponding size of immune system effectors $y(t)$. Time is measured in days. Other parameters: $\gamma_0 = \eta_0 = 0$ $z(t) = 1$ and $p = 0.9997$

The estimated probability that a tumor cell dies after an encounter with a CTL is $p^0 = 0.9997$ [74]. We simulated tumor dynamics for $t = 1{,}000$ days ($\approx$3 years), which is in line with the lifespan of chimeric mice. In Fig. 7.1, we set the tumor cells–CTLs binding rate to $k^0 = 0.4\,\text{day}^{-1}$. With a constant tumor cell death rate after de-complexification $\eta_0 = 0$, tumor size as well as immune effectors exhibited damped oscillations around a small value indicative of population-level tumor dormancy. However, for a non-null decrease rate of the "probability of dying" $\eta_0 = 4.5 * 10^{-3}$, a sudden onset of immuno-evasion at $t \approx 850$ days can be observed (Fig. 7.2). An initially oscillating immune response disappears after the tumor escape immune surveillance. At the onset of immuno-evasion, $p \approx 0.82$, which is reduced from its initial value p^0 but still remarkably large. Figure 7.3 shows the dependence of immuno-evasion onset on η_0. One can see that for $4.5 * 10^{-3} < \eta < 3.6 * 10^{-2}$ the onset of escape starts after adequate delays, whereas for $\eta_0 < 0.86 * 4.5 * 10^{-3}$ tumor escape is seen only at times greater than 1,000 days. In Fig. 7.4, we considered the effects of a smaller binding rate: $k = 0.25\ \text{day}^{-1}$. In contrast to the previous simulations, tumor size (as well as the immune effectors) exhibited sustained oscillations spacing from relatively small tumor sizes up to quite large sizes with $\eta = 0$. For a non-null decrease rate $\eta = 1.1 * 10^{-3}$, however, sudden immuno-evasion occurred at $t \approx 850$ days (second panel).

We now focus on the role of the probability $z(t)$ that an immune cell can recognize the tumor cell, and report baseline simulations with constant $p(t) = 0.997$ (i.e., $\eta = 0$), $\gamma_0 = \text{const} > 0$ and $z(0) = 0.9999$. Including the dynamics of $z(t)$ revealed that the system is more robust with respect to this time-varying parameter than with respect to $p(t)$. Indeed, immuno-evasion occurs only for small values of the "recognition probability" $z(t)$. Moreover, the dynamics of $z(t)$ were slower than those of $p(t)$ obtained in the previous set of simulations. We conclude that a larger (constant) "decrease rate" γ_0 is required to enable immuno-evasion than the "decrease rate" of the probability of dying (η_0). For example, to reproduce a

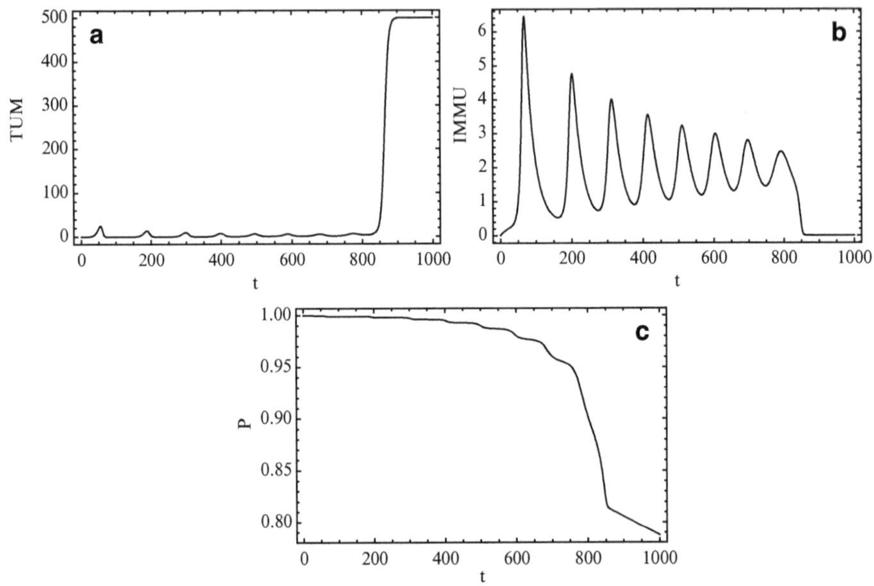

Fig. 7.2 *Immunoevasion after dynamical changes in the probability of cancer cell dying* $p(t)$ *for* $\eta = 4.5 * 10^{-3}$ *and a strong binding rate* $k = 0.4\,day^{-1}$. **(a)** behavior of $x(t)$: immuno-evasion onset at $t \approx 850$. **(b)** Behavior of the CTLs. **(c)** Time changes of $p(t)$. Other parameters are as in [74]. Time is measured in days. Other parameters: $\gamma_0 = \eta_0 = 0$ $z(t) = 1$ and $p = 0.9997$

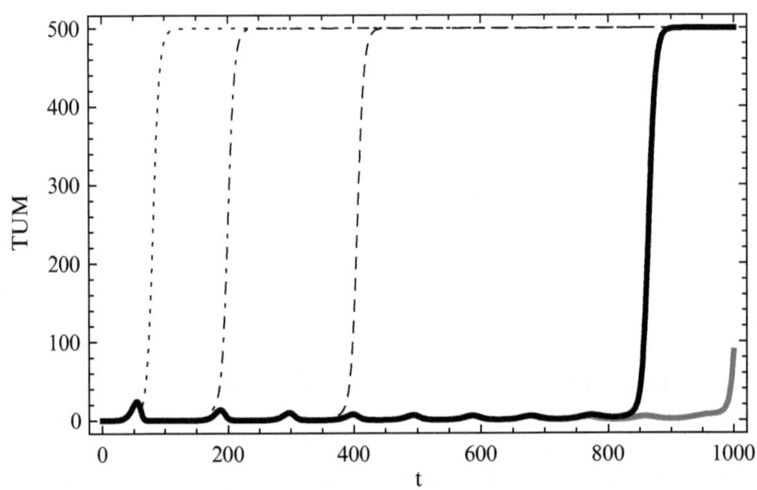

Fig. 7.3 *Effect on* $x(t)$ *of varying the decrease rate* η_0 *of the probability of dying for a strong binding rate* $k = 0.4\,day^{-1}$. *Solid gray line* $\eta_0 = 0.86 * 4.5 * 10^{-3}$; *solid black line*: $\eta_0 = 4.5 * 10^{-3}$; *dashed line*: $\eta_0 = 2 * 4.5 * 10^{-3}$; *dot-dashed line*: $\eta_0 = 4 * 4.5 * 10^{-3}$; *dotted line*: $\eta_0 = 8 * 4.5 * 10^{-3}$. Other parameters are as in [74]. Time is measured in days. Other parameters: $\gamma_0 = 0$ $z(t) = 1$ and $p(0) = 0.9997$

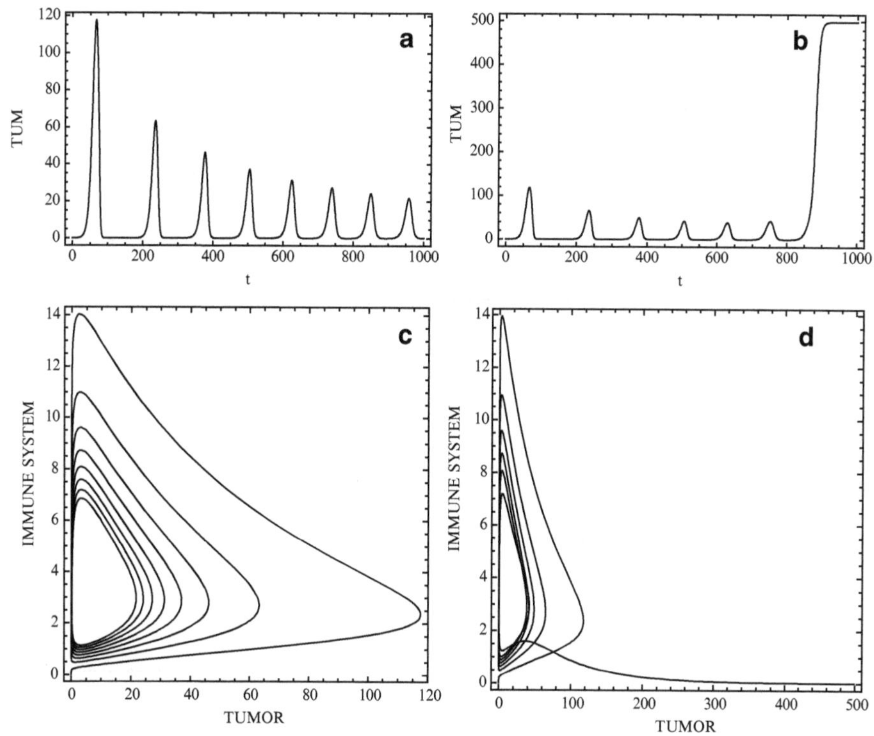

Fig. 7.4 *Immuno-evasion due to dynamical changes in the probability of dying* $p(t)$ *, for a weaker binding rate* $k = 0.25\, day^{-1}$. Tumor size $x(t)$, and phase plane $(x(t), y(t))$ in the absence of dynamical changes in p and z are shown, respectively, in panels (**a**) and (**c**). Panels (**b**) and (**d**) effects of dynamical changes in $p(t)$ in the case $\eta = 1.1 * 10^{-3}$. Panel (**b**) behavior of $x(t)$: immuno-evasion onsets at $t \approx 850$. Panel (**d**) corresponding phase-plane plot. Time is measured in days. Other parameters: $\gamma_0 = \eta_0 = 0$ $z(t) = 1$ and $p^0 = 0.9997$, and as in [74]

behavior similar to that reported in Fig. 7.1, we had to set $\gamma_0 = 0.013$. In Fig. 7.5, we show the simulations for the case $k^0 = 0.25$ (as in Fig. 7.1, where we recall $z(t) = 1$). Immuno-evasion is triggered at $t \approx 800$ for $\gamma_0 \approx 0.0030$, when $z(800) \approx 0.2$. Figure 7.6 shows how the onset of immuno-evasion depends on γ_0, suggesting that γ_0 should range in $3.47 * 10^{-3} < \gamma_0 < 1.56 * 10^{-2}$ for successful escape.

To investigate immuno-evasion through tumor-produced factors that are toxic to CTLs, we performed numerical simulations with varying ξ_0. The system produced a behavior that is qualitatively similar to those we reported above (c.f. Fig. 7.7).

In the previous examples, we focused on the case where, in the absence of immuno-evasive mechanisms, the tumor remains dormant. Here, we shall consider the impact of the proposed mechanisms on an "artificial" tumor dormancy induced by a highly idealized and efficient immunotherapy. We consider a tumor growing in a microenvironment characterized by a small baseline influx rate of CTLs,

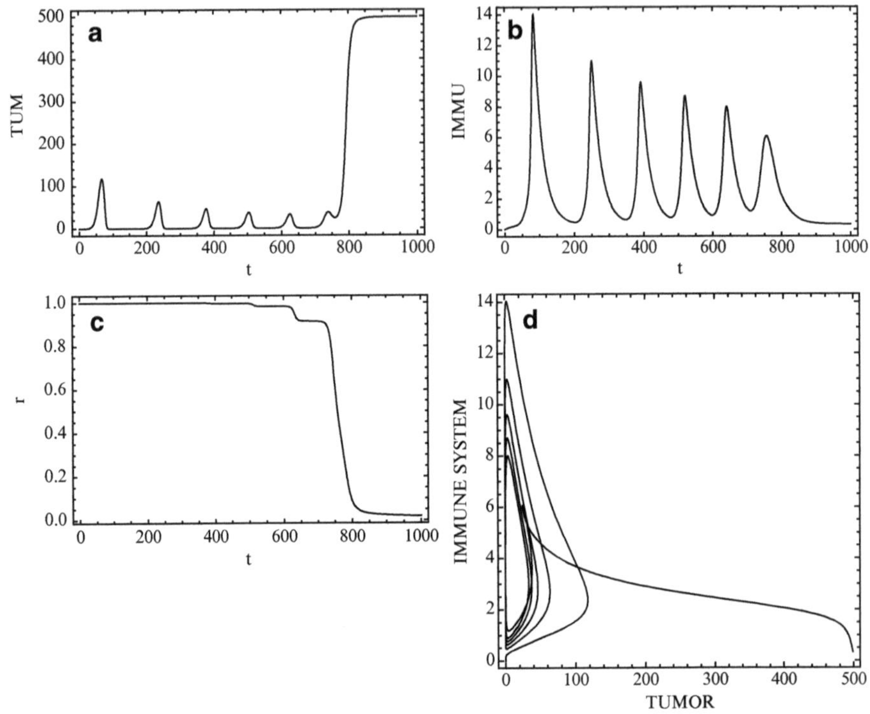

Fig. 7.5 *Behavior of tumor–CTL interplay in the presence of dynamical changes in the probability that a tumor cell is recognized by CTLs* $z(t)$ *in the case with constant* $\gamma_0(z) = 0.0039$, $k = 0.25\,day^{-1}$, $z(0) = 0.9999$, $\eta_0 = 0$ *and* $p^0 = 0.9997$. (**a**) Behavior of $x(t)$: immuno-evasion onsets at $t \approx 800$. (**b**) Behavior of the CTLs. (**c**) Time changes of $p(t)$. (**d**) (x,y) phase plane plot. Other parameters are as in [74]. Time is measured in days

$\sigma^* = 0.00136\,\text{day}^{-1}$, which is an order of magnitude less than the value reported in Tables 7.1 and 7.2 and used in the above simulations. In the absence of immune-evasion mechanisms, the tumor grows intrinsically rapidly and reaches 80% of its theoretical carrying capacity in 80 days, which is likely lethal for the host. By the inclusion of immuno-evasion mechanisms characterized by a time-varying probability of dying with decreased rate $\eta = 0.0005$, the full carrying capacity is reached at $t = 90$ days. In both cases, we simulated a highly idealized adoptive cellular immunotherapy, (Fig. 7.8) whose effect is modeled by increasing by 20-fold the rate σ up to the value $\sigma = 0.0272\,\text{day}^{-1}$. The therapy starts continuously at $t = 14\,\text{day}$, which is, of course, an idealized scenario. In both cases, therapy is initially very effective, and in the case with constant probability of dying ($\eta = 0$), the tumor remains controlled at a small size. In an intrinsically immuno-evasive tumor with $\eta = 0.0005$, however, tumor progression is observed at $t \approx 150$ days.

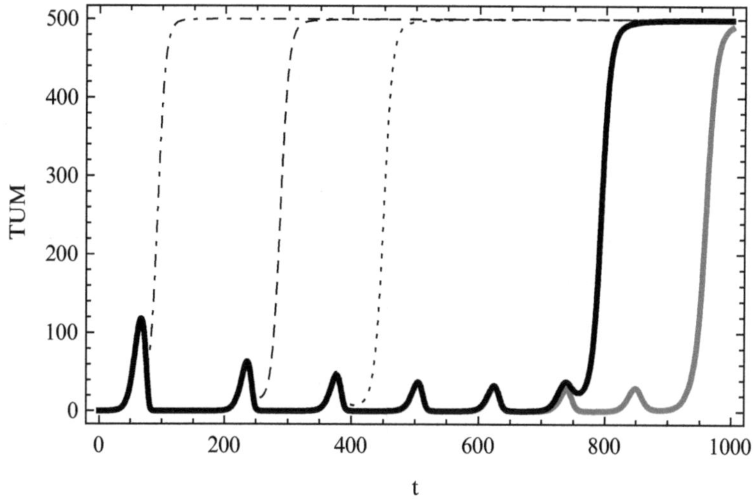

Fig. 7.6 *Immuno-evasion through decreased tumor detection rate* γ_0 or a weaker binding rate $k = 0.25$ day^{-1}, $\eta_0 = 0$ $z(t) = 0.9999$ and $p(0) = 0.9997$. *Solid gray line*: $\gamma_0 = 0.89 * 3.9 * 10^{-3}$; *solid line*: $\gamma_0 = 3.9 * 10^{-3}$; *dotted line*: $\gamma_0 = 1.5 * 3.9 * 10^{-3}$; *dashed line*: $\gamma_0 = 2 * 3.9 * 10^{-3}$; *dot-dashed line*: $\gamma_0 = 4 * 3.9 * 10^{-3}$. Other parameters are as in [74]

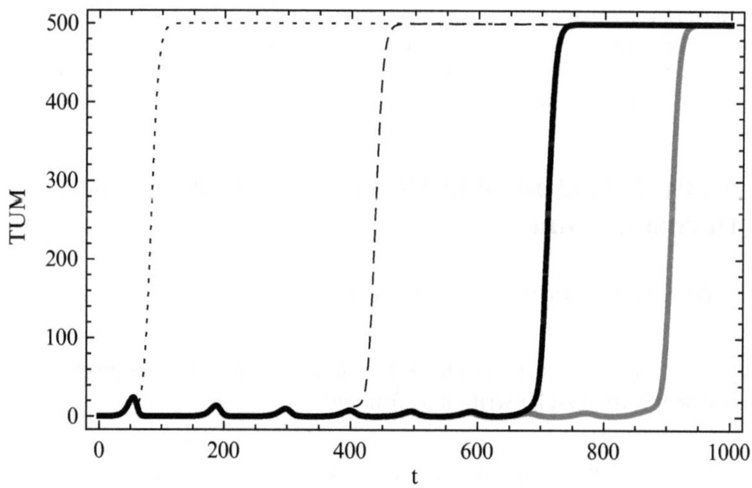

Fig. 7.7 *Immuno-evasion through the production of factors toxic for CTLs, in the case of constant* $p(t)$ *and* $k(t)$. Effect on $x(t)$ of varying ξ_0 in the case of $\varepsilon = 0.01$, $b_M = k = 0.4$ day^{-1} and $p(0) = 0.9997$. *Solid gray line*: $\xi_0 - 4 * 10^{-3}$, *solid line*: $\zeta_0 = 5 * 10^{-3}$; *dashed line*: $\xi_0 = 7.5 * 10^{\ 3}$; *dotted line*: $\gamma_0 = 3.5 * 10^{-2}$. Other parameters are as in [74]

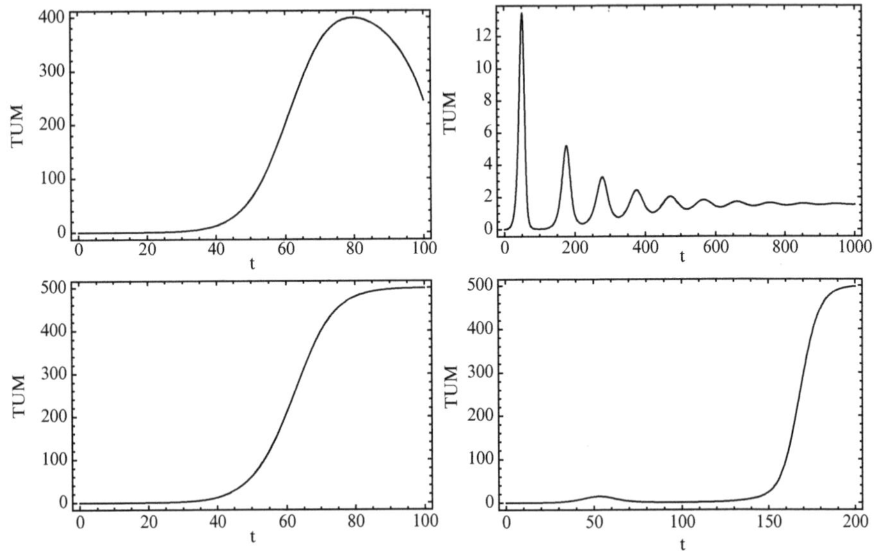

Fig. 7.8 *Behavior of tumor–CTLs under an ideal adoptive cellular immunotherapy* in the absence or presence of dynamical changes of *the probability of dying* $p(t)$. Baseline influx rate of CTLs: $\sigma = 0.00136\,\text{day}^{-1}$, $k = 0.25\,\text{day}^{-1}$, $p(0) = 0.9997$. (**a**) Tumor growth in the absence of therapy and constant $p(t)$. (**b**) Controlled growth of $x(t)$ in the presence of therapy such that for $t \geq 14$day it is $\sigma = 0.0272\,\text{day}^{-1}$, but constant $p(t)$. (**c**) Tumor growth in the absence of therapy and with time-varying $p(t)$ with $\eta = 0.0005$. (**d**) Growth of $x(t)$ in the presence of therapy such that for $t \geq 14$day it is $\sigma = 0.0272\,\text{day}^{-1}$, but time-varying $p(t)$ with $\eta = 0.0005$. Other parameters are as in [74]. Time is measured in days

Case Study 2: Modeling the Mesoscale Immuno-Evasion of a "Dormant" Tumor

A Toy Model of Tumor-Immune System Interplay

In certain studies[14,94], a toy model of tumor–IS interactions was proposed, which, in dimensional form, can be written as follows:

$$X' = (p_0 - \delta_0)X - jX^2 - m_0X - \frac{\beta_0 X^2}{1 + \left(\frac{X}{c}\right)^2}, \tag{7.15}$$

where $X(t)$ is the size of the tumor at time t, p_0X and δ_0X are respectively the baseline proliferation and apoptotic rates, jX^2 accounts for intercellular competition (e.g., for nutrients), and $\phi(X)X := \beta_0X^2/(1+(X/c)^2)$ is the rate of tumor cell lysis by the IS. Let β be the baseline IS strength. The term $-m_0X$ models the interplay of tumor cells with the innate immune defense [57]. This also allows us to identify $\phi(X)X$ as the contribution due to specific immune defenses.

Despite its simplicity, this model reproduces some of the basic properties of the interactions between tumor cells and the IS, including first and foremost multistability [14, 94]. This toy model is interesting because the specific rate of lysing of tumor cells by the immune effectors $\phi(X)$ is non-monotonic, which correlates with the fact that small tumors might produce an insufficient amount of antigens whereas large tumors decrease the ability of the IS to react due to immunosuppressors [35, 86]. One must note that if $m_0 < p_0 - \delta_0$, the model is underestimating the effectiveness of the immune response toward tumor cells, as (7.15) becomes $x' = (p_0 - \delta_0 - m_0)x > 0$ for small tumors. Since immune surveillance for small tumors is excluded, IS-induced tumor eradication cannot be achieved in the absence of therapies, unless the innate defenses are so strong that $m_0 > p_0 - \delta_0$. This limitation of the model must be stressed, but the model is applicable for tumors whose immunogenic activity is intermediate or low for low levels of X.

As in many tumor-growth models, this model also does not capture the myriad of complex dynamic molecular mechanisms that underlie the processes leading to proliferation, programmed cell death, senescence, and, of course, interaction with the IS. Thus, virtually all parameters appearing in (7.15) can be considered variable and affected by a major or minor extent of noise. Here, we shall focus on noisy variations of the baseline IS interaction rate β_1.

If $m_0 > p_0 - \delta_0$, it is convenient to nondimensionalize (7.15) with time unit $\tau_u = (p_0 - \delta_0 - m_0)^{-1}$ and tumor size $X_u = c$, which yields:

$$x' = x - \frac{x^2}{K} - \frac{\beta x^2}{1 + x^2}, \tag{7.16}$$

where $K = cj\tau_u$ and $\beta = \beta_0 c \tau_u^{-1}$. Note that τ_u, assumed here as reference time for nondimensionalizing the model, would be the characteristic time of exponential growth of the tumor (determined by the balance of tumor cells growth, apoptosis and other loss causes, and cell loss due to the interplay with innate immunity effectors) in the ideal case of absence of nonlinear competition of cells.

Let us assume that K is sufficiently "large", i.e., neither the self-competition term $-jX^2$ nor the adaptive IS-related term $-m_0 X$ induces dormancy. The deterministic behavior of the solutions of (7.16) is simple. There exist a β^* and $\beta^{**} > \beta^*$, such that, (1) if $0 < \beta < \beta^*$ there exists a unique and globally attractive macroscopic equilibrium that is near the carrying capacity K; (2) if $\beta > \beta^{**}$ there exists a small globally attractive equilibrium; and (3) if $\beta^* < \beta < \beta^{**}$ there exists a central unstable equilibrium and two equilibria, one microscopic and the other macroscopic. As a consequence at $\beta = \beta^*$ and at $\beta = \beta^{**}$, there is a hysteresis bifurcation.

Including Stochastic Fluctuations

Let the baseline IS strength, β, be subject to stochastic varying of the IS by adding a white noise:

$$\beta(t) = \beta + \sigma \xi(t),$$

where $\xi(t)$ is a white noise of unitary intensity. Therefore, $X(t)$ becomes a stochastic process, whose dynamics are ruled by the following Langevin–Ito stochastic differential equation (SDE) [56]:

$$x' = x - \frac{x^2}{K} - \frac{\beta x^2}{1+x^2} - \xi(t)\sigma\frac{x^2}{1+x^2}. \tag{7.17}$$

In a Langevin–Ito SDE, $x' = f(x) + \sigma g(x)\xi(t)$, the probability density of the random variable X at time t, denoted as $P(x,t)$, is obtained by solving the Fokker–Planck equation (FPE). It is possible to show that the FPE in the unidimensional case, such as ours, has a unique stationary solution of the form:

$$P_{\mathrm{st}}(x) = M\mathrm{Exp}\left(-\frac{2}{\sigma^2}U_{\mathrm{eff}}(x)\right), \tag{7.18}$$

where the function $U_{\mathrm{eff}}(x)$ (referred to as "effective potential" or "probability potential" in statistical physics [56]) is such that:

$$U_{\mathrm{eff}}(x) = \sigma^2 \log(g(x)) - \int_x \frac{f(z)}{g^2(z)}.$$

Note that the stationary probability density, whenever it exists, is globally attractive, i.e., for all initial conditions $P(x,0) = \rho_0(x)$ (in our case for any initial probability density of the tumor size), it is:

$$\lim_{t\to+\infty} P(x,t;\rho_0(x)) = P_{\mathrm{st}}(x). \tag{7.19}$$

In our case, it is easy to show for (7.17) that:

$$U_{\mathrm{eff}}(x) = \frac{x^3}{3K} + x\left(\frac{2}{K} + \beta\right) - \frac{K\beta + 1}{Kx} \tag{7.20}$$

$$-\sigma^2 \log(x^2+1) - \frac{x^2}{2} + \frac{1}{2x^2} - 2(1-\sigma^2)\log(x).$$

Although the above-outlined Gaussian noise approach allows these interesting analytical results, it has an inherent pitfall that strongly limits its biological applicability. In an infinitesimal interval $(t, t+\mathrm{d}t)$, the IS contribution to the change of the tumor size x is:

$$\mathrm{PROB}\left(-\frac{\beta x^2}{1+x^2}\mathrm{d}t - W(t)\sqrt{\mathrm{d}t}\sigma\frac{x^2}{1+x^2} > 0\right) > 0,$$

which means that the IS killer cells may actually generate tumor cells instead of killing them. A Gaussian perturbation of a positive parameter π is a good approximation if its standard deviation is far smaller than the average value of π, as in this case the probability of negativity of the now randomly varying π is very small and can be tolerated. In all other cases, the use of Gaussian noises is questionable.

As a consequence, it is more appropriate to perform an analysis based on the introduction of bounded noises $\nu(t)$, with $|\nu(t)| \leq B < \beta$ (so that $\beta + \nu(t) > 0$), such that:

$$x' = x - \frac{x^2}{K} - (\beta + \nu(t))\frac{x^2}{1+x^2}. \tag{7.21}$$

Let us choose β to ensure that the deterministic model has three equilibria, and B such that $\beta^* < \beta - B$ and $\beta + B < \beta^{**}$. This implies that, in the unperturbed case, there are three equilibria at $\beta - B$ and three at $\beta + B$. Let us call these equilibria a_{L}, b_{L}, and c_{L} for the lower bound and a_{U}, b_{U}, and c_{U} for the upper bound. Of course, $a_{\mathrm{U}} < a_{\mathrm{L}}$, $b_{\mathrm{U}} > b_{\mathrm{L}}$, and $c_{\mathrm{U}} < c_{\mathrm{L}}$. It follows from the differential inequalities

$$x - \frac{x^2}{K} - (\beta + B)\frac{x^2}{1+x^2} \leq x' \leq x - \frac{x^2}{K} - (\beta - B)\frac{x^2}{1+x^2} \tag{7.22}$$

that if $x(0) < b_2$, then $x(t) \in (a_1, a_2)$ for large times, whereas if $x(0) \in (b_1, +\infty)$, then $x(t) \in (c_1, c_2)$ for large times. In principle, for sufficiently large times, the probability density $\rho(x,t)$ is non-null only in $(a_1,a_2) \cup (c_1,c_2)$.

More interestingly, the fact that two initial distributions of $x(0)$ lead to two different and mutually exclusive asymptotic behaviors means that the asymptotical probability distribution, if it exists, depends on the initial conditions, i.e., there are multiple equilibria in the space $\mathfrak{I}$ of the probability measures. If the equilibrium does not exist, however, there are multiple attracting sets in $\mathfrak{I}$. This behavior is markedly different from the above-discussed case of Gaussian noise, where, as represented in (7.19), there is a unique and globally attractive stationary density.

Since we are only interested in the natural behavior of the tumor–IS interplay in the absence of human intervention, we limited our focus to random small or moderate initial values $x(0)$. Thus, we omitted the general assessment of the influence of B on the samples $x(t)$ for large times, and only focused on the case where the initial condition was suitably small. More formally, we are interested in the assessment of the qualitative changes of the probability density corresponding to the dynamics of a tumor, whose size $x(0)$ at the initial observation time was such that $x(0) \in A = (0, a^*)$, where a^* is a suitable small value (e.g., 1% of the carrying capacity).

Finally, we briefly mention that, in the case where the innate system is sufficiently reactive to have $m_0 > p_0 - \delta_0$, it follows from

$$x' < -(m_0 - (p_0 - \delta_0))x \tag{7.23}$$

that $P_{\mathrm{st}}(x) = \delta(x)$ independent of the bounded noise type.

Models of Bounded Noise

Since noise-induced transitions are dependent on the kind of adopted noise density [33], we considered two kinds of bounded noise. We introduced bounded noises by reviewing the best-studied non-dichotomic bounded stochastic processes.

The easiest—and apparently crudest—way of defining a bounded noise is to apply a bounded function to a random walk:

$$\xi(t) = h(W)$$

with

$$W'(t) = +\eta(t), \tag{7.24}$$

where $\eta(t)$ is a unitary intensity white noise, and $-B \leq h(W) \leq B$. The effect of the truncation of the tails of a random walk induced by the approach illustrated here is that, due to this "compression," the stationary probabilities $P(\nu)$ of this class of processes are such that

$$P(|B|) = +\infty.$$

This property makes this class of noises particularly useful for modeling bounded stochastic processes that extend the dichotomous noise with the stationary density

$$P_d(\xi) = \frac{1}{2}\delta(\xi - |B|).$$

Probably the best-studied bounded stochastic process obtained using this method is the so-called sine-Wiener noise [11, 12, 34] given by

$$\nu(t) = B\mathrm{Sin}\left(\sqrt{\frac{2}{\tau}}W(t)\right),$$

where $W(t)$ is a white noise. The sine-Wiener noise is such that $< \nu(t) >= 0$,$< \nu^2(t) >= B^2/2$ and

$$P(\xi) = \frac{A}{\sqrt{B^2 - \xi^2}}.$$

Moreover,

$$< \nu(t)\nu(t+z) >= \frac{B^2}{2}\mathrm{Exp}\left(-\frac{z}{\tau}\right)\left(1 - \mathrm{Exp}\left(-4\frac{t}{\tau}\right)\right),$$

where $z \geq 0$.

A different approach consists of using Tsallis noises [13, 87] $\nu(t)$, which are derived with the following Langevin equation [51, 85, 95]:

$$\nu'(t) = \tau^{-1}\left(-\frac{\nu}{1-\frac{\tau(1-q)}{D}\frac{\nu^2}{2}} + \sqrt{2D}\xi(t)\right), \tag{7.25}$$

where $-\infty < q < 1$ and $\xi(t)$ is a white Gaussian noise with zero mean and unitary intensity. Thus, $\nu(t)$ is a non-Gaussian noise with an average of zero and the following bounds:

$$-B < \nu(t) < B\,,\; B = \sqrt{\frac{2D}{\tau(1-q)}}. \tag{7.26}$$

The stationary density of ν is:

$$P_{\mathrm{st}}(\nu) = A_q\left(1-\frac{\nu^2}{B^2}\right)_+^{\frac{1}{1-q}},$$

where A_q is a normalization constant and $(z)_+ = \mathrm{Max}(z,0)$. Finally, the autocorrelation of $\nu(t)$ is approximately given by [51]:

$$\frac{\langle \nu(t)\nu(t+s)\rangle}{\langle \nu^2(t)\rangle} \approx \mathrm{Exp}\left(-|s|\frac{5-3q}{2\tau}\right).$$

In addition to previous work where only sine-Wiener and Tsallis noises were analyzed [41], here also consider the Cai–Lin noise [18, 19], which is obtained with the following Langevin–Ito equation:

$$\nu'(t) = -\frac{1}{\tau}\nu(t) + \sqrt{\frac{1}{\tau(\delta+1)}(B^2-\nu^2)}\eta(t), \tag{7.27}$$

where $\eta(t)$ is a white noise with unitary intensity.

The bounded nature of the above noise easily follows from the fact that at $\nu = +B$ it is $\nu' < 0$, whereas at $\nu = -B$ it is $\nu' > 0$.

The stationary density of ν is

$$P(\nu) = A(B^2-\nu^2)_+^{\delta}.$$

An interesting property of the Cai–Lin noise is that the process $\nu(t)$ has *exactly* the same autocorrelation of the Ornstein–Uhlenbeck process [18], i.e.,

$$\frac{\langle \nu(t)\nu(t+s)\rangle}{\langle \nu^2(t)\rangle} = \mathrm{Exp}\left(-\frac{|s|}{\tau}\right).$$

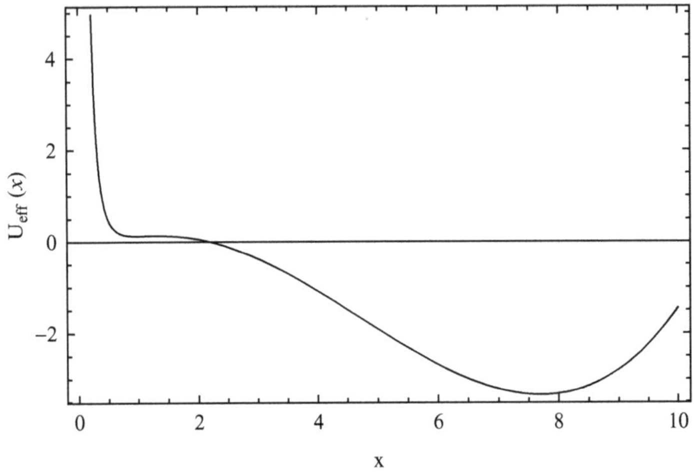

Fig. 7.9 *White noise perturbations of immune system levels. Plot of the "probability potential"* $U_{\text{eff}}(x)$ for $\sigma = 0.1$, $\beta = 1.8$, $K = 10$

Numerical Simulations

In this section, we assess the role of noise by simulating various values of our bifurcation parameter B. Since at $x = 1$ the maximum killing rate is equal to $\beta/2$, we assumed that $K > 10$. Thus, we used $K = 10$ as a test value. We note that there is a critical value $\beta_c = 2(1 - 1/K)$, such that the microscopic equilibrium is at $x = 1$, i.e., in correspondence to the maximum of $\phi(x)$. We scaled our model with $\tau_u \approx 5.56$ days obtained by fitting [62, 71] the above-mentioned data concerning experiments on chimeric mice [62, 71].

As for the autocorrelation time τ_c, it must summarize the temporal scale of most common random perturbations in the immune strength. We may assume that these fluctuations are of the order of a few of days, since they may reflect various phenomena related to the behavior of the patient and to external additional conditions, such as short infections (which induce temporary increase of immune system strength) or temporary short-lived immunodepressions [57]. Thus, we used $\tau_c = 0.2$ (slightly more than 1 day) and $\tau_c = 1$.

In the previous section, we analytically obtained the steady-state probability density corresponding to white-noise perturbations and emphasized that it is globally attractive. Figure 7.9 shows the potential $U_{\text{eff}}(x)$ for $\beta = 1.8$ and $\sigma = 0.1$. Since $\sigma^{-2} = 100$ and $U_{\text{eff}}(x)$ at the two minima is ≈ 0 and ≈ -3, it follows that the asymptotic distribution is very near to a Dirac's delta centered at x_{M}. The other peak may be neglected even in case of large σ, as shown in Fig. 7.10, which refers to $\sigma = 1.8/1.96$. To have equal values at the two minima of the potential, one must increase the value of up to approximately $\sqrt{2.2}$. The fact that also when σ is very small the probability density is largely centered close to the larger equilibrium state

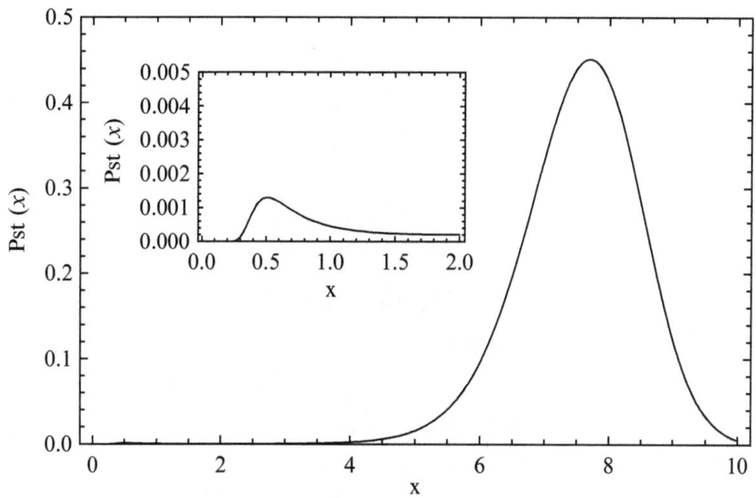

Fig. 7.10 *White noise perturbations of immune system levels. Plot of the stationary density* $P_{st}(x)$ for $\sigma = 1.8/1.96$, $\beta = 1.8$, $K = 10$. The inset shows a zoom for $0 < x < 2$, of the *left* part of the plot

of the deterministic system paradoxically suggests that the hypothetical gaussian statistical fluctuation affecting the immune system can enable the tumor to evade immune control in all cases.

From a biophysical point of view, however, this answer is neither complete nor satisfactory, since it is an asymptotic result that requires an assessment of the velocity of convergence. The density of x at $t = 1{,}000$ for $\sigma = 0.1$ is shown in Fig. 7.11, and we can see that the steady distribution has been fully reached. In dimensional time, however, this corresponds to 5,560 days ($t = 1{,}000 \times \tau_u$), or ≈ 15.21 years, which is an interval of time far longer than the average life of a chimeric mouse, which ranges from approximately 2–3 years [54], and, of course, it is also longer than the characteristic times of the evolutionary escape of the tumor from the IS.

As the velocity of convergence is low, we must assess transient values of $P(x,t)$ for clinically applicable values of $t = 3, 6$, and 12 months. We performed simulations assuming bounded noises of sine-Wiener, Tsallis, or Cai–Lin Cai type and under the condition that the initial tumor size $X(0)$ was uniformly distributed in a small range equal to $(0, 0.1)$. For the sake of simplicity, the probability density under these conditions will be henceforth called "conditional density."

In the case of sine-Wiener noise with $\tau = 1$ and $B = 0.2$, we obtained that at $T = 3$ months there is no probability or only a very low probability of escape with extreme values equal to $x = 2$ (upper panel of Fig. 7.12), whereas at $T = 6$ months the probability of escape is significant with some tumors reaching sizes near carrying capacity (lower panel of Fig. 7.12). Finally, for $T = 1$ year and $B = 0.2$, the probability of escape and macroscopic growth is very large and the density

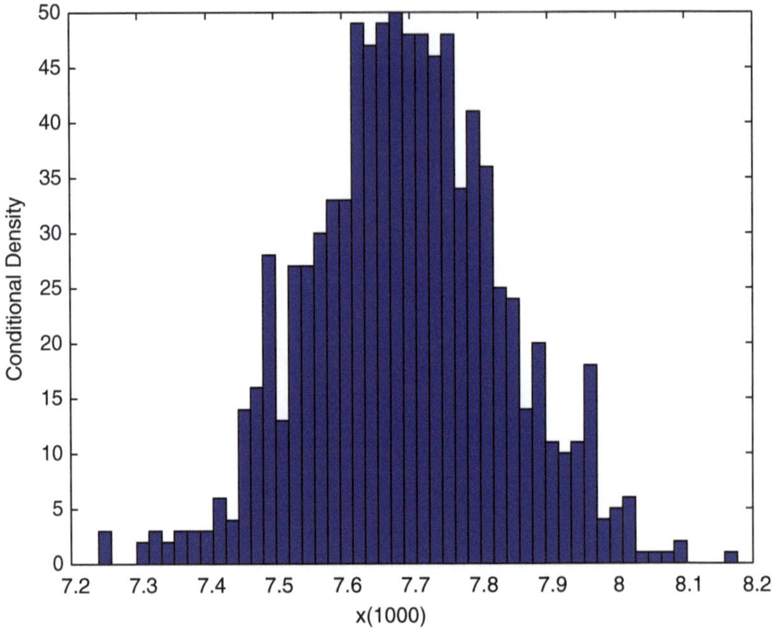

Fig. 7.11 *White noise perturbations of fluctuations of immune system levels. Plot of the* probability density of x for $t =\approx 15.2$ years induced by a Gaussian perturbation of $\beta = 1.8$ with $\sigma = 0.15$. Initially $x(0) \in (0, 0.1)$

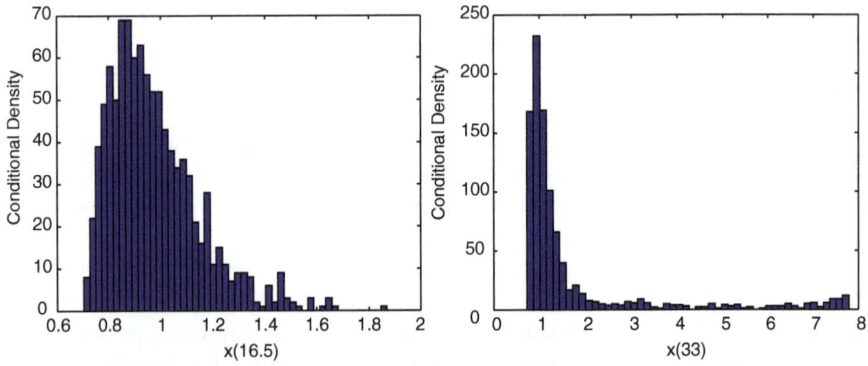

Fig. 7.12 Simulation of the effects that a bounded noise of the sine-Wiener type with $\tau = 1$ and $B = 0.2$ acts on the probability density $P(x,t)$ at finite times T with $x(0) \in (0, 0.1)$. (**a**) $T = 3$ months; (**b**) $T = 6$ months

is bimodal, whereas it is unimodal for $B = 0.04$. We conclude that an increase in B causes a qualitative change in $P(x,T)$ (Fig. 7.13). For smaller $\tau = 0.2$ (i.e., approximately 1 day) and $B = 0.2$ at $T = 6$ months, there is a small nonzero probability of escape, and a transition to bimodality with a considerable probability

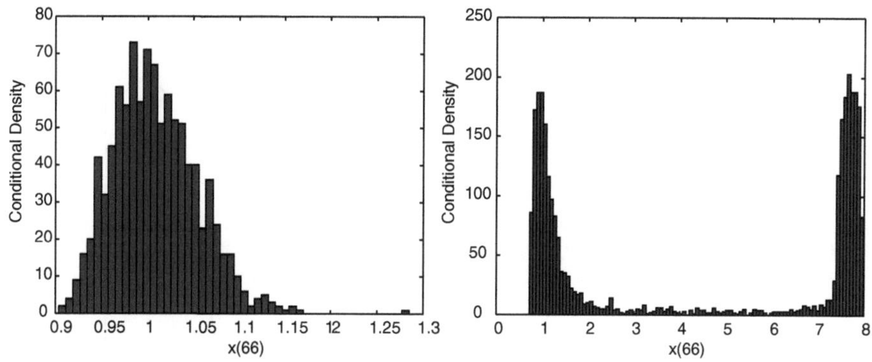

Fig. 7.13 Transitions induced to $P(x,T)$ with $T = 66 = 1$ years by a sine-Wiener noise with $\tau = 1$ and $x(0) \in (0, 0.1)$. (**a**) At $B = 0.04$ the density is unimodal. (**b**) At $B = 0.2$ the density is bimodal

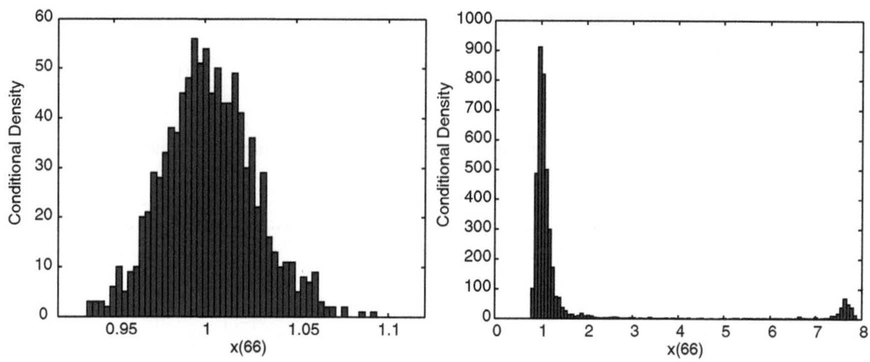

Fig. 7.14 Transitions induced to $P(x,T)$ with $T = 66 = 1$ years by a sine-Wiener noise with $\tau = 0.2$ and $x(0) \in (0, 0.1)$. (**a**) At $B = 0.04$ the density is unimodal. (**b**) At $B = 0.2$ the density is bimodal

of tumor explosion is seen at $T = 1$ year (Fig. 7.14). A representative time series of immuno-evasion with $\tau = 0.2$ and $B = 0.3$ is shown in Fig. 7.15.

In the case of Tsallis noise with $\tau = 1$ and $B = 0.2$, we obtained transitions to bimodality at $T = 1$ year, for both $q = 0.5$ (see Fig. 7.16) and $q = 0.1$. No transitions were observed at 3 or 6 months, when the density was unimodal centered at lower values of x. By decreasing τ to 0.2, we obtained that, at 1 year, for $q = 0.1$ there is a very small (one case is a thousand) probability of $x(66)$ reaching large values, whereas for $q = 0.5$ there is macroscopic progression in 11 cases out of 1,000. By increasing B to 0.4, we observed transitions at 1 year for $q = 0.5$, similar to that described above (Fig. 7.16).

In the case of Cai–Lin noise, we performed simulations to show the relevance of the noise density on simulation output. In all simulations, the bound B was constant but parameter δ was variable. For $\tau = 0.2$ and $\delta = 2$, there was no transition

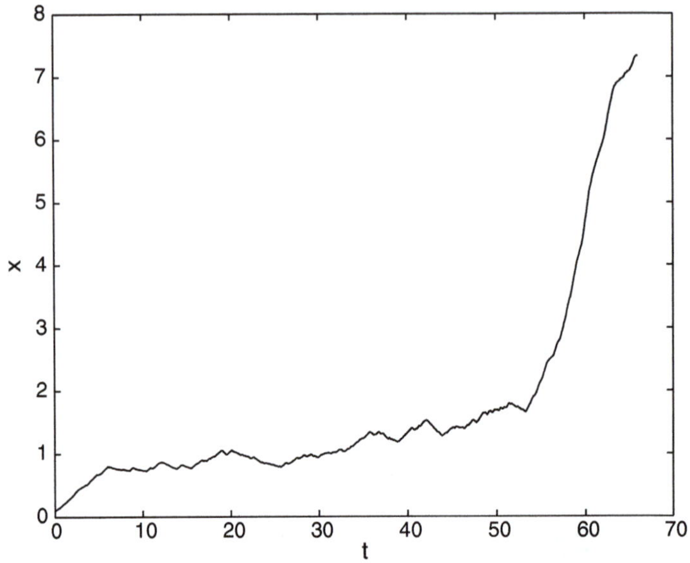

Fig. 7.15 Simulation of a tumor immuno-evasion induced by a sine-Wiener noise. Parameters: $B = 0.3$ and $\tau = 0.2$

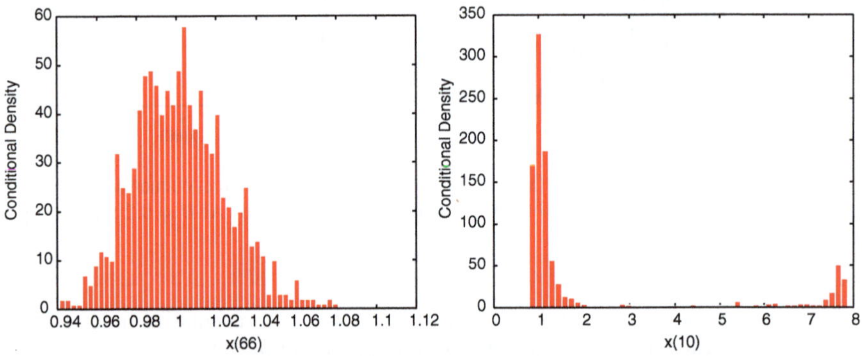

Fig. 7.16 Transitions induced to $P(x,T)$ with $T = 66 = 1$ years by a Tsallis noise with $\tau = 1$ and $q = 0.5$. $x(0) \in (0, 0.1)$. (**a**) At $B = 0.04$ the density is unimodal. (**b**) At $B = 0.2$ the density is bimodal

(Fig. 7.17) in contrast to $\delta = 0$ (corresponding density is constant in $(-B, B)$). Transitions were observed for $\tau = 1$ for both $\delta = 2$ and $\delta = 0$ (Fig. 7.18), with the revealed tumor sizes larger for $\delta = 0$.

Finally, we stress that other fast or slow fluctuations may be present, the latter for example due to psychological depressions [84]. In line with biological intuition, our simulations suggest that the first type of fluctuations is filtered out or have minimal effects, whereas the second type easily induces immuno-evasion.

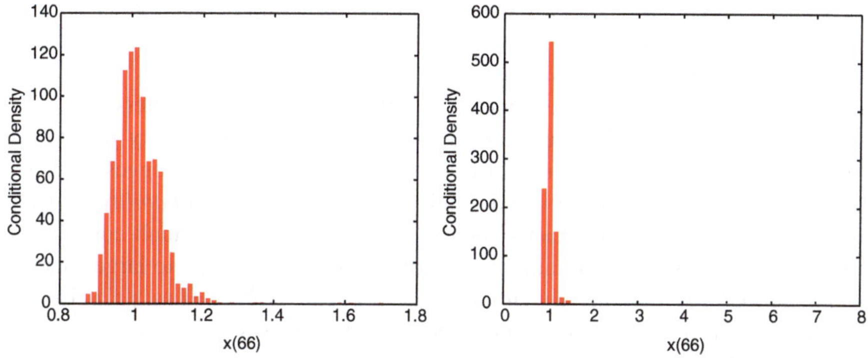

Fig. 7.17 Transitions induced to $P(x,T)$ with $T = 66 = 1$ years by a Cai noise with $B = 0.2$, $\tau = 0.2$ and $x(0) \in (0, 0.1)$. Effects of varying the shape parameter δ. (**a**) At $\delta = 2$ the density is unimodal. (**b**) At $\delta = 0$ the density is bimodal

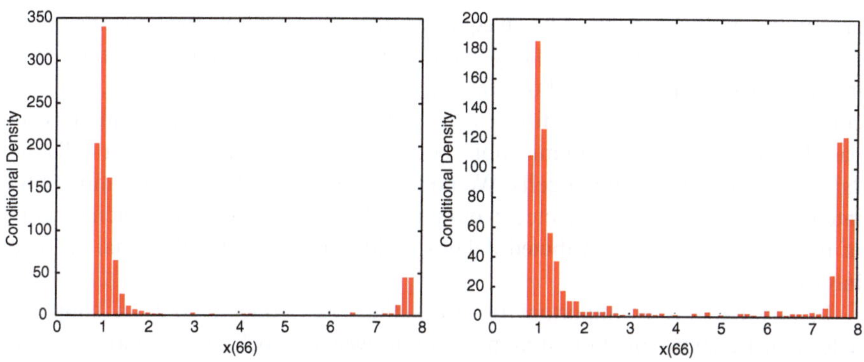

Fig. 7.18 Transitions induced to $P(x,T)$ with $T = 66 = 1$ years by a Cai noise with $B = 0.2$, $\tau = 1$, and $x(0) \in (0, 0.1)$. Effects of varying the shape parameter δ. (**a**) At $\delta = 2$ the density is bimodal. (**b**) At $\delta = 0$ the density again is bimodal, but the peak at large value of tumor size is higher than the other

Concluding Remarks

The seminal model by Kunznetsov [71, 74, case study 1] uniquely focuses on the interplay of tumor cells with CTLs, but is nevertheless able to qualitatively reproduce the phenomenon of immuno-evasion. It is an epigenetic model that is based on intercellular communication and the ability of tumor cells to adapt to an immune response. The interplay between tumor cells and specific immunity was chosen because of experimental evidence supporting the relevance of CTLs in determining dormancy or evasion of many important tumors, such as melanomas, ovarian carcinomas, and colorectal carcinomas [96], where the presence of infiltrating lymphocytes serves as a prognostic marker. Embedding a proposed evolutionary mechanism into a more complex setting of this model with detailed description of both adaptive and innate immunity should lead to results qualitatively similar to those illustrated here.

The model discussed in this chapter features two crucial parameters: the probability z that a tumor cell is recognized by a CTL and the probability p of tumor cells being killed by the immune effector cells. Our simulations suggest that: (1) the onset of immuno-evasion is extremely dependent on the *probability of dying* $p(t)$. Small changes in $p(t)$ can switch the tumor state from a small equilibrium under control to escape; (2) immuno-evasion is not sensitive even to large changes in *the probability that the tumor cells are recognized by CTLs* ($z(t)$). Escape is triggered in our simulations only when $z(t)$ becomes sufficiently small (e.g., $z \approx 0.2$); and (3) the dynamics of $z(t)$ are slower than those of $p(t)$ when $\gamma_0 = \eta$. Thus, we speculate that the prevalent mechanism underlying immuno-evasion is a reduction of immune-induced killing of tumor cells and not a reduced probability of recognizing tumor cells, at least in some cases. In the limit where the tumor cell–CTL complexes are at quasi-steady state, we showed that the dynamics of probability of dying are ruled by an equation that, from a purely mathematical point of view, is an imitation evolutionary game. Tomlinson and Bodmer were the first to introduce game theoretic methods in oncology [91]. Quite interestingly, although considering signals that are different from those studied here, their models are based on intercellular communications between "signal producer" tumor cells and "signal non-producer" tumor cells. We note that the timescale of propagation of the intercellular signals was neglected in our model for clarity. A simple implementation of this as a lumped or distributed delay should not contribute additional physical information.

In summary, it is important to stress that our model is only a simplistic deterministic approximation of numerous interweaved stochastic evolutionary scenarios leading to immuno-evasion. For the sake of simplicity, let us utilize the modeling framework proposed by Bazzani and Freguglia [4,5], in which phenotypes are modeled through a finite number of parameters. Thus, by understanding the probabilities of recognition and dying as phenotypic traits, we can say the tumor immunophenotype has an average given by the vector $\phi(t) = (p(t), k(t))$ with a very small variance. This is due to our hypothesis that a very efficient and rapid intercellular communication exists between tumor cells. Future research should include more appropriate frameworks with spatial dynamics that are particularly important in immuno-evasion of dormant tumors.

Our case study 2, which was mathematically based on multiplicative-bounded noises, showed that the unavoidable extrinsic fluctuations affecting the IS level (due to the multitude of its interactions) might be an important factor in determining evasion from dormancy at a mesoscale temporal range. However, we also showed tumor escape in the more realistic non-Gaussian setting. We have analytically shown that, with a bounded noise (and independent from the specific choice of it) with small amplitude, there is no global convergence toward a unique stationary density as observed with Gaussian noise.

Furthermore, the velocity of convergence toward a (unique or nonunique) steady-state probability density is a key parameter. We showed that if limiting the analysis at finite significant times, then the transition to large values is not reached if the oscillation amplitude B of the noise is too small, or, for the Tsallis and Cai–Lin noise, if the autocorrelation time τ_{corr} is small. Note that transition depends somewhat on the adopted noise model. Interestingly, both correlation time and the stationary density of the noise are key factors in triggering or preventing transition from tumor dormancy to malignant growth. Finally, we briefly note that, to the best of our knowledge, the model discussed herein included the first application of Cai–Lin noise in quantitative biology.

The two case studies reviewed herein involve nonspatial models. Spatial information, however, is very important in immuno-evasion of dormant tumors for various reasons. First, immuno-editing may modulate the spatial patterning and invasiveness of a tumor. Second, the diffusion of signals is instrumental in enacting both the immunosuppressive and intercellular communication networks. Third, and probably most interestingly, tumors may evolve evasion strategies that use exquisitely spatial mechanisms that cannot be represented by a modeling framework, as discussed herein. Vianello and colleagues [92] recently showed experimentally that tumor cells can produce chemicals that act as chemorepellents of CTLs. This is not surprising, since chemotactic motion of IS cells is a hallmark of the defense of the human body against *non-self* entities. Cellular effectors of both the innate and adaptive IS are able to reach their targets in response to gradients of various kinds of chemicals [29]. The potential role of immune effector-induced chemorepulsion was recently incorporated in our mathematical approach [42]. These and other spatial scenarios are currently under consideration [3].

Now, we describe one important observation. All models, including the discussed immuno-evasion model, although based on biological and biophysical information, require experimental validation. To the best of our knowledge, the existing literature illustrates immuno-editing only by means of qualitative clinical findings or experimental molecular findings. A complete quantitative study of the adaptive evasion from tumor dormancy allowing, for example, the plotting of tumor growth curves would be a remarkable resource. Thus, we hope that this theoretical work will trigger such experimental investigations, which would allow us to validate our model.

In this chapter, we considered only a few of the currently known biomechanisms of evasion of dormant tumors from immune control. In addition to cellular immunity, other mechanisms such as humoral immunity are likely to play a substantial role in determining immuno-evasion.

In summary, all discussed models focused on kinetic aspects, and thus partially disregarded a width of biological details. However, multiple distinct microscopic mechanisms often result in similar macroscopic behavior and can thus be neglected in first-order model approximation. Notwithstanding all these limitations, myriads of different aspects of immuno-evasive strategies have been illustrated. On a personal note and to conclude this work, it is worth noting that the term *dormancy* as often applied to micro-tumors and interpreted as *non-active* might be inappropriate. Micro-tumors, as we have seen, may have a very active life (not only due to noise).

Acknowledgments I very much thank Heiko Enderling, Nava Almog, and Lynn Hlatky for inviting me to contribute to this book! I also extend my thanks to the anonymous referees for their very useful suggestions.

This work was performed in the framework of the Integrated Project "P-medicine—from data sharing and integration via VPH models to personalized medicine" (project ID: 270089), which is partially funded by the European Commission under the 7th framework program.

References

1. Agarwal SA (Guest Editor) (2003) Seminars in oncology. I madew reference to a special issue of a journal. 29-3 (Suppl 7)
2. Ai BQ, Wang XJ, Liu GT, Liu LG (2003) Correlated noise in a logistic growth model. Phys Rev E 67:022903
3. Al Taamemi M, Chaplain M, d'Onofrio A (2011) Evasion of tumours from the control of the immune system: consequences of brief encounters. Biology Direct (in press)
4. Bazzani A, Freguglia P (2003) Evolution: Geometrical and dynamical aspects. Biol Forum 96:123–136
5. Bazzani A, Freguglia P (2004) An evolution model of phenotipic characters. WSEAS Trans Biol Biomed 4:369–373
6. Behera A, O'Rourke S (2008) Comment on "correlated noise in a logistic growth model". Phys Rev E 77:013901
7. Bellomo N (2010) Modeling the hiding-learning dynamics in large living systems. Appl Math Lett 23:907–911
8. Bellomo N, Delitala M (2008) From the mathematical kinetic stochastic game theory to modelling mutations, onset, progression and immune competition of cancer cells. Phys Life Rev 5:183–206
9. Bellomo N, Forni G (2008) Complex multicellular systems and immune competition: New paradigms looking for a mathematical theory. Curr Topics Dev Biol 81:485–502
10. Bleumer I, Oosterwijk E, de Mulder P, Mulders PF (2003) Immunotherapy for renal cell carcinoma. Eur Urol 44:65–75
11. Bobryk RB, Chrzeszczyk A (2005) Transitions induced by bounded noise. Physica A 358:263–272
12. Bobryk RB, Chrzeszczyk A (2009) Stability regions for Mathieu equation with imperfect periodicity. Phys Lett A 373:3532–3535
13. Borland L (1998) Ito-Langevin equations within generalized thermostatistics. Phys Lett A 245:67–72
14. Bose T, Trimper S (2009) Stochastic model for tumor growth with immunization. Phys Rev E 79:051903
15. Burnet FM (1957) Cancer – a biological approach. Br Med J 1:841–847
16. Burnet FM (1964) Immunological factors in the process of carcinogenesis. Br Med Bull 20:154–158
17. Burnet FM (1971) Immunological surveillance in neoplasia. Transplant Rev 7:3–25
18. Cai GQ, Lin YK (1996) Generation of non-Gaussian stationary stochastic process. Phys Rev E 54: 299–303
19. Cai GQ, Suzuki Y (2005) Response of systems under non-gaussian random excitation. Nonl Dyn 45:95–108
20. Cappuccio A, Elishmereni M, Agur Z (2006) Cancer immunotherapy by interleukin-21 treatment strategies evaluated in a mathematical model. Canc Res 66:7293–7300
21. Caravagna G, d'Onofrio A, Milazzo P, Barbuti R (2010) Antitumor immune surveillance through stochastic oscillations. J Th Bio 265:336–345
22. Cattani C, Ciancio A, d'Onofrio A (2010) Metamodeling of the learning-hiding competition between tumors and the immune system: A kinematic approach. Math Comp Model 52:62–69

23. Chakraborty AK, Kosmrlj A (2010) Statistical mechanical concepts in immunology. Annu Rev Phys Chem 61:283–303
24. Chaplain M, Matzavinos A (2006) Mathematical modelling of spatio-temporal phenomena in tumor immunology. Lect Notes Math Sci 1872:131–183
25. Chaplain M, Kuznetsov VA, James ZH, Stepanova LA (1998) Spatiotemporal dynamics of the immune system response to cancer. In: Horn MA, Simonett G, Webb G (eds) Mathematical models in medical and health sciences. Vanderbilt University Press, Nashville, pp 1–20
26. Cheon T (2003) Evolutionary stability of ecological hierarchy. Phys Rev Lett 90:258105
27. O'Connell J, Bennett MW, O'Sullivan GC, Collins JK, Shanahan F (1999) The Fas counter-attack: Cancer as a site of immune privilege. Immunol Today 20:46–50
28. DeBoer RJ, Hogeweg P, Hub F, Dullens J, DeWeger R, DenOtter W (1985) Macrophage T lymphocyte interactions in the anti-tumor immune response: A mathematical model. J Immunol 134:2748–2758
29. Delves P, Martin S, Burton D, Roitt I (2006) Essential immunology. Wiley-Blackwell, New York
30. De Pillis LG, Radunskaya AE, Wiseman CL (2005) A validated mathematical model of cell-mediated immune response to tumor growth. Canc Res 65:7950–7958
31. De Vito VT Jr, Hellman J, Rosenberg SA (eds) (2005) Cancer: Principles and practice of oncology. J P Lippincott, Philadelphia
32. De Vladar HP, Gonzalez JA (2004) Dynamic response of cancer under the influence of immunological activity and therapy. J Theor Biol 227:335–348
33. Deza R, Wio HS, Fuentes MA (2007) Noise-induced phase transitions: Effects of the noises, statistics and spectrum. AIP Conf. Proc. 913:62–67
34. Dimentberg MF (1988) Statistical dynamics of nonlinear and time-varying systems. Wiley, New York
35. d'Onofrio A (2005) A general framework for modeling tumor-immune system competition and immunotherapy: Analysis and medical inferences. Phys D 208:220–235
36. d'Onofrio A (2006) Tumor-immune system interaction: Modeling the tumor-stimulated proliferation of effectors and immunotherapy. Math Models Meth Appl Sci 16:1375–1401
37. d'Onofrio A (2007) Noisy oncology. In: Venturino E, Hoskins RH (eds) Aspects of mathematical modeling. Birkhauser, Basel, pp 229–234
38. d'Onofrio A (2007) Tumor evasion from immune system control: Strategies of a MISS to become a MASS. Chaos, Solitons and Fractals 31:261–268
39. d'Onofrio A (2008) Fuzzy oncology. Appl Math Lett 21:662–668
40. d'Onofrio A (2009) Fractal growth of tumors and other cellular populations: Linking the mechanistic to the phenomenological modeling and vice versa. Chaos, Solitons and Fractals 41:875–880
41. d'Onofrio A (2010) Bounded-noise-induced transitions in a tumor-immune system interplay. Phys Rev E 81:021923
42. d'Onofrio A (2011) Spatiotemporal effects of a possible chemorepulsion of tumor cells by immune system effectors. J Theor Biol 296:41–48
43. d'Onofrio A, Ciancio A (2011) Simple biophysical model of tumor evasion from immune system control. Phys Rev E 84:031910
44. d'Onofrio A, Tomlinson IPM (2007) A nonlinear mathematical model of cell turnover, differentiation and tumorigenesis in the intestinal crypt. J Theor Biol 224:367–374
45. d'Onofrio A, Gatti F, Cerrai P, Freschi L (2010) Delay-induced oscillatory dynamics of tumor-immune system interaction. Math Comp Mod 51:572–591
46. Dunn GP, Bruce AT, Ikeda H, Old LJ, Schreiber RD (2002) Cancer immunoediting: From immunosurveillance to tumor escape. Nat Immunol 3:991–998
47. Dunn GP, Old LJ, Schreiber RD (2004) The three Es of cancer immunoediting. Annu Rev Immunol 22:322–360
48. Ehrlich P (1909) Ueber den jetzigen Stand der Karzinomforschung. Ned Tijdschr Geneeskd 5:273–290

49. Fishelson Z, Berke G (1981) Tumor cell destruction by cytotoxic T lymphocytes: The basis of reduced antitumor cell activity in syngeneic hosts. J Immunol 25:2048–2052
50. Frey E (2010) Game theory: Theoretical concepts and applications to microbial communities. Physica A 389:3265–3298
51. Fuentes MA, Toral R, Wio HS (2001) Enhancement of stochastic resonance: The role of non Gaussian noises. Physica A 295:114–122
52. Gabrilovich DI, Hurwitz AA (eds) (2008) Tumor-induced immune suppression. Springer, Heidelberg
53. Gatti R, Robinson WA, Deinard AS, Nesbit M, McCullough JJ, Ballow M, Good RA (1973) Cyclic leukocytosis in chronic myelogenous leukemia: New perspectives on pathogenesis and therapy. Blood 41:771–783
54. Hedrich H (ed) The laboratory mouse. Elsevier, Amsterdam
55. Horsthemke W, Lefever R (1977) Phase transition induced by external noise. Phys Lett A 64:19–21
56. Horsthemke W, Lefever R (2007) Noise-induced transitions in physics, chemistry and biology. Springer, Heidelberg
57. Janeway CA Jr, Travers P, Walport M, Shlomchik MA (2001) Immunobiology, 5th edn. Garland, New York
58. Kaminski JM, Summers JB, Ward MB, Huber MR, Minev B (2004) Immunotherapy and prostate cancer. Canc Treat Rev 29:199–209
59. Kennedy BJ (1970) Cyclic leukocyte oscillations in chronic myelogenous leukemia during hydroxyurea therapy. Blood 35:751–760
60. Kim R, Emi M, Tanabe K (2007) Cancer immunoediting from immune surveillance to immune escape. Immunology 121:1–14
61. Kindt TJ, Osborne BA, Goldsby RA (2006) Kuby immunology. W H Freeman, New York
62. Kirschner D, Panetta JC (1998) Modeling immunotherapy of the tumor - immune interaction. J Math Biol 37:235–252
63. Kirschner D, Tsygvintsev A (2009) On the global dynamics of a model for tumor immunotherapy. Math Biosci Eng 6:573–583
64. Koebel CM, Vermi W, Swann JB, Zerafa N, Rodrig SJ, Old LJ, Smyth MJ, Schreiber RD (2007) Adaptive immunity maintains occult cancer in an equilibrium state. Nature 450:903–907
65. Kogan Y, Forys U, Shukron O, Kronik N, Agur Z (2010) Cellular immunotherapy for high grade gliomas: Mathematical analysis deriving efficacious infusion rates based on patient requirements. SIAM J Appl Math 70:1953–1976
66. Kondoh CM (2003) Foraging adaptation and the relationship between food-web complexity and stability. Science 299:5611–5613
67. Kronik N, Kogan Y, Vainstein V, Agur Z (2008) Improving alloreactive CTL immunotherapy for malignant gliomas using a simulation model of their interactive dynamics. Canc Immunol Immunother 57:425–439
68. Kurnick JT, Ramirez-Montagut T, Boyle LA, Andrews DM, Pandolfi F, Durda PJ, Butera D, Dunn IS, Benson EM, Gobin SJ, van den Elsen PJ (2011) A novel autocrine pathway of tumor escape from immune recognition: Melanoma cell lines produce a soluble protein that diminishes expression of the gene encoding the melanocyte lineage melan-A/MART-1 antigen through down-modulation of its promoter. J Immunol 167:1204–1211
69. Kuznetsov VA (1979) Dynamics of cellular immune anti-tumor reactions I. Synthesis of a multi-level model mathematical methods. In: Fedorov (ed) The theory of systems. Kyrghyz State University Press, Frunze, pp 57–71 (in Russian)
70. Kuznetsov VA, Knott GD (2001) Modeling tumor regrowth and immunotherapy. Math Comp Mod 33:1275–1287
71. Kuznetsov VA, Makalkin IA, Taylor MA, Perelson AS (1994) Nonlinear dynamics of immunogenic tumors: Parameter estimation and global bifurcations analysis. Bull Math Biol 56:295–321

72. Lefever R, Horsthemke W (1979) Bistability in fluctuating environments Implications in tumor immunology. Bull Math Biol 41:469–490
73. De Lisi C, Rescigno A (1977) Immune surveillance and neoplasia: A minimal mathematical model. Bull Math Biol 39:201–221
74. Matzavinos A, Chaplain M, Kuznetsov VA (2004) Mathematical modelling of the spatio-temporal response of cytotoxic T-lymphocytes to a solid tumor. Math Med Biol 21:1–34
75. Mehta BC, Agarwal MB (1980) Cyclic oscillations in leukocyte count in chronic myeloid leukemia. Acta Hematol 63:68–70
76. Meloni G, Trisolini SM et al (2002) How long can we give interleukin-2? Clinical and immunological evaluation of AML patients after 10 or more years of IL2 administration. Leukemia 16:2016–2018
77. Owen M, Sherratt J (1997) Pattern formation and spatiotemporal irregularity in a model for macrophage-tumor interaction. J Theor Biol 189:63–80
78. Owen M, Sherratt J (1999) Mathematical modelling of macrophage dynamics in tumors. Math Meth Mod Appl Sci 9:513–539
79. Pappalardo F, Lollini PL, Castiglione F, Motta S (2005) Modeling and simulation of cancer immunoprevention vaccine. Bioinformatics 21:2891–2897
80. Pardoll D (2003) Does the immune system see tumors as foreign or self? Annu Rev Immunol 21:807–839
81. Parisi G (1990) A simple model for the immune network. Proc Nat Acad Sci USA 87:429–433
82. Pennisi M, Pappalardo F, Palladini A, Nicoletti G, Nanni P, Lollini PL, Motta S (2010) Modeling the competition between lung metastases and the immune system using agents. BMC Bioinformatics 11(Suppl 7):S13
83. Perelson AS, Weisbuch G (1997) Immunology for physicists. Rev Mod Phys 69:1219–1268
84. Reiche EM, Nunes SO, Morimoto HK (2004) Stress, depression, the immune system, and cancer. Lancet Oncol 5:617–625
85. Revelli JA, Sanchez AD, Wio HS (2002) Effect of non-Gaussian noises on the stochastic resonance-like phenomenon in gated traps. Physica D 168–169:165–170
86. Schmielau J, Finn OJ (2001) Activated granulocytes and granulocyte-derived hydrogen peroxide are the underlying mechanism of suppression of t-cell function in advanced cancer patients. Canc Res 61:4756–4760
87. Stariolo DA (1994) The Langevin and Fokker-Planck equations in the framework of a generalized statistical mechanics. Phys Lett A 185:262–264
88. Stepanova NV (1980) Course of the immune reaction during the development of a malignant tumor. Biophizika 24:917–923
89. Stewart TJ, Abrams SI (2008) How tumors escape mass destruction. Oncogene 27:5894–5903
90. Thomas L (1959) Discussion. In: Lawrence HS (ed) Cellular and humoral aspects of the hypersensitive states. Hoeber-Harper, New York, pp 529–532
91. Tomlinson IPM, Bodmer WF (1997) Modelling the consequences of interactions between tumor cells. Br J Canc 75:157–160
92. Vianello F, Papeta N, Chen T, Kraft P, White N, Hart WK, Kircher MF, Swart E, Rhee S, Palù G, Irimia D, Toner M, Weissleder R, Poznansky MC (2006) Murine B16 melanomas expressing high levels of the chemokine stromal-derived factor-1/CXCL12 induce tumor-specific T cell chemorepulsion and escape from immune control. J Immunol 176(5):2902–2914
93. Vodopick H, Rupp EM, Edwards CL, Goswitz FA, Beauchamp JJ (1972) Spontaneous cyclic leukocytosis and thrombocytosis in chronic granulocytic leukemia. New Engl J Med 286:284–290
94. Wie RZ, Shao YZ, He YZ (2006) Pure multiplicative stochastic resonance of a theoretical anti-tumor model with seasonal modulability. Phys Rev E 73:060902
95. Wio HS, Toral R (2004) Effect of non-Gaussian noise sources in a noise-induced transition. Physica D 193:161–168
96. Zitvogel L, Tesniere A, Kroemer G (2006) Cancer despite immunosurveillance: Immunoselection and immunosubversion. Nat Rev Imm 6:715–727

Chapter 8
Tumor Dormancy and Cancer Stem Cells: Two Sides of the Same Coin?

Sonja Kleffel and Tobias Schatton

Abstract Increasing evidence suggests that tumor dormancy represents an important mechanism underlying the observed failure of existing therapeutic modalities to fully eradicate cancers. In addition to its more established role in maintaining minimal residual disease after treatment, dormancy might also critically contribute to early stages of tumor development and the formation of clinically undetectable micrometastatic foci. There are striking parallels between the concept of tumor dormancy and the cancer stem cell (CSC) theory of tumor propagation. For instance, the CSC hypothesis similarly predicts that a subset of self-renewing cancer cells—that is CSCs—is responsible for tumor initiation, bears the preferential ability to survive tumor therapy, and persists long term to ultimately cause delayed cancer recurrence and metastatic progression. Additionally, many of the biological mechanisms involved in controlling the dormant state of a tumor can also govern CSC behavior, including cell cycle modifications, alteration of angiogenic processes, and modulation of antitumor immune responses. In fact, quiescence and immune escape are emerging hallmark features of at least some CSCs, indicating significant overlap between dormant cancer populations and CSCs. Herein, we crucially dissect whether CSCs occupy specific roles in orchestrating the switch between dormancy and exuberant tumor growth. We elucidate how recently uncovered CSC biological features could enable these cells to evade immunologic clearance and regulate cancer expansion, relapse, and progression. We propose that the study of CSC immunobiological pathways holds the promise to critically advance our understanding of the processes mediating tumor dormancy. Ultimately, such research endeavors could unravel novel therapeutic avenues that efficiently target both proliferating and dormant CSCs to minimize the risk of tumor recurrence in cancer patients.

S. Kleffel, M.Sc. • T. Schatton, Pharm.D., Ph.D. (✉)
Harvard Skin Disease Research Center, Department of Dermatology, Brigham and Women's Hospital, Harvard Institutes of Medicine, Room 673B, 77 Avenue Louis Pasteur, Boston, MA 02115, USA
e-mail: tobias.schatton@tch.harvard.edu

H. Enderling et al. (eds.), *Systems Biology of Tumor Dormancy*, Advances in Experimental Medicine and Biology 734,
DOI 10.1007/978-1-4614-1445-2_8, © Springer Science+Business Media New York 2013

Keywords Cancer stem cell • Tumor dormancy • Minimal residual disease • Disseminated tumor cell • Circulating tumor cell • Metastasis • Tumor initiation • Tumorigenicity • Progression • Recurrence • Relapse • Therapy resistance • Epidemiology • Animal models • Antitumor immunity • Immune evasion • Immunoescape • Angiogenesis • Angiogenic switch • Apoptosis • Quiescence

Tumor Dormancy: Definition and Clinical Evidence

Tumor dormancy describes a protracted phase of cancer progression, in which single tumor cells or microscopic tumor cell clusters remain clinically occult and show no evidence of growth, yet retain their capacity for malignant progression [1, 2]. The possibility that cancers may lie dormant in the body for prolonged periods of time without causing overt neoplasia has been recognized for many decades [3]. Historically, "tumor dormancy" has been defined as the reappearance of a cancer many years after clinical remission (Fig. 8.1a) [3]. Although there is no strict definition regarding the time interval between initial treatment and disease recurrence, many publications on tumor dormancy have referred to a "disease-free" period in excess of 5 years. Indeed, extensive epidemiologic studies of recurrence patterns in large cohorts of patients with various types of cancer support the existence of dormant tumors [4–10]. For example, it has been reported that 20% of patients with breast cancer who were clinically disease-free upon surgical removal of their primary tumor developed recurrences 5–25 years later [4–6]. Similarly, a meta-analysis revealed that the mean disease-free interval for patients with primary cutaneous melanoma was 14.3 years after diagnosis and 22.3 years for patients with primary ocular melanoma [10]. Comparative genomic hybridization (CGH) analyses of primary vs. patient-matched recurrent breast carcinomas and melanomas revealed that most recurrences were clonally related to their primary lesions [11, 12], further supporting the existence of dormant tumor cells or tumor populations that are ultimately responsible for delayed relapses. Furthermore, direct evidence for the existence of dormant tumor cells comes from studies in patients with hematologic malignancies [7, 8]. For example, in the case of chronic myeloid leukemia

Fig. 8.1 (continued) Metastasis describes a process during which cancer cells disseminate from the primary tumor to invade the surrounding tissue, intravasate into the blood or lymphatic microcirculation, transit through the vasculature and/or lymphatic system, and exit into the parenchyma of distant tissues to form secondary lesions. Dormant tumor cells may exist at various stages of the metastatic cascade, including as circulating tumor cells (CTCs) during the translocation phase and as single disseminated tumor cells (DTCs) or micrometastatic foci upon extravasation into the metastatic organ. Analogous to the prominent role of microenvironmental cues in governing the tumor dormant state of primary disease, specific niche environments can also orchestrate dormancy vs. exuberant proliferation of metastatic tumors. In particular, microenvironments conducive for quiescence can induce metastatic tumor cell dormancy. Furthermore, the establishment of an equilibrium between pro- and anti-angiogenic signals, antitumor immunity, and immune evasion can promote tumor mass dormancy

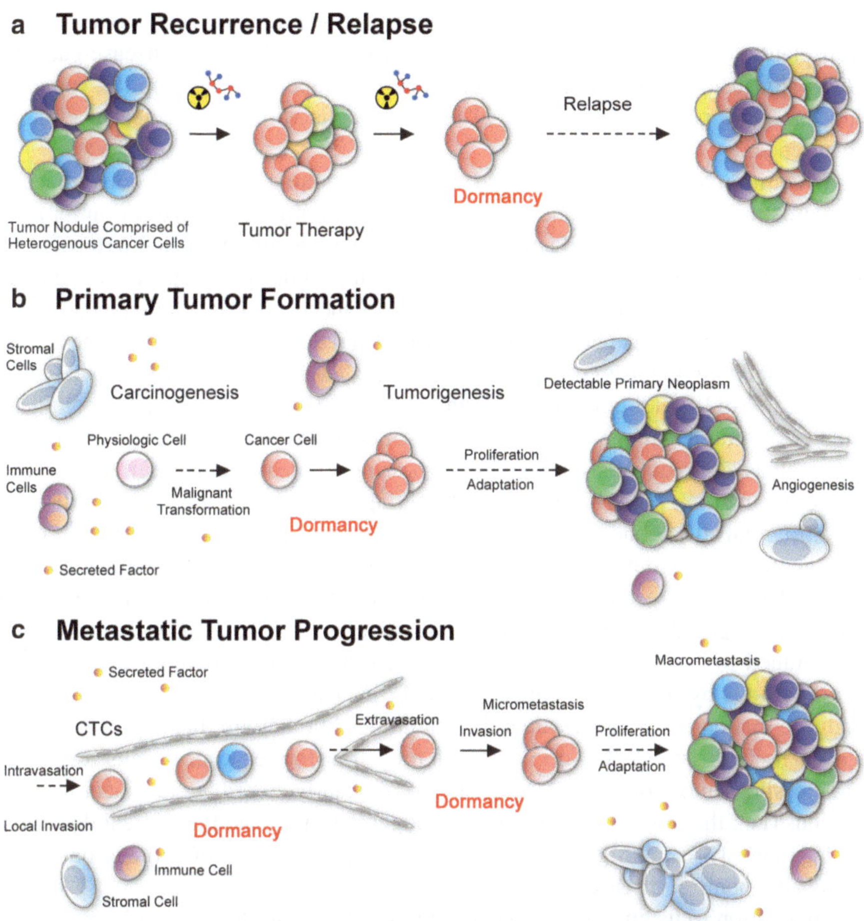

Fig. 8.1 Cancer dormancy occurrence during various phases of tumor propagation. (**a**). Current treatment modalities, including chemotherapy and ionizing radiation, preferentially target the proliferating bulk of tumor cells. However, they often fail to fully eradicate the malignant mass. Increasing evidence suggests that a subpopulation of dormant cancer cells bears the preferential ability to survive tumor therapy and persist long term, ultimately causing tumor relapse. (**b**). In addition to the more established role of this refractory cell population in maintaining minimal residual disease after treatment, dormancy might also occur during early phases of tumorigenesis. Malignant transformation might result both from the cumulative acquisition of oncogenic mutations in physiologic cells and from the dysregulation of niche elements, including stromal and immune compartments, extracellular matrix (ECM), and growth factors. Microenvironmental cues are also important for preventing initial tumor outgrowth by maintaining the dormant state of solitary cells and clinically undetectable lesions over extended periods of time. For instance, niche milieus characterized by hypoxia, nonphysiologic pH and metabolite concentrations, and low nutrient, ECM and growth factor availability can cause cancer cell quiescence and tumor mass dormancy. Release from microenvironmental growth constraints ultimately triggers overt neoplasia. Tumor angiogenesis and immune evasion represent essential mediators of virulent tumor expansion. Consistently, they have been implicated in the switch from cancer dormancy to tumor proliferation. (**c**). Dormancy may also happen during metastatic disease progression.

(CML), nearly 85% of patients in sustained remission after allogeneic bone marrow transplantation remain polymerase chain reaction-positive for disease-causing *BCR-ABL* fusion gene transcripts after years of therapeutic follow-up [8], suggesting that most patients carry residual leukemic cells without evidence of clinical recurrence.

In addition to its more established role in maintaining minimal residual disease after treatment, tumor dormancy might also occur during early stages of tumor progression (Fig. 8.1b) [1]. In fact, epidemiologic studies in unselected autopsies of trauma victims revealed that dormant tumors are highly prevalent in the general population [13–18]. For instance, while thyroid cancers are only clinically apparent in about 0.1% in adults between the ages of 50 and 70 years [19], routine necropsy of deceased individuals of comparable ages revealed that more than 35% of asymptomatic thyroid glands contained malignant nodules [14, 17]. Similarly, histological studies of routine autopsy tissues identified microscopic foci of prostate and breast carcinomas at frequencies markedly surpassing clinical incidence rates for the respective cancers [16–18], further highlighting that the reservoir of latent disease is substantial. It is conceivable that the prevalence and incidence of cancer could rise continually as detection thresholds are lowered by advances in imaging and molecular biology techniques [13]. In support of this notion, the prevalence of histologically verifiable papillary carcinomas smaller than 0.5 mm in diameter was determined to be close to 100% in unselected thyroid specimens in a systematic autopsy study by Harach and colleagues [14]. Taken together, these autopsy findings indicate that dormant tumors are quite ubiquitous in otherwise healthy individuals. Given its high prevalence, dormancy of fully transformed cancer cells and non-progressing tumor populations could thus significantly contribute to cancer latency periods (i.e., the time separating the carcinogenic insult from the clinical detection of the tumor [1]) and might represent a common process preceding the initiation of clinically manifest disease.

In addition to its emergence in early phases of primary tumor development and delayed cancer recurrence after initial treatment, tumor dormancy can also occur during the metastatic process (Fig. 8.1c) [20]. The formation and growth of tumor metastases that impair the function of vital organs is the leading cause of cancer-related deaths [21]. During the metastatic cascade, cancer cells from a primary neoplasm invade the surrounding tissue, enter the blood or lymphatic microcirculation at the primary tissue site (intravasation), survive transit within the vasculature and/or lymphatic system, translocate to distant organs or tissues where they exit the bloodstream and/or lymphatic vessel (extravasation), and adapt to and survive the foreign microenvironment of the distant site in ways that facilitate the formation of a proliferating secondary tumor (colonization) [22]. Conventional thinking holds that metastasis happens during late stages of tumor progression [23]. However, recent evidence suggests that tumor cell dissemination might occur early in the tumorigenic process—that is in parallel with primary tumor development [24–26]. Clinical findings supporting this view come from studies on disseminated tumor cells (DTCs) and circulating tumor cells (CTCs) [25, 27, 28]. For instance, CGH analysis of single DTCs isolated from the bone marrow of node-negative breast cancer patients after curative resection of the primary neoplasm revealed similar genetic abnormalities between DTCs and matched primary tumor samples [27]. In contrast, DTCs from patients with detectable metastatic disease demonstrated more pronounced chromosomal aberrations than primary tumor cells [27], suggesting the genetic alterations that enable the progression of primary lesions are insufficient for establishing metastatic tumor foci. Rather, the acquisition of additional genomic abnormalities at the secondary site appears to be required for robust metastatic tumor growth. Given the marked prevalence of DTCs and/or occult metastatic disease, even among node-negative patients [28, 29], the hematogenous spread of tumor cells and their successful colonization of distant tissues may represent inefficient processes [30]. In support of this notion, an increasing body of literature indicates that metastases form from only a fraction of cells that populate secondary sites, whereas many DTCs remain as either solitary dormant cells or dormant micrometastases [20, 31]. While the detection of DTCs during asymptomatic tumor stages and/or upon successful resection of the primary neoplasm is being increasingly recognized as an early determinant of subsequent metastasis formation, many patients with measurable DTC levels do not develop clinically apparent metastases [32–35]. Similarly, CTCs can be detected in cancer patients in sustained remission even after years of follow-up [36]. For example, Meng et al. found that more than 30% of breast cancer survivors without evidence of clinical disease had CTCs 7–22 years after mastectomy [28]. While the relatively short half-life of CTCs (about 1–2 h [28]) indicates that CTCs themselves might not be exclusively dormant, the findings by Meng and colleagues suggest the presence of clinically undetectable, yet replicating tumor masses in patients with cancer dormancy that keep replenishing the pool of CTCs for many years. Together, these clinical findings corroborate the presence of dormant tumor cells at various stages of the metastatic cascade, including CTCs during the translocation phase and/or single dormant DTCs or micrometastatic foci upon extravasation into the metastatic organ (see Fig. 8.1c).

In aggregate, a plethora of clinical observations emphasizes a significant role for tumor dormancy in human disease. However, many of the clinical studies outlined above are based on epidemiologic studies and thus cannot provide direct mechanistic insights into the molecular networks and cellular population dynamics that orchestrate the state of tumor dormancy.

Mechanistic Insights into the Maintenance of Tumor Dormancy

Significant efforts in the field have been made to understand the control mechanisms underlying cancer dormancy [37]. To that end, a number of sophisticated experimental systems have been developed that successfully model dormancy and have allowed researchers to further characterize the molecular machinery regulating this process [1]. These models have not only enhanced our understanding of the biology underlying cancer dormancy, but have also substantiated the presence of dormant tumor cells in the course of both primary and metastatic tumor development, as well as during neoplastic recurrence after treatment. Insights from experimental dormancy models could ultimately be translated into the clinic upon validation in patient biospecimens [38].

Mechanistically, tumor dormancy can be subdivided into (1) "cellular dormancy" or "tumor cell dormancy," i.e., cancer cell-intrinsic programs that induce growth arrest at the single-cell level; and (2) "tumor mass dormancy" or "tumor population dormancy," which describes cellular interaction dynamics that antagonize the expansion of a tumor population consisting of actively dividing cells [1].

Tumor Cell Dormancy

Cellular dormancy can occur when tumor cells enter a state of quiescence through a G_0/G_1 arrest of the cell cycle [1]. Experimentally, this quiescent state is commonly defined by (1) absence of proliferation, (2) lack of programmed cell death, and (3) retention of fluorescent tracers that are typically diluted to undetectable levels after a few cycles of cell division [39, 40]. Evidence of cellular dormancy comes from various tumor model systems [39, 41–43]. In a murine breast cancer metastasis model, Naumov and colleagues found large numbers of viable non-cycling solitary DTCs that were negative for the proliferation marker Ki67 and demonstrated in vivo retention of a nanofluorescence dye, indicating lack of cell division [39]. Similarly, a study by Luzzi et al. revealed that 36% of intraperitoneally injected murine B16 melanoma cells remained as solitary DTCs in vivo, showing no evidence of proliferation in secondary tissues [42]. These studies highlight the crucial importance of the given microenvironment in a non-orthotopic site in governing the fate of a metastasized tumor cell [31]. Specifically, a niche that is conducive for quiescence prevents DTCs from actively cycling, thereby disallowing the formation of a metastatic nodule [44]. Indeed, numerous experimental metastasis studies have unraveled niche signals that can induce and maintain the dormant state of a tumor by promoting cell cycle arrest, including activated stress signaling, reduced urokinase receptor (uPAR) expression, and loss of integrin and/or extracellular matrix (ECM) function [41, 43, 45–47].

In addition to its role in metastatic tumor dormancy, cell cycle arrest has also been implicated in the maintenance of the dormant state of a tumor in response to various forms of cancer therapy, including chemotherapy and ionizing radiation [1, 2]. For example, while treatment of breast cancer-bearing mice with the alkylating agent doxorubicin significantly reduced metastatic tumor burden, the chemotherapeutic regimen spared solitary, nondividing DTCs [48]. Comparably, treatment with the proteasome inhibitor bortezomib eradicated the majority of replicating multiple myeloma cells but invariably left behind a surviving cell fraction that had entered a prolonged state of quiescence in response to this form of therapy [49]. This growth arrest was linked to eIF-2alpha phosphorylation and the induction of endoplasmic reticulum (ER) stress apoptotic signaling [49]. Increased activation of eIF-2alpha was also reported in the context of chemotherapy-induced squamous cell carcinoma dormancy [50]. Consistent with findings generated with these tumor dormancy models, the timing of late recurrence in breast cancer patients undergoing mastectomy was unaffected by prolonged adjuvant chemotherapy, further emphasizing that growth-arrested tumor subpopulations may constitute the pool of chemoresistant cancer cells that could ultimately drive delayed post-resection recurrence [51].

Overall, the studies discussed above highlight tumor therapy resistance as a critical biological feature of dormant cancer cells (see Fig. 8.1a). Additionally, both animal studies modeling the tumor dormant state and clinical observations emphasize the importance of the microenvironment in orchestrating the balance between cellular dormancy and neoplastic proliferation. Specifically, a niche environment fostering quiescence of DTCs or primary solitary tumor cells might prevent dormant cancer cells from reawakening. Niche environments restraining tumor cell growth may include angiostatic milieus that result in hypoxia and/or reduced nutrient, growth and ECM factor availability, as well as inflammatory and apoptosis-conducive conditions [44]. This suggests additional biological traits pertinent to dormant tumor cells: (1) their ability to survive an environment devoid of blood vessels, (2) their competence to evade immunologic clearance, and (3) their preferential endowment with anti-apoptotic mechanisms (see Fig. 8.1b, c) [37].

Tumor Population Dormancy: Angiogenesis

Both data emerging from the clinic and experimental dormancy models suggest critical roles for arrested angiogenesis and the establishment of equilibrium between immune escape and rejection in maintaining the dormant state of a tumor [52, 53], particularly in the setting of tumor mass dormancy. In contrast to cellular dormancy, tumor mass dormancy is not characterized by the absence of proliferation and apoptosis on the cellular level. Rather, it describes the balance of the two, which prevents a neoplasm from increasing in size [54].

The vast majority of solid cancers rely on the recruitment of functional blood vessels—a process termed angiogenesis [55]—to support the continuous expansion of a growing tumor mass [56, 57]. Although tumors are proliferation-competent, they are often unable to induce angiogenic sprouting of blood vessels and/or remodel preexisting tumor vasculature, leading to limited cancer perfusion and nutrient supply [52], and, as a result, hypoxia and programmed cancer cell death [54]. Such avascular tumors are growth-arrested and remain microscopic in size because their marked proliferation rate is balanced by enhanced apoptosis [58–61]. This angiogenesis-dependent dynamic equilibrium between tumor proliferation and cancer cell death is referred to as "angiogenic dormancy" [62], and the transition from a pre-vascular lesion to a highly vascularized and progressively expanding tumor mass is defined as the "angiogenic switch" [63]. Angiogenic tumor dormancy and, correspondingly, the angiogenic switch are controlled by the integrated action between pro-angiogenic signals, such as VEGF (vascular endothelial growth factor) or PDGF (platelet-derived growth factor), and anti-angiogenic factors, including angiostatin, endostatin, and thrombospondin-1 (Tsp-1) [64–66]. Primary and metastatic tumor environments devoid of angiogenesis-promoting factors and/or milieus in which anti-angiogenic signaling molecules are more prevalent will cause cancer dormancy, as demonstrated in numerous tumor models [1, 60, 65–68]. For example, exogenous addition of the anti-angiogenic factor angiostatin potently inhibited both murine and human primary carcinoma growth in mice by inducing dormant microscopic tumor foci, in which proliferation was balanced by apoptosis [65]. Similarly, elevated Tsp-1 expression levels mediated dormancy maintenance in models of murine hematopoietic tumors, mammary, and kidney carcinoma [66, 67]. Tsp-1 repression, on the other hand, was sufficient to cause cancer progression and was associated with activation of the *MYC* oncogene [66, 67], a well-established modulator of the dormant state of a tumor [69]. Comparably, potent angiogenic stimuli are sufficient to revert cancer dormancy by triggering the angiogenic switch [70]. For instance, it has been demonstrated that poorly tumorigenic human leukemia cells that exclusively form dormant lesions in immunocompromised mice resumed growth following both transient VEGF stimulation and in vivo administration of pro-angiogenic cells [68]. Taken together, these experimental findings highlight that angiogenesis induction represents a fundamental event underlying the shift from tumor dormancy to progressive cancer outgrowth.

Tumor Population Dormancy: Immune Surveillance

In addition to angiogenesis suppression, alternative control mechanisms have been described that prevent escape from cancer dormancy. For instance, the immune system may also help contain tumor outgrowth and metastatic progression [71, 72]. In fact, numerous reports have suggested a pertinent role for the antitumor immune response in dormancy maintenance [53, 73, 74]. Direct evidence of immunity-induced tumor dormancy almost exclusively comes from animal models [1] that typically involve immunization of mice with irradiated cancer cells or tumor-associated antigen (TAA) peptides followed by challenge with viable tumor cells [53]. For example, in the DBA/2 murine lymphoma model, dormancy can be induced by subcutaneous injection of syngeneic L5178Y lymphoma cells prior to intraperitoneally rechallenging mice with the same cell line [74]. Similarly, in the BCL-1 lymphoma model, immunization of mice with BCL1-derived immunoglobulin (Ig) can trigger anti-idiotype immune responses that induce cancer dormancy [75, 76]. Furthermore, vaccination of mice with DA1-3b leukemia cell line variants following subsequent challenge with viable DA1-3b cells can cause leukemia dormancy [77, 78]. Taken together, these models clearly support that the immune system has an important role in inducing a tumor dormant state, particularly in the context of vigorous antitumor immune responses. Thus, immune-mediated dormancy in humans might apply primarily to highly immunogenic cancers, such as malignant melanoma [79]. In fact, melanomas with a donor origin have been reported in kidney and bone marrow transplant recipients receiving immunosuppressive medication [80, 81]. Strikingly, some donors had been in complete melanoma remission for decades [80], indicating that the donor immune response may have prevented the outgrowth of DTCs that ultimately escaped from dormancy in the immunosuppressed transplant recipient. Despite these findings, whether antitumor immunity is generally applicable to clinical cancer dormancy maintenance remains a matter of considerable debate [38]. In spite of this controversy, the aforementioned experimental models and clinical observations support the notion that dormant tumor cells employ immuno-escape strategies that enable their persistence in the context of robust adaptive and innate immune responses. Immunological mechanisms facilitating the establishment of an equilibrium between cancer elimination and immune evasion may thus promote tumor mass dormancy [53, 71]. Furthermore, escape from cytotoxic immune effectors might also represent a critical survival mechanism for quiescent tumor cells [53]. Mechanistically, dormant cancer subsets have been found to both passively evade immunologic clearance (e.g., by downregulating immunogenic TAAs [73]) and actively suppress immune effector functions (e.g., through preferential expression of immune-inhibitory members of the B7/CD28 costimulatory superfamily) [82, 83].

In summary, a number of experimental animal models have been developed that have allowed researchers to gain mechanistic insights into the immunobiological processes governing cancer dormancy vs. tumorigenic outgrowth, including cell cycle modifications, alteration of angiogenic responses, and modulation of antitumor immunity [1, 37]. However, further study is required to determine exactly how these regulatory mechanisms act in concert to sustain the tumor dormant state and why some dormant tumors are released from growth restraints. Beyond this, it remains unclear what mechanisms underlie clinical tumor dormancy and whether experimental dormancy models accurately reflect human disease [38]. Importantly, detailed biological characterization of dormant tumor populations from clinical material has been limited by the absence of reliable biomarkers and methodologies capable of sufficiently isolating these cells.

Recent advances in research on physiologic and malignant stem cell systems could potentially alleviate this shortcoming. In fact, there are striking parallels between dormant tumor cells, physiologic stem cells, and cancer stem cells (CSCs), which can likewise undergo prolonged periods of quiescence [84], preferentially resist current forms of tumor therapy [85], survive immune attack [86], and selectively withstand anti-angiogenic and other hostile microenvironments [87]. The emerging literature on the biological functions of CSCs and their potential relevance to clinical tumor dormancy will be the focus of the following sections of this chapter.

The Cancer Stem Cell Paradigm: A New Perspective on Tumor Progression

Tumors, like physiologic tissues, are complex structures composed of genetically, phenotypically, and functionally heterogeneous cell populations that differ in their ability to initiate and maintain tumor growth [88]. The prevailing stochastic model of neoplastic progression postulates that the accumulation of genetic mutations followed by clonal selection induces tumor heterogeneity [89]. According to this theory, all cancer cells—regardless of their phenotype—possess equivalent intrinsic capacities to (1) proliferate, (2) initiate tumor growth, and (3) cause relapse [90]. Phenotypic and functional differences of cancer cells, however, can be explained only in part by sequential acquisition of genetic variations, which has led to the development of the CSC hypothesis of tumor growth [91, 92]. The CSC model posits that only a subpopulation of cancer cells within a tumor (i.e., CSCs) can proliferate extensively and give rise to the morphologically and functionally diverse cancer cell progeny that comprises the malignant lesion [93]. Pursuant to the CSC concept, tumors are organized as defined hierarchies with CSCs at their apex. It should be noted, however, that these two models of tumorigenicity are not necessarily mutually exclusive, as stochastic processes may be at play within the CSC population [94].

CSCs have been operationally defined by their (1) preferential ability to initiate tumor growth, (2) capacity to self-renew, and (3) competence to differentiate into various non-self renewing tumor bulk populations [95], thus providing an explanation for functional differences between tumor subpopulations [93]. Experimentally, verification of these defining CSC traits requires serial xenotransplantation at limiting dilution of marker-defined clinical cancer subpopulations into an orthotopic site of immunocompromised animals (typically NOD/SCID (non-obese diabetic/severe combined immunodeficiency) mice) [95, 96]. Using this approach, the preferential ability to initiate and maintain tumor growth is ascribed to a putative CSC population. More importantly, the CSCs' capacity to self-renew and differentiate also requires experimental confirmation in serial in vivo passaging studies. Typically, this is done by demonstrating reestablishment of the original patient tumor heterogeneity in primary and secondary cancer xenografts upon inoculation of immunocompromised hosts with purified CSCs [95]. Recent studies employing lineage-tracking methodologies have allowed for a side-by-side comparison of CSC-defining traits by utilizing distinct genetic labels to trace CSCs vs. differentiated tumor cells in vivo [96]. These experiments have further substantiated the selective ability of CSC fractions to both undergo cell divisions that expand the CSC pool and generate differentiated cancer cell progeny, whereas non-CSC subsets exclusively give rise to differentiated tumor populations [97–99]. Other reports suggest a more dynamic regulation of CSC phenotype and function, allowing for a bidirectional interconversion between the CSC compartment and non-stem cancer cells in response to contextual signals from the microenvironment [100, 101]. Such findings of CSC plasticity highlight the critical importance of the host immune system and the stromal microenvironment in governing tumor growth rates as well as CSC and non-stem cancer cell fate [86, 102]. These considerations further emphasize that variations in the methodologies and experimental model systems used to assess cancer "stemness" require careful scrutiny, as they might influence experimental outcomes [94].

Despite the fundamental advances that CSC studies have yielded regarding our understanding of functional tumor heterogeneity, the CSC model remains a topic of considerable controversy [103, 104]. The disagreements surrounding the CSC theory have partly resulted from the implication of a direct relationship between

CSCs and physiologic stem cells as the cellular origin for cancer [105]. While some experimental tumor model systems have indeed identified adult tissue stem cells as the source of malignant transformation [106, 107], other models have indicated that progenitor or transient amplifying (TA) cells [108, 109] and perhaps even terminally differentiated cells [110] could acquire CSC-like properties through a series of mutagenic events. Additionally, the commonly accepted definition of a CSC does not infer that transdifferentiation plasticity (i.e., differentiation along a particular lineage) associated with both physiologic stem cells and cancers [111] is intrinsic to the CSC pool [95, 96]. Thus, the term CSC refers to the functional traits of the cancer cell, rather than its cellular origin and biological properties within normal tissues [103].

Disparity also exists regarding the very definition of a CSC, given recent findings of CSC biomarker plasticity [100, 101]. If CSCs were defined to be a distinct, unchangeable subpopulation that resides at the apex of a hierarchically organized tumor, then such data could shed doubt on the validity of the classical CSC model. Certainly, if the majority of cancer cells are in a constant state of flux with regard to CSC phenotype and function, it is arduous to conceptualize a hierarchy of stemness. As outlined above, however, CSC behavior critically depends on the composition and configuration of the surrounding niche environment [87, 112]. Similarly, in physiologic tissues, stem cell competence is linked to the contextual signals originating from specialized niches [113, 114]. Accordingly, stem cell competence is not restricted to discrete cell subsets characterized by a defined expression pattern [115]. Rather, the vast majority of cells not irreversibly committed to differentiation harbors stem cell potential [116]. Indeed, genetic lineage tracing studies revealed that TA populations primed for terminal differentiation can "de-differentiate" into physiologic stem cells, such as in response to in vivo stem cell loss [117]. However, this observed reversibility of stem cell potential as a function of the microenvironment or epigenetic factors does not invalidate the hierarchical organization of physiologic tissues. Therefore, the perception that changing microenvironments dynamically modulate CSC properties does not preclude that the conglomerate tumor populations adhere to defined hierarchies. Findings of bidirectional interconversion between CSCs and non-stem cancer cells thus do not undermine the CSC hypothesis, as the distinct tumor populations retain their identities in the sense that they can be distinguished phenotypically and functionally at any given time point within a tumor [103].

Additional dissonance regarding the CSC model has resulted from the assumption that CSCs should represent only a small fraction of cancer cells [118, 119], paralleling low frequencies of physiologic stem cells in healthy normal tissues [116]. However, "rareness" is not a defining trait of CSCs [93], and relative CSC frequencies—analogous to physiologic stem cell systems [120]—likely rely on the complex interplay between stem and non-stem cells, host and microenvironmental factors, differentiation, and apoptosis kinetics. Indeed, recent data suggest a considerable variability of relative CSC numbers depending on tumor type, stage of malignant progression, niche constitution (e.g., growth and ECM factor availability and degree of vascularization), and host immunocompetence [87, 103]. Hence, CSC frequencies cannot be stated in absolute numbers, but rather relative to the specific experimental system used [96, 121]. Given the potential misconceptions associated with the term "CSC," many investigators in the field refer to CSCs as tumor-initiating cells (TICs) or tumor-propagating cells [95].

Pioneering work in the CSC field originated almost two decades ago, when John Dick and colleagues reported that human acute myeloid leukemia (AML) demonstrates hierarchical organization, with a population of leukemic stem cells (LSCs) at its apex [122, 123]. Subsequent studies have further revealed that CSCs are integral to cancer propagation in a number of solid tumor entities, including those of the bladder, bone, breast, central nervous system, colon, head and neck, liver, ovaries, pancreas, and skin [97, 124–137]. Meanwhile, the importance of CSCs in experimental tumor initiation and primary tumor growth has been firmly established [121]. However, the role of CSCs in multistage cancer progression, especially with respect to metastasis formation, requires further study. Recent evidence indicates that the CSC compartment might indeed critically contribute to the metastatic process [22, 138]. For instance, in several tumor models, efficient metastatic colonization, i.e., the initial expansion of metastatic foci at the secondary site, critically depends on CSCs [135, 138, 139]. Moreover, CSC frequencies correlate with disease progression in clinical cancer specimens [97, 140], further highlighting the potential translational relevance of the CSC paradigm. CSC–niche interactions and, in particular, a process described as the epithelial to mesenchymal transition (EMT) represent potential mechanisms by which CSCs located at the invading front of the tumor might acquire migratory traits to colonize secondary sites [141, 142].

Beyond Self-Renewal and Differentiation: Further Attributes of CSCs

Accumulating evidence demonstrates preferential CSC resistance to current standard-of-care cancer therapeutic regimens, including chemotherapy and ionizing radiation [85, 143, 144], which target specific phases of the cell cycle based on the understanding that cancer cells are highly proliferative in contrast to normal tissues. Complete eradication of primary and especially disseminated disease, however, is often limited by the preferential ability of tumor cells to activate DNA damage recognition and repair pathways, alter cell cycle checkpoints, inhibit apoptotic signaling, and decrease intracellular drug accumulation [85]. Strikingly, several studies have provided experimental evidence that the aforementioned survival mechanisms are preferentially operative in CSCs vs. tumor bulk populations [97, 99, 145–148]. Their enhanced ability to survive cytotoxic therapy identifies CSCs as the likely cause of disease relapse. Consistently, recent studies revealed increased CSC frequencies in cancer patients following therapeutic intervention [145, 147, 149–151].

Accordingly, novel treatment strategies designed to target chemoresistant CSCs, as well as the majority of non-stem tumor cells, could facilitate eradication of cancers resistant to current forms of therapy [152, 153].

Beyond therapy resistance, many of the biological mechanisms underlying CSC-driven tumorigenesis and neoplastic progression are also beginning to be unraveled, including their preferential ability to (1) induce angiogenic responses, (2) interpret contextual niche signals in favor of tumor growth, (3) selectively survive hostile microenvironments, and (4) evade immunological clearance [86, 87, 143]. As already outlined in more detail above, tumor initiation, growth, and metastasis crucially depend on the creation of a vascular network that provides tumor cells with nutrients and oxygen, while removing metabolic wastes and carbon dioxide [56, 57]. Several recent reports have identified specific roles for CSCs in tumor vascularization. For example, in comparison with patient-matched non-stem tumor cells, CSCs bear the preferential ability to promote tumor angiogenesis through enhanced secretion of VEGF [154]. Other studies have found that CSCs may preferentially reside within perivascular niches, where they can stimulate vascularization by triggering angiogenic growth factor production in niche-constituent cells [112]. This interplay between CSCs and their microenvironment is bidirectional, given that the surrounding pro-angiogenic niche may in turn promote CSC maintenance and self-renewal [155]. An additional mechanism by which CSCs may promote tumor vascularization is by their ability to transdifferentiate into ECM-rich, vessel-like structures [156–158]. This process, termed “vasculogenic mimicry” [111], might enable CSCs to generate their own vascular network independently of true cancer angiogenesis in order to perfuse tumor tissue. Furthermore, CSCs have mechanisms in place that enable their preferential survival in hypoxic milieus [159, 160] and microenvironments characterized by non-physiological pH levels and toxic metabolite concentrations [161].

Accumulating evidence indicates that, in addition to conferring competitive advantages to a progressing neoplasm by promoting cancer vascularization, CSC populations also have a preferential ability to both evade and actively modulate antitumor immune responses [86, 124, 162–167]. The concept of “tumor surveillance” advanced by Drs. Burnet and Thomas [168, 169] posits that the immune system can eliminate cancerous clones, thereby acting as a major safeguard from the development of malignant disease. While this hypothesis remains the topic of considerable debate [71], both clinical findings and experimental results generated over the past decades support the notion that a functional immune system can in some cases prevent tumor initiation and growth [170, 171]. The recently uncovered ability of CSCs to actively thwart antitumor immune responses might thus confer selective growth advantages to these virulent cancer subsets. CSC immunological functions include evasion from immune clearance, induction of clonal anergy or deletion, and activation of regulatory immune cells (discussed in more detail below) [86]. Accordingly, future cancer immunotherapeutic protocols that consider CSCs and their immunomodulatory functions hold promise toward achieving more durable responses in tumor patients.

Regulation of the Tumor Dormant State: Clues from CSC Biology?

There are striking parallels between the CSC theory of tumor development and the concept of cancer dormancy [1, 172]. Dormant tumor cells, like CSCs, bear the preferential competence to survive both cytotoxic treatment modalities and less fertile microenvironments incapable of fostering tumor propagation [37]. Analogous to the involvement of CSCs in multiple facets of tumorigenesis, cancer dormancy can occur during early phases of tumor development and metastatic tumor spread, and is especially relevant to delayed cancer recurrence [1]. Furthermore, both dormant cancer cells and CSCs preferentially evade antitumor immune responses [53]. Also, contextual signals from the microenvironment, as well as angiogenic factors, can govern both the dormant state of a tumor and CSC fate [52, 61].

At the very least, the commonalities between CSCs and dormant tumor cells point to a partial overlap of these two cancer subpopulations. At present, we do not know whether CSCs undergo prolonged phases of dormancy, and only few studies have addressed the question of CSC quiescence. The very definition of CSCs is based on their preferential ability to initiate robust tumor outgrowth, which seems incompatible with a cancer dormant state. In light of the intriguing relationship between common CSC attributes and the mechanisms controlling the tumor dormant state, however, insights from CSC biology could help guide future research on tumor dormancy. In fact, recent evidence suggests that the CSC compartment itself might consist of heterogeneous subpopulations [173, 174], including a predominantly quiescent, slow-cycling fraction [175, 176]. The CSC pool, or subsets thereof, might therefore exploit phases of cellular dormancy to ensure long-term tumor maintenance and survival in harmful tumor environments, including growth-inhibitory niches and cytotoxic milieus. Finally, given the capacity of CSCs to cause delayed relapse and drive rapid tumor expansion, CSC immunobiological pathways might play critical roles in the switch from tumor dormancy to vigorous tumor outgrowth. Future dormancy studies should perhaps consider dormant CSCs as opposed to differentiated dormant tumor foci as a likely source of ultimate disease recurrence. CSC biology, and CSC markers in particular, could thus critically advance our understanding of the mechanisms underlying clinical tumor dormancy, by enabling the prospective isolation and characterization of dormant tumor cells. In the following sections, we dissect whether CSCs, or subsets thereof, might indeed represent dormant tumor populations, by juxtaposing mechanistic insights into CSC behavior with processes regulating maintenance of cancer dormancy.

Cancer Recurrence: Are Dormant Tumors Comprised of Therapy-Resistant Cancer Stem Cells?

Perhaps the most obvious link between CSCs and dormant tumor populations is their converging ability to survive cancer therapy. Indeed, dormant lesions are often spared by current treatment modalities, as determined both in animal models and clinical disease [48, 51]. Similarly, CSC frequencies are markedly enhanced in local and metastatic tumor recurrences post therapy compared to those in pre-therapy samples [145–147, 149, 150]. Importantly, therapeutic refractoriness of dormant tumor cell populations is attributable to several resistance mechanisms also intrinsic to the CSC compartment, including impairment of cancer apoptotic pathways, alteration of cell cycle checkpoints, and reduced drug accumulation. One mechanism of particular relevance to CSC resistance is decreased intracellular buildup of chemotherapeutics accomplished by energy-dependent efflux pumps known as ABC (ATP binding cassette) transporters [85]. For instance, LSCs express marked levels of several ABC transporter proteins implicated in leukemia resistance [177, 178]. In human malignant melanoma, CSCs can be prospectively isolated based on their selective expression of the ABC superfamily member, ABCB5 [97]. In a series of colorectal cancer patients, 5-FU-based chemoradiation therapy revealed a markedly enhanced abundance of $ABCB5^+$ tumor cells when residual disease was detected [147]. Moreover, ABCB5 frequency inversely correlated with recurrence-free survival in a large cohort of liver cancer patients [179]. Importantly, ABCB5 inhibition sensitized human melanoma [180], colorectal carcinoma [147], and hepatocellular carcinoma [179] CSCs to chemotherapy-induced cell obliteration. Whether dormant cancer cells express elevated levels of ABC transporters is currently unknown. However, investigating the ABC transporter repertoire of dormant tumor subsets might represent an important line of investigation, given the aforementioned associations of distinct ABC superfamily members with CSC drug resistance and clinical cancer recurrence.

Direct evidence for a potential overlap between dormant cancer fractions and CSCs is given by the propensity of both populations to undergo cell cycle arrest in response to various forms of therapy. As outlined above, mitotic quiescence is a hallmark feature of the dormant state of a tumor [1], especially in the context of chemotherapy and radiation-induced cellular dormancy [2]. Strikingly, relative quiescence represents an important mechanism underlying preferential CSC resistance to various forms of tumor therapy. For instance, LSCs isolated from AML patients showed a preferential G_0/G_1 arrest of the cell cycle [181], suggesting resistance to therapies targeted at the proliferating bulk of tumor cells. In CML, resistance to the BCR-ABL tyrosine kinase inhibitor (TKI) imatinib has been linked to the failure of the chemotherapeutic regimen to deplete the quiescent LSC compartment [182, 183]. Similarly, second generation TKIs failed to induce apoptosis in the highly quiescent fraction of LSCs [184, 185]. Evidence for CSC-associated quiescence and resultant therapy resistance has also been generated in solid tumors. For example, in glioblastoma and breast cancers, CSC frequency was enriched after ionizing radiation [145, 146, 186]. The preferential radioresistance of both breast and brain CSCs was mechanistically linked to activation of DNA damage repair pathways and to significantly reduced rates of proliferation and apoptosis induction through the involvement of DNA checkpoint kinases [145, 146, 187]. A direct link between tumor dormancy and CSCs is further reinforced by findings of semiquiescent cancer initiators in hepatocellular carcinoma [176]. Specifically, this study revealed that a subset of slow cycling liver CSCs preferentially survived genotoxic chemoradiation by reversing DNA damage in a CD13 (also known as aminopeptidase N)-dependent manner [176]. Additional insights into the mechanisms of CSC cell cycle arrest have been generated in model organisms. For instance, in a mouse model of CML, maintenance of LSC quiescence and resistance was functionally related to the *PML* (promyelocytic leukemia) tumor suppressor gene [188]. In an AML model, cell cycle restriction mediated by p21 activation was critical in preventing excess accumulation of DNA damage and functional exhaustion of LSCs [148]. Thus, certain chemotherapeutic agents not only spare resistant CSCs and dormant tumor populations, but also specifically enrich for CSCs by inducing a state of quiescence.

Similarly, dormancy induction models harness chemotherapeutic effects to promote tumor cell quiescence [49, 50, 189]. Analogous to the p21-dependent cell cycle arrest observed in CSCs [148], p21 activation also induced reversible quiescence and prevention of apoptosis in dormant tumor populations [190]. Taken together, these findings clearly demonstrate significant overlap of dormant cancer cells and CSCs, particularly in the context of cytotoxic treatment modalities. Therefore, CSC quiescence might critically contribute to latent disease and delayed cancer recurrences caused by dormant tumor subsets. Accordingly, biomarkers and signaling pathways intrinsic to the CSC compartment could prove useful in enhancing our understanding of the relationship between tumor cell dormancy and prolonged time to cancer relapse.

How CSC Signaling and Niche Interactions May Govern Tumor Dormancy

As outlined above, quiescence is a key regulator of the tumor dormant state, particularly in response to various forms of tumor therapy. However, cell cycle arrest also critically contributes to the maintenance of tumor dormancy in the absence of genotoxic insults. Not surprisingly, several molecules implicated in cell cycle control, either directly or through the involvement of distinct signaling networks, can foster tumor cell dormancy. Indeed, the cell cycle checkpoint modulator, p38, has been extensively associated with dormancy perpetuation in various tumor types [1]. For instance, high p38-MAPK (mitogen-activated protein kinase) vs. ERK (extracellular signal-regulated kinase)-MAPK signaling ratios induced tumor growth arrest in disease models of fibrosarcoma, melanoma, breast cancer, and prostate carcinoma [41, 191]. Similarly, prostate CSCs undergo phases of tumor dormancy upon activation of p38-MAPK signaling [192]. Another important regulator of tumor cell dormancy is the *MYC* oncogene, the inactivation of which elicits cellular senescence in diverse tumor types [193]. Strikingly, in a mouse model of hepatocellular carcinoma, *MYC* inactivation resulted in the generation of stem-like cells that could differentiate into normal cellular lineages while retaining their latent potential to become cancerous, hence demonstrating a direct link between oncogene-dependence, CSCs, and preservation of the tumor dormant state [69]. Several downstream effectors of the MYC oncoprotein, including NDRG (N-Myc downstream-regulated gene) family members, have also been implicated in cell cycle arrest of human prostate cancer and glioma CSCs [192, 194].

Additional molecular networks integral to both stem cell quiescence and tumor dormancy maintenance include the Sonic hedgehog (Shh), Notch, and Wnt/β(beta)-catenin signaling pathways [195–198]. For example, Shh-mediated upregulation of the polycomb group (PcG) protein, Bmi1, sustained CSC quiescence and self-renewal [133, 199, 200]. Increased Bmi1 levels also correlated with late (>10 years) metastatic relapse in breast cancer patients [201], indicating that expression of this PcG member might coincide with long-term dormant cancer populations responsible for delayed recurrences. Similarly, Wnt and Notch activity levels control the balance between quiescence and self-replication of both CSCs [202–205] and dormant tumor cells [206–209].

In addition to tumor cell-intrinsic mechanisms, cues from the niche environment can prominently regulate cancer dormancy and CSC quiescence alike. Interestingly, microenvironments conducive for dormancy upkeep have also been found to enable CSC preservation. For example, microenvironmental milieus characterized by transforming growth factor-beta (TGF-β) signaling can trigger both CSC immaturity and tumor dormancy maintenance [210, 211]. In particular, the TGF pathway members, TGFB2 and TGFB3, were determined to be intrinsically associated with both the cancer dormant state and CSCs [37, 164]. Furthermore, uPAR, a prominent downstream target of TGF-β [212], can likewise modulate both CSC biology and tumor cell dormancy. Indeed, high uPAR expression conferred CSC-like properties to breast cancer cells [213] and sustained dormancy in various cancer models (e.g., by activation of the cell cycle inhibitor, p38) [45, 191]. The transcription factor NFkappaB represents an upstream activator of TGF-β signaling and resultant uPAR expression levels [212] that has also been implicated in CSC quiescence [214]. An additional link between dormant tumor subsets and CSCs is their overlapping use of the BMP (bone morphogenetic protein) and mTOR (mammalian target of rapamycin) pathways to balance quiescence vs. proliferation. For example, BMP7 secreted by bone stromal cells induced CSC dormancy in prostate cancer [192], and exogenous addition of BMP4 to glioblastoma-bearing mice resulted in stunted CSC proliferation [215]. Similarly, niche environments that promote mTOR signaling enable in vivo survival of both dormant tumor cells and CSCs by inhibiting their proliferation [189, 216]. Finally, the ECM and integrin composition of the tumor environment might regulate both CSC behavior and the tumor dormant state [43, 46, 211, 217, 218].

In conclusion, the convergence of niche-constituent factors and signaling pathways integral to CSC biology and dormancy maintenance suggests significant overlap between both tumor subpopulations. However, while dampening the aforementioned signaling cascades typically favors the tumor dormant state, CSCs frequently demonstrate aberrant pathway activation during phases of robust tumor outgrowth [195]. Thus, CSCs do not always coincide with dormant cancer populations, but appear to play crucial roles in the shift from dormancy to proliferation, given that the intricate balance between pro- and anti-proliferative CSC-intrinsic signals can govern tumor dormancy. Prolonged phases of quiescence could be particularly relevant to the prevention of premature exhaustion of the CSC pool.

Angiogenic Control of Tumor Dormancy: Is There a Role for CSCs?

Many of the molecular pathways and niche-constituent factors discussed above not only control cancer dormancy and CSC quiescence but also orchestrate the tumor angiogenic process. For example, activated BMP signaling can stimulate tumor angiogenesis [219, 220] and CSC-driven tumor growth [215]. Similarly, endothelial cell-mediated Notch activation releases dormant tumor cells from growth constraints by inducing the angiogenic switch [208] and also enables CSC-mediated tumor angiogenesis in glioblastoma [158]. As outlined in more detail above, CSCs promote tumor vascularization and subsequent cancer expansion in several complementary ways: through (1) preferential secretion of VEGF into the surrounding environment [154], (2) stimulation of angiogenic growth factor production by niche-constituent cells [112, 155], and (3) transdifferentiation into vessel-like structures that might perfuse tumor tissue in the absence of true angiogenesis [156–158], via a process termed "vasculogenic mimicry" [111], which could ensure minimal perfusion and hence sustained viability of a critical tumor mass ultimately responsible for delayed cancer relapse. Accordingly, in scenarios of tumor emergence and growth, CSCs might counteract the dormant state of a tumor by fueling the creation of a vascular network that provides the expanding cancer with nutrients and oxygen.

Nevertheless, CSCs still might undergo phases of dormancy, especially when exposed to niche environments conducive for angiostasis. Indeed, whereas CSCs trigger robust tumor outgrowth when exposed to perivascular niches [112, 155], hypoxic environments were found to maintain an undifferentiated, quiescent state of CSCs [159, 160]. Importantly, hypoxia induces apoptotic cell death in the bulk of proliferating tumor cells [159, 160]. CSCs, on the other hand, preferentially survive hypoxic conditions by expressing the so-called hypoxia-inducible factors (HIFs) [159, 160] also involved in cell cycle arrest (e.g., through p21 activation) [221]. A significant overlap of CSCs and dormant tumor populations in angiostatic microenvironments is also suggested by findings in model organisms that have revealed tumor regression upon *MYC* inactivation, concomitant with enhanced niche expression of the antiangiogenic protein, Tsp-1 [67], and selective survival of stem-like cancer cells capable of causing disease relapse upon oncogene reactivation [69].

In summary, several studies have unraveled intriguing parallels between the mechanisms regulating CSC behavior and angiogenic control of tumor dormancy. While distinct CSC-specific functions might critically promote cancer vascularization in settings of robust tumor outgrowth, CSCs also bear the preferential capacity to survive angiostatic environments associated with dormancy maintenance. Together, these findings suggest that CSCs might represent major culprits in the switch from cancer dormancy to virulent tumor outgrowth. Because CSCs consolidate both the selective competence to persist prolonged phases of nutrient and oxygen deprivation and the preferential ability to aggressively promote tumor progression by triggering robust angiogenic responses, CSCs, or subsets thereof, are likely to represent the dormant tumor populations ultimately responsible for delayed cancer recurrences.

Finally, the results reviewed herein highlight the critical importance of the microenvironment, especially with respect to its angiogenic status and resultant oxygen and nutrient levels, in governing CSC quiescence vs. proliferation.

CSC Escape from Immune Surveillance: Relationship to Tumor Dormancy?

The idea that a functional immune system can eliminate transformed cells before they progress into clinically manifest disease was first proposed by Ehrlich over a century ago [222]. In 1957, Drs. Burnet and Thomas advanced the concept of "tumor surveillance," which postulates that cancerous cells can be recognized and cleared by the immune system during early stages of tumorigenesis, based on their altered expression of self-antigens [168, 169]. While the immunosurveillance hypothesis remains a topic of active controversy [71], both clinical and experimental findings generated over the past decades indeed support a role for the immune system in eradicating cancer cells. For example, several animal models of immunodeficiency, including NOD/SCID, RAG2$^{-/-}$ (recombination activation gene 2 knockout), IFN$\gamma^{-/-}$ (interferon-gamma knockout), and Pfp$^{-/-}$ (perforin knockout) mice, showed increased incidence rates of spontaneous and carcinogen-induced sarcomas, lymphomas, and epithelial carcinomas [71, 171, 223]. In addition, epidemiological studies revealed that immunocompromised patients exhibit markedly enhanced risks for developing a wide range of cancers [170]. Meanwhile, numerous antigens have been characterized that demonstrate selective or elevated expression levels in cancers vs. physiologic tissues [79, 224], empowering the immune system to recognize malignant cells as foreign [225, 226]. Two distinct types of such tumor antigens can be distinguished [79]: (1) tumor-specific antigens (TSAs), which describe aberrant gene products originating from chromosomal mutations and/or genomic rearrangements (e.g., leukemia-specific *BCR-ABL* fusion transcripts [227]), and (2) TAAs that result from abnormal transcriptional and/or translational activation in cancers vis-à-vis normal tissues, including lineage-specific differentiation antigens (e.g., melanoma antigen recognized by T-cells, MART-1) [228]). Although TSA- and TAA-specific immune responses can be detected in cancer patients, they often fail to fully inhibit tumor outgrowth [225, 229].

One possible explanation for the limited efficacy of immune cells to reject the tumor is that highly aggressive, rapidly dividing cancer cells might simply exceed the capacity of the antitumor immune response [223]. An alternative explanation is that transformed cells evolve from a host's own tissues and therefore predominantly express self-antigens to which immune effector cells have been "tolerized" [223]. "Immunogenic tolerance" describes a process that prevents immunogenic clearance of tissue cells expressing a particular set of antigens [230, 231]. Tolerance to self-antigens is achieved by processes that result in either elimination of autoreactive lymphocytes through apoptotic cell death (clonal deletion) [232] or induction of functional unresponsiveness of antigen-reactive cells (clonal anergy) [233]. Self-tolerance refers to the ability of the immune system to recognize and protect self major histocompatibility complex (MHC)/*self* peptide complex-bearing cells from immune-mediated rejection while simultaneously retaining its capacity to launch an immune response against cells expressing MHC/*foreign* peptide complexes [230, 231]. Under physiological conditions, self-tolerance prevents the emergence of autoimmune disorders [230, 231]. However, in the context of cancer, it may facilitate tumor progression by protecting transformed cells from immune-mediated rejection [223]. Taken together, findings of enhanced tumor incidence rates in immunocompromised patients and animal models and the characterization of TAA-specific immune responses support the notion that a functional immune system can recognize and destroy nascent transformed cells, thereby impeding tumor development [71]. At the same time, this potential tumor suppressor role of antitumor immunity might also promote the selection of cancer cells capable of immune evasion [71, 223].

Both dormant tumor subsets and CSCs can actively escape immune-mediated elimination [53, 86], indicating that these virulent tumor fractions might coincide with immune-privileged cancer subpopulations that survive immune attack to ultimately drive neoplastic progression. One mechanism by which both tumor populations can escape immunologic clearance involves downmodulation of TAAs. For instance, compared to tumor bulk populations, both melanoma and glioblastoma CSCs demonstrated low to absent expression of TAAs, including MART-1, gp100, NY-ESO-1, and MAGE-A [86, 163, 164]. Comparably, decreased expression of TAAs also distinguished the pool of dormant tumor cells in various disease models [73, 234]. Consequently, CSCs and dormant cancer subsets might not be recognized by the antitumor immune response, which may render them resistant to rejection by immune effector populations. Together, these findings might thus provide a novel explanation for the relative ineffectiveness of cytotoxic T lymphocytes (CTLs) to fully eradicate cancers [229].

Another mechanism through which CSCs and dormant tumor cells could evade rejection by CTLs is through downregulation or absence of MHC class I antigens. CTLs and, to a lesser degree, $CD4^+$ T cells critically rely on the surface expression of class I MHC molecules to recognize cancer cells as non-self [235, 236]. Hence, partial or complete loss of surface MHC class I may protect tumor target cells from CTL-mediated lysis [229, 237, 238]. Of note, CSCs have demonstrated reduced MHC class I expression compared to tumor bulk populations [163, 164]. Similarly, dormant cancer fractions resistant to vaccine-induced immunity showed decreased class I MHC levels [53], suggesting significant overlap of CSCs and dormant tumor populations. Importantly, decreased MHC class I expression has been associated with neoplastic progression, therapeutic unresponsiveness, and adverse clinical outcomes in cancer patients [237, 239–241], which may at least in part relate to the inability of the antitumor immune response to eliminate the MHC class I pool of dormant CSCs.

In addition to their preferential ability to withstand CTL cytolytic activity, CSCs and dormant cancer cells might also engage in active tolerance induction in order to modulate the host antitumor immune response. One major mechanism of tolerance induction involves functional inactivation of antigen-reactive lymphocytes, also referred to as clonal anergy [233]. An immunological process that has been implicated in the induction of anergy is direct inhibition of antigen-reactive cells via the secretion of immunosuppressive factors, including TGF-β or PGE-2 (prostaglandin

E2) [242, 243]. Elevated levels of both TGF-β(beta) pathway members and PGE-2 were detected in melanoma and glioma CSCs, respectively [164, 166]. Correspondingly, TGF-β signaling is activated in dormant tumor subsets compared to that in proliferative cancer cells [37], and PGE-2 activity orchestrated the tumor dormant state in the murine L5178Y lymphoma model [244].

Induction of anergy can further result from impaired stimulation of antigen-specific immune cells [245]. According to the "two-signal paradigm," antigen-dependent T cell activation requires two distinct but complementary signals: naïve T cells receive signal 1 through T cell receptor (TCR) engagement with the MHC/antigenic peptide complex presented by an antigen presenting cell (APC) [245]. Signal 2 describes an antigen-independent stimulus triggered through ligation of an APC-expressed positive costimulatory ligand to its respective receptor present on T cells [245]. Positive costimulatory signaling is required for full T cell activation leading to interleukin-2 (IL-2) production and clonal expansion [246]. The so-called negative or inhibitory costimulatory signals may function to downregulate immune responses [245, 247]. Of note, costimulatory signaling events are not restricted to T cell–APC interactions, but might also encompass interactions between T cells and non-lymphoid cells, including cancer cells [248]. Remarkably, CSC fractions displayed selective expression for the negative costimulator programmed death (PD)-1 [163, 164] and its ligand, PD-L1 [163, 166, 167], suggesting an important role for the PD-1–PD-L1/PD-L2 signaling axis in CSC-driven tumor immune evasion. Likewise, in the DA1-3b leukemia model, dormant tumor cells were found to progressively overexpress PD-L1, thereby inhibiting T-cell activation and CTL-mediated tumor eradication [82, 83]. The involvement of PD pathway members in the escape of host antitumor immunity and neoplastic progression is well established [249–253], indicating a significant involvement of both CSCs and dormant tumor fractions in PD-mediated immune evasion.

B7 costimulatory ligands represent another class of molecules associated with both CSCs and dormant tumor populations. In fact, B7.1 (CD80) expression was activated in dormant tumor subsets and blockade of the molecule reversed their protection from CTL-induced lysis [83]. Similarly, B7.2 (CD86) expression was restricted to the CSC compartment in melanomas [164], and inhibition of B7.2 on melanoma CSCs stunted T cell activation via the induction of regulatory T cells (Tregs) [164]. Tregs are important mediators of immunological self-tolerance that can potently inhibit the activation and cytokine production of other immune compartments, including tumor-reactive CTLs [254–256]. Consistently, Treg induction has been implicated in tumor immune evasion in cancer patients [257–261]. The induction and/or active recruitment of Tregs might thus represent an additional mechanism by which both CSCs and dormant tumor cells could thwart anticancer immune responses.

In sum, several CSC-specific immunomodulatory functions parallel tolerogenic properties of dormant tumor fractions, suggesting significant overlap between both cancer subsets. However, dormancy maintenance requires the establishment of an intricate balance between the anticancer immune response and tumor immune escape. CSCs might indeed play pivotal roles in maintaining the dormant state of cancer by circumventing immunogenic clearance during phases of robust antitumor immunity. At the same time, CSC-mediated tumor immune escape might promote virulent cancer propagation. Accordingly, the immunologic privileges of CSCs could orchestrate the switch from sustained tumor dormancy to cancer expansion in the context of functional antitumor immune responses.

Conclusions

Tumor dormancy represents an important mechanism underlying the observed failure of current therapeutic modalities to fully eradicate cancers. In addition to its more established role in maintaining minimal residual disease after treatment, dormancy might also critically contribute to early stages of tumorigenesis and the formation of micrometastatic disease. There are intriguing parallels between the cancer dormancy hypothesis and the CSC model of tumor propagation. Both CSCs and dormant tumor populations engage in several survival-promoting and virulence-conferring mechanisms, including therapy resistance and evasion of antitumor immune responses. Moreover, distinct sets of molecules and signaling networks shown to orchestrate the tumor dormant state are likewise operative in CSCs. In particular, cancer cell-intrinsic pathways and niche environments involved in tumor angiogenesis and induction of quiescence can govern both CSC fate and dormancy maintenance.

Detailed biological characterization of dormant tumor populations has been limited by the absence of reliable biomarkers and methodologies capable of sufficiently isolating these cells. Given the significant overlap of CSCs with dormant tumor fractions, biomarkers and immunobiological functions specific to the CSC compartment could shed light on the molecular events driving the cancer dormant state. Accordingly, insights into CSC pathobiology could lead to improved experimental systems that might further our understanding of the regulatory elements underlying clinical tumor dormancy and resultant cancer recurrences. Ultimately, such research endeavors may well translate into strategies that successfully eradicate dormant CSCs to minimize the risk of tumor relapse in cancer patients.

Acknowledgments We apologize to those colleagues whose studies were not cited in this review due to space limitations. We thank Dr. Quinn Charbonneau for her review of the manuscript during various stages of its evolution. Our research is supported by a Research Career Development Award from the Dermatology Foundation and a Melanoma Research Scholar Award from the Outrun the Sun Melanoma Foundation (to T.S.).

References

1. Aguirre-Ghiso JA (2007) Models, mechanisms and clinical evidence for cancer dormancy. Nat Rev Cancer 7:834–846
2. Goss PE, Chambers AF (2010) Does tumour dormancy offer a therapeutic target? Nat Rev Cancer 10:871–877
3. Hadfield G (1954) The dormant cancer cell. Br Med J 2:607–610
4. Karrison TG, Ferguson DJ, Meier P (1999) Dormancy of mammary carcinoma after mastectomy. J Natl Cancer Inst 91:80–85

5. Saphner T, Tormey DC, Gray R (1996) Annual hazard rates of recurrence for breast cancer after primary therapy. J Clin Oncol 14:2738–2746
6. Demicheli R, Terenziani M, Valagussa P, Moliterni A, Zambetti M, Bonadonna G (1994) Local recurrences following mastectomy: support for the concept of tumor dormancy. J Natl Cancer Inst 86:45–48
7. Davis TA, Maloney DG, Czerwinski DK, Liles TM, Levy R (1998) Anti-idiotype antibodies can induce long-term complete remissions in non-Hodgkin's lymphoma without eradicating the malignant clone. Blood 92:1184–1190
8. Chomel JC, Brizard F, Veinstein A, Rivet J, Sadoun A, Kitzis A, Guilhot F, Brizard A (2000) Persistence of BCR-ABL genomic rearrangement in chronic myeloid leukemia patients in complete and sustained cytogenetic remission after interferon-alpha therapy or allogeneic bone marrow transplantation. Blood 95:404–408
9. Tsao H, Cosimi AB, Sober AJ (1997) Ultra-late recurrence (15 years or longer) of cutaneous melanoma. Cancer 79:2361–2370
10. Crowley NJ, Seigler HF (1990) Late recurrence of malignant melanoma. Analysis of 168 patients. Ann Surg 212:173–177
11. Waldman FM, DeVries S, Chew KL, Moore DH 2nd, Kerlikowske K, Ljung BM (2000) Chromosomal alterations in ductal carcinomas in situ and their in situ recurrences. J Natl Cancer Inst 92:313–320
12. Chin L, Garraway LA, Fisher DE (2006) Malignant melanoma: genetics and therapeutics in the genomic era. Genes Dev 20:2149–2182
13. Black WC, Welch HG (1993) Advances in diagnostic imaging and overestimations of disease prevalence and the benefits of therapy. N Engl J Med 328:1237–1243
14. Harach HR, Franssila KO, Wasenius VM (1985) Occult papillary carcinoma of the thyroid. A "normal" finding in Finland. A systematic autopsy study. Cancer 56:531–538
15. Montgomery TR, Whitlock GF, Nohlgren JE, Lewis AM (1961) What becomes of the patient with latent or occult carcinoma of the prostate. J Urol 86:655–658
16. Munsie WJ, Foster EA (1968) Unsuspected very small foci of carcinoma of the prostate in transurethral resection specimens. Cancer 21:692–768
17. Mortensen JD, Bennett WA, Woolner LB (1955) Incidence of carcinoma in thyroid glands removed at 1000 consecutive routine necropsies. Surg Forum 5:659–663
18. Nielsen M, Thomsen JL, Primdahl S, Dyreborg U, Andersen JA (1987) Breast cancer and atypia among young and middle aged women: a study of 110 medicolegal autopsies. Br J Cancer 56:814–819
19. Feldman AR, Kessler L, Myers MH, Naughton MD (1986) The prevalence of cancer. Estimates based on the Connecticut Tumor Registry. N Engl J Med 315:1394–1397
20. Pantel K, Brakenhoff RH, Brandt B (2008) Detection, clinical relevance and specific biological properties of disseminating tumour cells. Nat Rev Cancer 8:329–340
21. Pantel K, Brakenhoff RH (2004) Dissecting the metastatic cascade. Nat Rev Cancer 4:448–456
22. Chaffer CL, Weinberg RA (2011) A perspective on cancer cell metastasis. Science 331:1559–1564
23. Fidler IJ, Hart IR (1982) Biological diversity in metastatic neoplasms: origins and implications. Science 217:998–1003
24. Husemann Y, Geigl JB, Schubert F, Musiani P, Meyer M, Burghart E, Forni G, Eils R, Fehm T, Riethmuller G, Klein CA (2008) Systemic spread is an early step in breast cancer. Cancer Cell 13:58–68
25. Schardt JA, Meyer M, Hartmann CH, Schubert F, Schmidt-Kittler O, Fuhrmann C, Polzer B, Petronio M, Eils R, Klein CA (2005) Genomic analysis of single cytokeratin-positive cells from bone marrow reveals early mutational events in breast cancer. Cancer Cell 8:227–239
26. Weckermann D, Muller P, Wawroschek F, Harzmann R, Riethmuller G, Schlimok G (2001) Disseminated cytokeratin positive tumor cells in the bone marrow of patients with prostate cancer: detection and prognostic value. J Urol 166:699–703
27. Schmidt-Kittler O, Ragg T, Daskalakis A, Granzow M, Ahr A, Blankenstein TJ, Kaufmann M, Diebold J, Arnholdt H, Muller P, Bischoff J, Harich D, Schlimok G, Riethmuller G, Eils R,

Klein CA (2003) From latent disseminated cells to overt metastasis: genetic analysis of systemic breast cancer progression. Proc Natl Acad Sci USA 100:7737–7742
28. Meng S, Tripathy D, Frenkel EP, Shete S, Naftalis EZ, Huth JF, Beitsch PD, Leitch M, Hoover S, Euhus D, Haley B, Morrison L, Fleming TP, Herlyn D, Terstappen LW, Fehm T, Tucker TF, Lane N, Wang J, Uhr JW (2004) Circulating tumor cells in patients with breast cancer dormancy. Clin Cancer Res 10:8152–8162
29. Pantel K, Alix-Panabieres C, Riethdorf S (2009) Cancer micrometastases. Nat Rev Clin Oncol 6:339–351
30. Cameron MD, Schmidt EE, Kerkvliet N, Nadkarni KV, Morris VL, Groom AC, Chambers AF, MacDonald IC (2000) Temporal progression of metastasis in lung: cell survival, dormancy, and location dependence of metastatic inefficiency. Cancer Res 60:2541–2546
31. Chambers AF, Groom AC, MacDonald IC (2002) Dissemination and growth of cancer cells in metastatic sites. Nat Rev Cancer 2:563–572
32. Braun S, Pantel K, Muller P, Janni W, Hepp F, Kentenich CR, Gastroph S, Wischnik A, Dimpfl T, Kindermann G, Riethmuller G, Schlimok G (2000) Cytokeratin-positive cells in the bone marrow and survival of patients with stage I, II, or III breast cancer. N Engl J Med 342:525–533
33. Braun S, Cevatli BS, Assemi C, Janni W, Kentenich CR, Schindlbeck C, Rjosk D, Hepp F (2001) Comparative analysis of micrometastasis to the bone marrow and lymph nodes of node-negative breast cancer patients receiving no adjuvant therapy. J Clin Oncol 19:1468–1475
34. Cote RJ, Rosen PP, Lesser ML, Old LJ, Osborne MP (1991) Prediction of early relapse in patients with operable breast cancer by detection of occult bone marrow micrometastases. J Clin Oncol 9:1749–1756
35. Mansi JL, Gogas H, Bliss JM, Gazet JC, Berger U, Coombes RC (1999) Outcome of primary-breast-cancer patients with micrometastases: a long-term follow-up study. Lancet 354:197–202
36. Pantel K, Riethdorf S (2009) Pathology: are circulating tumor cells predictive of overall survival? Nat Rev Clin Oncol 6:190–191
37. Almog N (2010) Molecular mechanisms underlying tumor dormancy. Cancer Lett 294: 139–146
38. Uhr JW, Pantel K (2011) Controversies in clinical cancer dormancy. Proc Natl Acad Sci USA 108:12396–12400
39. Naumov GN, MacDonald IC, Weinmeister PM, Kerkvliet N, Nadkarni KV, Wilson SM, Morris VL, Groom AC, Chambers AF (2002) Persistence of solitary mammary carcinoma cells in a secondary site: a possible contributor to dormancy. Cancer Res 62:2162–2168
40. Heyn C, Ronald JA, Ramadan SS, Snir JA, Barry AM, MacKenzie LT, Mikulis DJ, Palmieri D, Bronder JL, Steeg PS, Yoneda T, MacDonald IC, Chambers AF, Rutt BK, Foster PJ (2006) In vivo MRI of cancer cell fate at the single-cell level in a mouse model of breast cancer metastasis to the brain. Magn Reson Med 56:1001–1010
41. Aguirre-Ghiso JA, Ossowski L, Rosenbaum SK (2004) Green fluorescent protein tagging of extracellular signal-regulated kinase and p38 pathways reveals novel dynamics of pathway activation during primary and metastatic growth. Cancer Res 64:7336–7345
42. Luzzi KJ, MacDonald IC, Schmidt EE, Kerkvliet N, Morris VL, Chambers AF, Groom AC (1998) Multistep nature of metastatic inefficiency: dormancy of solitary cells after successful extravasation and limited survival of early micrometastases. Am J Pathol 153:865–873
43. White DE, Kurpios NA, Zuo D, Hassell JA, Blaess S, Mueller U, Muller WJ (2004) Targeted disruption of beta1-integrin in a transgenic mouse model of human breast cancer reveals an essential role in mammary tumor induction. Cancer Cell 6:159–170
44. Psaila B, Lyden D (2009) The metastatic niche: adapting the foreign soil. Nat Rev Cancer 9:285–293
45. Aguirre Ghiso JA (2002) Inhibition of FAK signaling activated by urokinase receptor induces dormancy in human carcinoma cells in vivo. Oncogene 21:2513–2524
46. Chang SF, Chang CA, Lee DY, Lee PL, Yeh YM, Yeh CR, Cheng CK, Chien S, Chiu JJ (2008) Tumor cell cycle arrest induced by shear stress: roles of integrins and Smad. Proc Natl Acad Sci USA 105:3927–3932

47. Yu W, Kim J, Ossowski L (1997) Reduction in surface urokinase receptor forces malignant cells into a protracted state of dormancy. J Cell Biol 137:767–777
48. Naumov GN, Townson JL, MacDonald IC, Wilson SM, Bramwell VH, Groom AC, Chambers AF (2003) Ineffectiveness of doxorubicin treatment on solitary dormant mammary carcinoma cells or late-developing metastases. Breast Cancer Res Treat 82:199–206
49. Schewe DM, Aguirre-Ghiso JA (2009) Inhibition of eIF2alpha dephosphorylation maximizes bortezomib efficiency and eliminates quiescent multiple myeloma cells surviving proteasome inhibitor therapy. Cancer Res 69:1545–1552
50. Ranganathan AC, Zhang L, Adam AP, Aguirre-Ghiso JA (2006) Functional coupling of p38-induced up-regulation of BiP and activation of RNA-dependent protein kinase-like endoplasmic reticulum kinase to drug resistance of dormant carcinoma cells. Cancer Res 66:1702–1711
51. Demicheli R, Miceli R, Moliterni A, Zambetti M, Hrushesky WJ, Retsky MW, Valagussa P, Bonadonna G (2005) Breast cancer recurrence dynamics following adjuvant CMF is consistent with tumor dormancy and mastectomy-driven acceleration of the metastatic process. Ann Oncol 16:1449–1457
52. Gimbrone MA Jr, Leapman SB, Cotran RS, Folkman J (1972) Tumor dormancy in vivo by prevention of neovascularization. J Exp Med 136:261–276
53. Quesnel B (2008) Tumor dormancy and immunoescape. APMIS 116:685–694
54. Holmgren L, O'Reilly MS, Folkman J (1995) Dormancy of micrometastases: balanced proliferation and apoptosis in the presence of angiogenesis suppression. Nat Med 1:149–153
55. Folkman J (1971) Tumor angiogenesis: therapeutic implications. N Engl J Med 285:1182–1186
56. Folkman J (2002) Role of angiogenesis in tumor growth and metastasis. Semin Oncol 29: 15–18
57. Kerbel RS (2008) Tumor angiogenesis. N Engl J Med 358:2039–2049
58. Udagawa T, Fernandez A, Achilles EG, Folkman J, D'Amato RJ (2002) Persistence of microscopic human cancers in mice: alterations in the angiogenic balance accompanies loss of tumor dormancy. FASEB J 16:1361–1370
59. Achilles EG, Fernandez A, Allred EN, Kisker O, Udagawa T, Beecken WD, Flynn E, Folkman J (2001) Heterogeneity of angiogenic activity in a human liposarcoma: a proposed mechanism for "no take" of human tumors in mice. J Natl Cancer Inst 93:1075–1081
60. Almog N, Henke V, Flores L, Hlatky L, AKung AL, Wright RD, Berger R, Hutchinson L, Naumov GN, Bender E, Akslen LA, Achilles EG, Folkman J (2006) Prolonged dormancy of human liposarcoma is associated with impaired tumor angiogenesis. FASEB J 20: 947–949
61. Naumov GN, Bender E, Zurakowski D, Kang SY, Sampson D, Flynn E, Watnick RS, Straume O, Akslen LA, Folkman J, Almog N (2006) A model of human tumor dormancy: an angiogenic switch from the nonangiogenic phenotype. J Natl Cancer Inst 98:316–325
62. Naumov GN, Akslen LA, Folkman J (2006) Role of angiogenesis in human tumor dormancy: animal models of the angiogenic switch. Cell Cycle 5:1779–1787
63. Hanahan D, Folkman J (1996) Patterns and emerging mechanisms of the angiogenic switch during tumorigenesis. Cell 86:353–364
64. Folkman J, Merler E, Abernathy C, Williams G (1971) Isolation of a tumor factor responsible for angiogenesis. J Exp Med 133:275–288
65. O'Reilly MS, Holmgren L, Chen C, Folkman J (1996) Angiostatin induces and sustains dormancy of human primary tumors in mice. Nat Med 2:689–692
66. Watnick RS, Cheng YN, Rangarajan A, Ince TA, Weinberg RA (2003) Ras modulates Myc activity to repress thrombospondin-1 expression and increase tumor angiogenesis. Cancer Cell 3:219–231
67. Giuriato S, Ryeom S, Fan AC, Bachireddy P, Lynch RC, Rioth MJ, van Riggelen J, Kopelman AM, Passegue E, Tang F, Folkman J, Felsher DW (2006) Sustained regression of tumors upon MYC inactivation requires p53 or thrombospondin-1 to reverse the angiogenic switch. Proc Natl Acad Sci USA 103:16266–16271

68. Indraccolo S, Stievano L, Minuzzo S, Tosello V, Esposito G, Piovan E, Zamarchi R, Chieco-Bianchi L, Amadori A (2006) Interruption of tumor dormancy by a transient angiogenic burst within the tumor microenvironment. Proc Natl Acad Sci USA 103:4216–4221
69. Shachaf CM, Kopelman AM, Arvanitis C, Karlsson A, Beer S, Mandl S, Bachmann MH, Borowsky AD, Ruebner B, Cardiff RD, Yang Q, Bishop JM, Contag CH, Felsher DW (2004) MYC inactivation uncovers pluripotent differentiation and tumour dormancy in hepatocellular cancer. Nature 431:1112–1117
70. Bergers G, Benjamin LE (2003) Tumorigenesis and the angiogenic switch. Nat Rev Cancer 3:401–410
71. Dunn GP, Bruce AT, Ikeda H, Old LD, Schreiber RD (2002) Cancer immunoediting: from immunosurveillance to tumor escape. Nat Immunol 3:991–998
72. Koebel CM, Vermi W, Swann JB, Zerafa N, Rodig SJ, Old LJ, Smyth MJ, Schreiber RD (2007) Adaptive immunity maintains occult cancer in an equilibrium state. Nature 450:903–907
73. Weinhold KJ, Miller DA, Wheelock EF (1979) The tumor dormant state. Comparison of L5178Y cells used to establish dormancy with those that emerge after its termination. J Exp Med 149:745–757
74. Weinhold KJ, Goldstein LT, Wheelock EF (1979) The tumor dormant state. Quantitation of L5178Y cells and host immune responses during the establishment and course of dormancy in syngeneic DBA/2 mice. J Exp Med 149:732–744
75. Uhr JW, Tucker T, May RD, Siu H, Vitetta ES (1991) Cancer dormancy: studies of the murine BCL1 lymphoma. Cancer Res 51:5045s–5053s
76. Yefenof E, Picker LJ, Scheuermann RH, Tucker TF, Vitetta ES, Uhr JW (1993) Cancer dormancy: isolation and characterization of dormant lymphoma cells. Proc Natl Acad Sci USA 90:1829–1833
77. Saudemont A, Buffenoir G, Denys A, Desreumaux P, Jouy N, Hetuin D, Bauters F, Fenaux P, Quesnel B (2002) Gene transfer of CD154 and IL12 cDNA induces an anti-leukemic immunity in a murine model of acute leukemia. Leukemia 16:1637–1644
78. Vereecque R, Buffenoir G, Preudhomme C, Hetuin D, Bauters F, Fenaux P, Quesnel B (2000) Gene transfer of GM-CSF, CD80 and CD154 cDNA enhances survival in a murine model of acute leukemia with persistence of a minimal residual disease. Gene Ther 7:1312–1316
79. Rosenberg SA (1999) A new era for cancer immunotherapy based on the genes that encode cancer antigens. Immunity 10:281–287
80. MacKie RM, Reid R, Junor B (2003) Fatal melanoma transferred in a donated kidney 16 years after melanoma surgery. N Engl J Med 348:567–568
81. Curtis RE, Rowlings PA, Deeg HJ, Shriner DA, Socie G, Travis LB, Horowitz MM, Witherspoon RP, Hoover RN, Sobocinski KA, Fraumeni JF Jr, Boice JD Jr (1997) Solid cancers after bone marrow transplantation. N Engl J Med 336:897–904
82. Saudemont A, Jouy N, Hetuin D, Quesnel B (2005) NK cells that are activated by CXCL10 can kill dormant tumor cells that resist CTL-mediated lysis and can express B7-H1 that stimulates T cells. Blood 105:2428–2435
83. Saudemont A, Quesnel B (2004) In a model of tumor dormancy, long-term persistent leukemic cells have increased B7-H1 and B7.1 expression and resist CTL-mediated lysis. Blood 104:2124–2133
84. Lane SW, Scadden DT, Gilliland DG (2009) The leukemic stem cell niche: current concepts and therapeutic opportunities. Blood 114:1150–1157
85. Dean M, Fojo T, Bates S (2005) Tumour stem cells and drug resistance. Nat Rev Cancer 5:275–284
86. Schatton T, Frank MH (2009) Antitumor immunity and cancer stem cells. Ann N Y Acad Sci 1176:154–169
87. Hjelmeland AB, Lathia JD, Sathornsumetee S, Rich JN (2011) Twisted tango: brain tumor neurovascular interactions. Nat Neurosci 14:1375–1381
88. Hanahan D, Weinberg RA (2011) Hallmarks of cancer: the next generation. Cell 144:646–674
89. Michor F, Iwasa Y, Nowak MA (2004) Dynamics of cancer progression. Nat Rev Cancer 4:197–205

90. Nowell PC (1976) The clonal evolution of tumor cell populations. Science 194:23–28
91. Bruce WR, Van Der Gaag H (1963) A quantitative assay for the number of Murine Lymphoma Cells capable of proliferation in vivo. Nature 199:79–80
92. Hamburger AW, Salmon SE (1977) Primary bioassay of human tumor stem cells. Science 197:461–463
93. Reya T, Morrison SJ, Clarke MF, Weissman IL (2001) Stem cells, cancer, and cancer stem cells. Nature 414:105–111
94. Dirks P (2010) Cancer stem cells: invitation to a second round. Nature 466:40–41
95. Clarke MF, Dick JE, Dirks PB, Eaves CJ, Jamieson CH, Jones DL, Visvader J, Weissman IL, Wahl GM (2006) Cancer stem cells–perspectives on current status and future directions: AACR workshop on cancer stem cells. Cancer Res 66:9339–9344
96. Schatton T, Frank NY, Frank MH (2009) Identification and targeting of cancer stem cells. Bioessays 31:1038–1049
97. Schatton T, Murphy GF, Frank NY, Yamaura K, Waaga-Gasser AM, Gasser M, Zhan Q, Jordan S, Duncan LM, Weishaupt C, Fuhlbrigge RC, Kupper TS, Sayegh MH, Frank MH (2008) Identification of cells initiating human melanomas. Nature 451:345–349
98. Lathia JD, Gallagher J, Myers JT, Li M, Vasanji A, McLendon RE, Hjelmeland AB, Huang AY, Rich JN (2011) Direct in vivo evidence for tumor propagation by glioblastoma cancer stem cells. PLoS One 6:e24807
99. Vlashi E, Kim K, Lagadec C, Donna LD, McDonald JT, Eghbali M, Sayre JW, Stefani E, McBride W, Pajonk F (2009) In vivo imaging, tracking, and targeting of cancer stem cells. J Natl Cancer Inst 101:350–359
100. Chaffer CL, Brueckmann I, Scheel C, Kaestli AJ, Wiggins PA, Rodrigues LO, Brooks M, Reinhardt F, Su Y, Polyak K, Arendt LM, Kuperwasser C, Bierie B, Weinberg RA (2011) Normal and neoplastic nonstem cells can spontaneously convert to a stem-like state. Proc Natl Acad Sci USA 108:7950–7955
101. Quintana E, Shackleton M, Foster HR, Fullen DR, Sabel MS, Johnson TM, Morrison SJ (2010) Phenotypic heterogeneity among tumorigenic melanoma cells from patients that is reversible and not hierarchically organized. Cancer Cell 18:510–523
102. Scadden DT (2006) The stem-cell niche as an entity of action. Nature 441:1075–1079
103. Gupta PB, Chaffer CL, Weinberg RA (2009) Cancer stem cells: mirage or reality? Nat Med 15:1010–1012
104. Jordan CT (2009) Cancer stem cells: controversial or just misunderstood? Cell Stem Cell 4:203–205
105. Bjerkvig R, Tysnes BB, Aboody KS, Najbauer J, Terzis AJ (2005) Opinion: the origin of the cancer stem cell: current controversies and new insights. Nat Rev Cancer 5:899–904
106. Barker N, Ridgway RA, van Es JH, van de Wetering M, Begthel H, van den Born M, Danenberg E, Clarke AR, Sansom OJ, Clevers H (2009) Crypt stem cells as the cells-of-origin of intestinal cancer. Nature 457:608–611
107. Zhu L, Gibson P, Currle DS, Tong Y, Richardson RJ, Bayazitov IT, Poppleton H, Zakharenko S, Ellison DW, Gilbertson RJ (2009) Prominin 1 marks intestinal stem cells that are susceptible to neoplastic transformation. Nature 457:603–607
108. Jamieson CH, Ailles LE, Dylla SJ, Muijtjens M, Jones C, Zehnder JL, Gotlib J, Li K, Manz MG, Keating A, Sawyers CL, Weissman IL (2004) Granulocyte-macrophage progenitors as candidate leukemic stem cells in blast-crisis CML. N Engl J Med 351:657–667
109. Krivtsov AV, Twomey D, Feng Z, Stubbs MC, Wang Y, Faber J, Levine JE, Wang J, Hahn WC, Gilliland DG, Golub TR, Armstrong SA (2006) Transformation from committed progenitor to leukaemia stem cell initiated by MLL-AF9. Nature 442:818–822
110. Sun B, Chen M, Hawks CL, Pereira-Smith OM, Hornsby PJ (2005) The minimal set of genetic alterations required for conversion of primary human fibroblasts to cancer cells in the subrenal capsule assay. Neoplasia 7:585–593
111. Hendrix MJ, Seftor EA, Hess AR, Seftor RE (2003) Vasculogenic mimicry and tumour-cell plasticity: lessons from melanoma. Nat Rev Cancer 3:411–421
112. Calabrese C, Poppleton H, Kocak M, Hogg TL, Fuller C, Hamner B, Oh EY, Gaber MW, Finklestein D, Allen M, Frank A, Bayazitov IT, Zakharenko SS, Gajjar A, Davidoff A,

Gilbertson RJ (2007) A perivascular niche for brain tumor stem cells. Cancer Cell 11:69–82
113. Powell K (2005) Stem-cell niches: it's the ecology, stupid! Nature 435:268–270
114. Sato T, van Es JH, Snippert HJ, Stange DE, Vries RG, van den Born M, Barker N, Shroyer NF, van de Wetering M, Clevers H (2011) Paneth cells constitute the niche for Lgr5 stem cells in intestinal crypts. Nature 469:415–418
115. Klein AM, Simons BD (2011) Universal patterns of stem cell fate in cycling adult tissues. Development 138:3103–3111
116. Simons BD, Clevers H (2011) Strategies for homeostatic stem cell self-renewal in adult tissues. Cell 145:851–862
117. Nakagawa T, Sharma M, Nabeshima Y, Braun RE, Yoshida S (2010) Functional hierarchy and reversibility within the murine spermatogenic stem cell compartment. Science 328:62–67
118. Shackleton M, Quintana E, Fearon ER, Morrison SJ (2009) Heterogeneity in cancer: cancer stem cells versus clonal evolution. Cell 138:822–829
119. Quintana E, Shackleton M, Sabel MS, Fullen DR, Johnson TM, Morrison SJ (2008) Efficient tumour formation by single human melanoma cells. Nature 456:593–598
120. Morrison SJ, Kimble J (2006) Asymmetric and symmetric stem-cell divisions in development and cancer. Nature 441:1068–1074
121. Rosen JM, Jordan CT (2009) The increasing complexity of the cancer stem cell paradigm. Science 324:1670–1673
122. Bonnet D, Dick JE (1997) Human acute myeloid leukemia is organized as a hierarchy that originates from a primitive hematopoietic cell. Nat Med 3:730–737
123. Lapidot T, Sirard C, Vormoor J, Murdoch B, Hoang T, Caceres-Cortes J, Minden M, Paterson B, Caligiuri MA, Dick JE (1994) A cell initiating human acute myeloid leukaemia after transplantation into SCID mice. Nature 367:645–648
124. Chan KS, Espinosa I, Chao M, Wong D, Ailles L, Diehn M, Gill H, Presti J Jr, Chang HY, van de Rijn M, Shortliffe L, Weissman IL (2009) Identification, molecular characterization, clinical prognosis, and therapeutic targeting of human bladder tumor-initiating cells. Proc Natl Acad Sci USA 106:14016–14021
125. Suva ML, Riggi N, Stehle JC, Baumer K, Tercier S, Joseph JM, Suva D, Clement V, Provero P, Cironi L, Osterheld MC, Guillou L, Stamenkovic I (2009) Identification of cancer stem cells in Ewing's sarcoma. Cancer Res 69:1776–1781
126. Al-Hajj M, Wicha MS, Benito-Hernandez A, Morrison SJ, Clarke MF (2003) Prospective identification of tumorigenic breast cancer cells. Proc Natl Acad Sci USA 100:3983–3988
127. Singh SK, Hawkins C, Clarke ID, Squire JA, Bayani J, Hide T, Henkelman RM, Cusimano MD, Dirks PB (2004) Identification of human brain tumour initiating cells. Nature 432:396–401
128. O'Brien CA, Pollett A, Gallinger S, Dick JE (2007) A human colon cancer cell capable of initiating tumour growth in immunodeficient mice. Nature 445:106–110
129. Ricci-Vitiani L, Lombardi DG, Pilozzi E, Biffoni M, Todaro M, Peschle C, De Maria R (2007) Identification and expansion of human colon-cancer-initiating cells. Nature 445:111–115
130. Dalerba P, Dylla SJ, IPark IK, Liu R, Wang X, Cho RW, Hoey T, Gurney A, Huang EH, Simeone DM, Shelton AA, Parmiani G, Castelli C, Clarke MF (2007) Phenotypic characterization of human colorectal cancer stem cells. Proc Natl Acad Sci USA 104:10158–10163
131. Prince ME, Sivanandan R, Kaczorowski A, Wolf GT, Kaplan MJ, Dalerba P, Weissman IL, Clarke MF, Ailles LE (2007) Identification of a subpopulation of cells with cancer stem cell properties in head and neck squamous cell carcinoma. Proc Natl Acad Sci USA 104:973–978
132. Yang ZF, Ho DW, Ng MN, Lau CK, Yu WC, Ngai P, Chu PW, Lam CT, Poon RT, Fan ST (2008) Significance of CD90(+) cancer stem cells in human liver cancer. Cancer Cell 13:153–166
133. Zhang S, Balch C, Chan MW, Lai HC, Matei D, Schilder JM, Yan PS, Huang TH, Nephew KP (2008) Identification and characterization of ovarian cancer-initiating cells from primary human tumors. Cancer Res 68:4311–4320
134. Li C, Heidt DG, Dalerba P, Burant CF, Zhang L, Adsay V, Wicha M, Clarke MF, Simeone DM (2007) Identification of pancreatic cancer stem cells. Cancer Res 67:1030–1037

135. Hermann PC, Huber SL, Herrler T, Aicher A, Ellwart JW, Guba M, Bruns CJ, Heeschen C (2007) Distinct populations of cancer stem cells determine tumor growth and metastatic activity in human pancreatic cancer. Cell Stem Cell 1:313–323
136. Boiko AD, Razorenova OV, van de Rijn M, Swetter SM, Johnson DL, Ly DP, Butler PD, Yang GP, Joshua B, Kaplan MJ, Longaker MT, Weissman IL (2010) Human melanoma-initiating cells express neural crest nerve growth factor receptor CD271. Nature 466:133–137
137. Civenni G, Walter A, Kobert N, Mihic-Probst D, Zipser M, Belloni B, Seifert B, Moch H, Dummer R, van den Broek M, Sommer L (2011) Human CD271-positive melanoma stem cells associated with metastasis establish tumor heterogeneity and long-term growth. Cancer Res 71:3098–3109
138. Liu H, Patel MR, Prescher JA, Patsialou A, Qian D, Lin J, Wen S, Chang YF, Bachmann MH, Shimono Y, Dalerba P, Adorno M, Lobo N, Bueno J, Dirbas FM, Goswami S, Somlo G, Condeelis J, Contag CH, Gambhir SS, Clarke MF (2010) Cancer stem cells from human breast tumors are involved in spontaneous metastases in orthotopic mouse models. Proc Natl Acad Sci USA 107:18115–18120
139. Malanchi I, Santamaria-Martinez A, Susanto E, Peng H, Lehr HA, Delaloye JF, Huelsken J (2011) Interactions between cancer stem cells and their niche govern metastatic colonization. Nature 481(7379):85–89
140. Liu R, Wang X, Chen GY, Dalerba P, Gurney A, Hoey T, Sherlock G, Lewicki J, Shedden K, Clarke MF (2007) The prognostic role of a gene signature from tumorigenic breast-cancer cells. N Engl J Med 356:217–226
141. Polyak K, Weinberg RA (2009) Transitions between epithelial and mesenchymal states: acquisition of malignant and stem cell traits. Nat Rev Cancer 9:265–273
142. Thiery JP, Acloque H, Huang RY, Nieto MA (2009) Epithelial-mesenchymal transitions in development and disease. Cell 139:871–890
143. Eyler CE, Rich JN (2008) Survival of the fittest: cancer stem cells in therapeutic resistance and angiogenesis. J Clin Oncol 26:2839–2845
144. Vlashi E, McBride WH, Pajonk F (2009) Radiation responses of cancer stem cells. J Cell Biochem 108:339–342
145. Bao S, Wu Q, McLendon RE, Hao Y, Shi Q, Hjelmeland AB, Dewhirst MW, Bigner DD, Rich JN (2006) Glioma stem cells promote radioresistance by preferential activation of the DNA damage response. Nature 444:756–760
146. Diehn M, Cho RW, Lobo NA, Kalisky T, Dorie MJ, Kulp AN, Qian D, Lam JS, Ailles LE, Wong M, Joshua B, Kaplan MJ, Wapnir I, Dirbas FM, Somlo G, Garberoglio C, Paz B, Shen J, Lau SK, Quake SR, Brown JM, Weissman IL, Clarke MF (2009) Association of reactive oxygen species levels and radioresistance in cancer stem cells. Nature 458:780–783
147. Wilson BJ, Schatton T, Zhan Q, Gasser M, Ma J, Saab KR, Schanche R, Waaga-Gasser AM, Gold JS, Huang Q, Murphy GF, Frank MH, Frank NY (2011) ABCB5 identifies a therapy-refractory tumor cell population in colorectal cancer patients. Cancer Res 71:5307–5316
148. Viale A, De Franco F, Orleth A, Cambiaghi V, Giuliani V, Bossi D, Ronchini C, Ronzoni S, Muradore I, Monestiroli S, Gobbi A, Alcalay M, Minucci S, Pelicci PG (2009) Cell-cycle restriction limits DNA damage and maintains self-renewal of leukaemia stem cells. Nature 457:51–56
149. Li X, Lewis MT, Huang J, Gutierrez C, Osborne CK, Wu MF, Hilsenbeck SG, Pavlick A, Zhang X, Chamness GC, Wong H, Rosen J, Chang JC (2008) Intrinsic resistance of tumorigenic breast cancer cells to chemotherapy. J Natl Cancer Inst 100:672–679
150. Yu F, Yao H, Zhu P, Zhang X, Pan Q, Gong C, Huang Y, Hu X, Su F, Lieberman J, Song E (2007) Let-7 regulates self renewal and tumorigenicity of breast cancer cells. Cell 131:1109–1123
151. Gupta PB, Onder TT, Jiang G, Tao K, Kuperwasser C, Weinberg RA, Lander ES (2009) Identification of selective inhibitors of cancer stem cells by high-throughput screening. Cell 138:645–659
152. Zhou BB, Zhang H, Damelin M, Geles KG, Grindley JC, Dirks PB (2009) Tumour-initiating cells: challenges and opportunities for anticancer drug discovery. Nat Rev Drug Discov 8: 806–823
153. Frank NY, Schatton T, Frank MH (2010) The therapeutic promise of the cancer stem cell concept. J Clin Invest 120:41–50

154. Bao S, Wu Q, Sathornsumetee S, Hao Y, Li Z, Hjelmeland AB, Shi Q, McLendon RE, Bigner DD, Rich JN (2006) Stem cell-like glioma cells promote tumor angiogenesis through vascular endothelial growth factor. Cancer Res 66:7843–7848
155. Beck B, Driessens G, Goossens S, Youssef KK, Kuchnio A, Caauwe A, Sotiropoulou PA, Loges S, Lapouge G, Candi A, Mascre G, Drogat B, Dekoninck S, Haigh JJ, Carmeliet P, Blanpain C (2011) A vascular niche and a VEGF-Nrp1 loop regulate the initiation and stemness of skin tumours. Nature 478:399–403
156. Frank NY, Schatton T, Kim S, Zhan Q, Wilson BJ, Ma J, Saab KR, Osherov V, Widlund HR, Gasser M, Waaga-Gasser AM, Kupper TS, Murphy GF, Frank MH (2011) VEGFR-1 expressed by malignant melanoma-initiating cells is required for tumor growth. Cancer Res 71:1474–1485
157. Ricci-Vitiani L, Pallini R, Biffoni M, Todaro M, Invernici G, Cenci T, Maira G, Parati EA, Stassi G, Larocca LM, De Maria R (2010) Tumour vascularization via endothelial differentiation of glioblastoma stem-like cells. Nature 468:824–828
158. Wang R, Chadalavada K, Wilshire J, Kowalik U, Hovinga KE, Geber A, Fligelman B, Leversha M, Brennan C, Tabar V (2010) Glioblastoma stem-like cells give rise to tumour endothelium. Nature 468:829–833
159. Li Z, Bao S, Wu Q, Wang H, Eyler C, Sathornsumetee S, Shi Q, Cao Y, Lathia J, McLendon RE, Hjelmeland AB, Rich JN (2009) Hypoxia-inducible factors regulate tumorigenic capacity of glioma stem cells. Cancer Cell 15:501–513
160. Pietras A, Hansford LM, Johnsson AS, Bridges E, Sjolund J, Gisselsson D, Rehn M, Beckman S, Noguera R, Navarro S, Cammenga J, Fredlund E, Kaplan DR, Pahlman S (2009) HIF-2alpha maintains an undifferentiated state in neural crest-like human neuroblastoma tumor-initiating cells. Proc Natl Acad Sci USA 106:16805–16810
161. Vlashi E, Lagadec C, Vergnes L, Matsutani T, Masui K, Poulou M, Popescu R, Della Donna L, Evers P, Dekmezian C, Reue K, Christofk H, Mischel PS, Pajonk F (2011) Metabolic state of glioma stem cells and nontumorigenic cells. Proc Natl Acad Sci USA 108:16062–16067
162. Majeti R, Chao MP, Alizadeh AA, Pang WW, Jaiswal S, Gibbs KD Jr, van Rooijen N, Weissman IL (2009) CD47 is an adverse prognostic factor and therapeutic antibody target on human acute myeloid leukemia stem cells. Cell 138:286–299
163. Di Tomaso T, Mazzoleni S, Wang E, Sovena G, Clavenna D, Franzin A, Mortini P, Ferrone S, Doglioni C, Marincola FM, Galli R, Parmiani G, Maccalli C (2010) Immunobiological characterization of cancer stem cells isolated from glioblastoma patients. Clin Cancer Res 16:800–813
164. Schatton T, Schutte U, Frank NY, Zhan Q, Hoerning A, Robles SC, Zhou J, Hodi FS, Spagnoli GC, Murphy GF, Frank MH (2010) Modulation of T-cell activation by malignant melanoma initiating cells. Cancer Res 70:697–708
165. Todaro M, D'Asaro M, Caccamo N, Iovino F, Francipane MG, Meraviglia S, Orlando V, La Mendola C, Gulotta G, Salerno A, Dieli F, Stassi G (2009) Efficient killing of human colon cancer stem cells by gammadelta T lymphocytes. J Immunol 182:7287–7296
166. Wei J, Barr J, Kong LY, Wang Y, Wu A, Sharma AK, Gumin J, Henry V, Colman H, Priebe W, Sawaya R, Lang FF, Heimberger AB (2010) Glioblastoma cancer-initiating cells inhibit T-cell proliferation and effector responses by the signal transducers and activators of transcription 3 pathway. Mol Cancer Ther 9:67–78
167. Wei J, Barr J, Kong LY, Wang Y, Wu A, Sharma AK, Gumin J, Henry V, Colman H, Sawaya R, Lang FF, Heimberger AB (2010) Glioma-associated cancer-initiating cells induce immunosuppression. Clin Cancer Res 16:461–473
168. Burnet M (1957) Cancer: a biological approach. III. Viruses associated with neoplastic conditions. IV. Practical applications. Br Med J 1:841–847
169. Burnet M (1957) Cancer; a biological approach. I. The processes of control. Br Med J 1:779–786
170. Grulich AE, van Leeuwen MT, Falster MO, Vajdic CM (2007) Incidence of cancers in people with HIV/AIDS compared with immunosuppressed transplant recipients: a meta-analysis. Lancet 370:59–67

171. Shankaran V, Ikeda H, Bruce AT, White JM, Swanson PE, Old LJ, Schreiber RD (2001) IFNgamma and lymphocytes prevent primary tumour development and shape tumour immunogenicity. Nature 410:1107–1111
172. Jordan CT, Guzman ML, Noble M (2006) Cancer stem cells. N Engl J Med 355:1253–1261
173. Dalerba P, Kalisky T, Sahoo D, Rajendran PS, Rothenberg ME, Leyrat AA, Sim S, Okamoto J, Johnston DM, Qian D, Zabala M, Bueno J, Neff NF, Wang J, Shelton AA, Visser B, Hisamori S, Shimono Y, van de Wetering M, Clevers H, Clarke MF, Quake SR (2011) Single-cell dissection of transcriptional heterogeneity in human colon tumors. Nat Biotechnol 29:1120–1127
174. Pece S, Tosoni D, Confalonieri S, Mazzarol G, Vecchi M, Ronzoni S, Bernard L, Viale G, Pelicci PG, Di Fiore PP (2010) Biological and molecular heterogeneity of breast cancers correlates with their cancer stem cell content. Cell 140:62–73
175. Kusumbe AP, Bapat SA (2009) Cancer stem cells and aneuploid populations within developing tumors are the major determinants of tumor dormancy. Cancer Res 69:9245–9253
176. Haraguchi N, Ishii H, Mimori K, Tanaka F, Ohkuma M, Kim HM, Akita H, Takiuchi D, Hatano H, Nagano H, Barnard GF, Doki Y, Mori M (2010) CD13 is a therapeutic target in human liver cancer stem cells. J Clin Invest 120:3326–3339
177. de Grouw EP, Raaijmakers MH, Boezeman JB, van der Reijden BA, van de Locht LT, de Witte TJ, Jansen JH, Raymakers RA (2006) Preferential expression of a high number of ATP binding cassette transporters in both normal and leukemic CD34+CD38- cells. Leukemia 20:750–754
178. Wulf GG, Wang RY, Kuehnle I, Weidner D, Marini F, Brenner MK, Andreeff M, Goodell MA (2001) A leukemic stem cell with intrinsic drug efflux capacity in acute myeloid leukemia. Blood 98:1166–1173
179. Cheung ST, Cheung PF, Cheng CK, Wong NC, Fan ST (2011) Granulin-epithelin precursor and ATP-dependent binding cassette (ABC)B5 regulate liver cancer cell chemoresistance. Gastroenterology 140(1):344–355
180. Frank NY, Margaryan A, Huang Y, Schatton T, Waaga-Gasser AM, Gasser M, Sayegh MH, Sadee W, Frank MH (2005) ABCB5-mediated doxorubicin transport and chemoresistance in human malignant melanoma. Cancer Res 65:4320–4333
181. Guan Y, Gerhard B, Hogge DE (2003) Detection, isolation, and stimulation of quiescent primitive leukemic progenitor cells from patients with acute myeloid leukemia (AML). Blood 101:3142–3149
182. Graham SM, Jorgensen HG, Allan E, Pearson C, Alcorn MJ, Richmond L, Holyoake TL (2002) Primitive, quiescent, Philadelphia-positive stem cells from patients with chronic myeloid leukemia are insensitive to STI571 in vitro. Blood 99:319–325
183. Michor F, Hughes TP, Iwasa Y, Branford S, Shah NP, Sawyers CL, Nowak MA (2005) Dynamics of chronic myeloid leukaemia. Nature 435:1267–1270
184. Copland M, Hamilton A, Elrick LJ, Baird JW, Allan EK, Jordanides N, Barow M, Mountford JC, Holyoake TL (2006) Dasatinib (BMS-354825) targets an earlier progenitor population than imatinib in primary CML but does not eliminate the quiescent fraction. Blood 107:4532–4539
185. Jorgensen HG, Allan EK, Jordanides NE, Mountford JC, Holyoake TL (2007) Nilotinib exerts equipotent antiproliferative effects to imatinib and does not induce apoptosis in CD34+ CML cells. Blood 109:4016–4019
186. Phillips TM, McBride WH, Pajonk F (2006) The response of CD24(–/low)/CD44+ breast cancer-initiating cells to radiation. J Natl Cancer Inst 98:1777–1785
187. Hambardzumyan D, Becher OJ, Rosenblum MK, Pandolfi PP, Manova-Todorova K, Holland EC (2008) PI3K pathway regulates survival of cancer stem cells residing in the perivascular niche following radiation in medulloblastoma in vivo. Genes Dev 22:436–448
188. Ito K, Bernardi R, Morotti A, Matsuoka S, Saglio G, Ikeda Y, Rosenblatt J, Avigan DE, Teruya-Feldstein J, Pandolfi PP (2008) PML targeting eradicates quiescent leukaemia-initiating cells. Nature 453:1072–1078
189. Schewe DM, Aguirre-Ghiso JA (2008) ATF6alpha-Rheb-mTOR signaling promotes survival of dormant tumor cells in vivo. Proc Natl Acad Sci USA 105:10519–10524

190. Marches R, Hsueh R, Uhr JW (1999) Cancer dormancy and cell signaling: induction of p21(waf1) initiated by membrane IgM engagement increases survival of B lymphoma cells. Proc Natl Acad Sci USA 96:8711–8715
191. Aguirre-Ghiso JA, Estrada Y, Liu D, Ossowski L (2003) ERK(MAPK) activity as a determinant of tumor growth and dormancy; regulation by p38(SAPK). Cancer Res 63:1684–1695
192. Kobayashi A, Okuda H, Xing F, Pandey PR, Watabe M, Hirota S, Pai SK, Liu W, Fukuda K, Chambers C, Wilber A, Watabe K (2011) Bone morphogenetic protein 7 in dormancy and metastasis of prostate cancer stem-like cells in bone. J Exp Med 208(13): 2641–2655
193. Wu CH, van Riggelen J, Yetil A, Fan AC, Bachireddy P, Felsher DW (2007) Cellular senescence is an important mechanism of tumor regression upon c-Myc inactivation. Proc Natl Acad Sci USA 104:13028–13033
194. Schilling SH, Hjelmeland AB, Radiloff DR, Liu IM, Wakeman TP, Fielhauer JR, Foster EH, Lathia JD, Rich JN, Wang XF, Datto MB (2009) NDRG4 is required for cell cycle progression and survival in glioblastoma cells. J Biol Chem 284:25160–25169
195. Takebe N, Harris PJ, Warren RQ, Ivy SP (2011) Targeting cancer stem cells by inhibiting Wnt, Notch, and Hedgehog pathways. Nat Rev Clin Oncol 8:97–106
196. Taipale J, Beachy PA (2001) The Hedgehog and Wnt signalling pathways in cancer. Nature 411:349–354
197. Reya T, Clevers H (2005) Wnt signalling in stem cells and cancer. Nature 434:843–850
198. Ables JL, Breunig JJ, Eisch AJ, Rakic P (2011) Not(ch) just development: Notch signalling in the adult brain. Nat Rev Neurosci 12:269–283
199. Lessard J, Sauvageau G (2003) Bmi-1 determines the proliferative capacity of normal and leukaemic stem cells. Nature 423:255–260
200. Liu S, Dontu G, Mantle ID, Patel S, Ahn NS, Jackson KW, Suri P, Wicha MS (2006) Hedgehog signaling and Bmi-1 regulate self-renewal of normal and malignant human mammary stem cells. Cancer Res 66:6063–6071
201. Joensuu K, Hagstrom J, Leidenius M, Haglund C, Andersson LC, Sariola H, Heikkila P (2011) Bmi-1, c-myc, and Snail expression in primary breast cancers and their metastases–elevated Bmi-1 expression in late breast cancer relapses. Virchows Arch 459:31–39
202. Vermeulen L, De Sousa EMF, van der Heijden M, Cameron K, de Jong JH, Borovski T, Tuynman JB, Todaro M, Merz C, Rodermond H, Sprick MR, Kemper K, Richel DJ, Stassi G, Medema JP (2010) Wnt activity defines colon cancer stem cells and is regulated by the microenvironment. Nat Cell Biol 12:468–476
203. de Sousa EMF, Colak S, Buikhuisen J, Koster J, Cameron K, de Jong JH, Tuynman JB, Prasetyanti PR, Fessler E, van den Bergh SP, Rodermond H, Dekker E, van der Loos CM, Pals ST, van de Vijver MJ, Versteeg R, Richel DJ, Vermeulen L, Medema JP (2011) Methylation of cancer-stem-cell-associated Wnt target genes predicts poor prognosis in colorectal cancer patients. Cell Stem Cell 9:476–485
204. Sikandar SS, Pate KT, Anderson S, Dizon D, Edwards RA, Waterman ML, Lipkin SM (2010) NOTCH signaling is required for formation and self-renewal of tumor-initiating cells and for repression of secretory cell differentiation in colon cancer. Cancer Res 70:1469–1478
205. Fan X, Matsui W, Khaki L, Stearns D, Chun J, Li YM, Eberhart CG (2006) Notch pathway inhibition depletes stem-like cells and blocks engraftment in embryonal brain tumors. Cancer Res 66:7445–7452
206. Gattelli A, Cirio MC, Quaglino A, Schere-Levy C, Martinez N, Binaghi M, Meiss RP, Castilla LH, Kordon EC (2004) Progression of pregnancy-dependent mouse mammary tumors after long dormancy periods. Involvement of Wnt pathway activation. Cancer Res 64:5193–5199
207. Gestl SA, Leonard TL, Biddle JL, Debies MT, Gunther EJ (2007) Dormant Wnt-initiated mammary cancer can participate in reconstituting functional mammary glands. Mol Cell Biol 27:195–207
208. Indraccolo S, Minuzzo S, Masiero M, Pusceddu I, Persano L, Moserle L, Reboldi A, Favaro E, Mecarozzi M, Di Mario G, Screpanti I, Ponzoni M, Doglioni C, Amadori A (2009) Cross-talk between tumor and endothelial cells involving the Notch3-Dll4 interaction marks escape from tumor dormancy. Cancer Res 69:1314–1323

209. Masiero M, Minuzzo S, Pusceddu I, Moserle L, Persano L, Agnusdei V, Tosello V, Basso G, Amadori A, Indraccolo S (2011) Notch3-mediated regulation of MKP-1 levels promotes survival of T acute lymphoblastic leukemia cells. Leukemia 25:588–598
210. Gohongi T, Fukumura D, Boucher Y, Yun CO, Soff GA, Compton C, Todoroki T, Jain RK (1999) Tumor-host interactions in the gallbladder suppress distal angiogenesis and tumor growth: involvement of transforming growth factor beta1. Nat Med 5:1203–1208
211. Schober M, Fuchs E (2011) Tumor-initiating stem cells of squamous cell carcinomas and their control by TGF-beta and integrin/focal adhesion kinase (FAK) signaling. Proc Natl Acad Sci USA 108:10544–10549
212. Tobar N, Villar V, Santibanez JF (2010) ROS-NFkappaB mediates TGF-beta1-induced expression of urokinase-type plasminogen activator, matrix metalloproteinase-9 and cell invasion. Mol Cell Biochem 340:195–202
213. Jo M, Eastman BM, Webb DL, Stoletov K, Klemke R, Gonias SL (2010) Cell signaling by urokinase-type plasminogen activator receptor induces stem cell-like properties in breast cancer cells. Cancer Res 70:8948–8958
214. Guzman ML, Neering SJ, Upchurch D, Grimes B, Howard DS, Rizzieri DA, Luger SM, Jordan CT (2001) Nuclear factor-kappaB is constitutively activated in primitive human acute myelogenous leukemia cells. Blood 98:2301–2307
215. Piccirillo SG, Reynolds BA, Zanetti N, Lamorte G, Binda E, Broggi G, Brem H, Olivi A, Dimeco F, Vescovi AL (2006) Bone morphogenetic proteins inhibit the tumorigenic potential of human brain tumour-initiating cells. Nature 444:761–765
216. Gaur P, Sceusi EL, Samuel S, Xia L, Fan F, Zhou Y, Lu J, Tozzi F, Lopez-Berestein G, Vivas-Mejia P, Rashid A, Fleming JB, Abdalla EK, Curley SA, Vauthey JN, Sood AK, Yao JC, Ellis LM (2010) Identification of cancer stem cells in human gastrointestinal carcinoid and neuroendocrine tumors. Gastroenterology 141:1728–1737
217. Lathia JD, Gallagher J, Heddleston JM, Wang J, Eyler CE, Macswords J, Wu Q, Vasanji A, McLendon RE, Hjelmeland AB, Rich JN (2010) Integrin alpha 6 regulates glioblastoma stem cells. Cell Stem Cell 6:421–432
218. Vaillant F, Asselin-Labat ML, Shackleton M, Forrest NC, Lindeman GJ, Visvader JE (2008) The mammary progenitor marker CD61/beta3 integrin identifies cancer stem cells in mouse models of mammary tumorigenesis. Cancer Res 68:7711–7717
219. Rothhammer T, Bataille F, Spruss T, Eissner G, Bosserhoff AK (2007) Functional implication of BMP4 expression on angiogenesis in malignant melanoma. Oncogene 26:4158–4170
220. Cunha SI, Pardali E, Thorikay M, Anderberg C, Hawinkels L, Goumans MJ, Seehra J, Heldin CH, ten Dijke P, Pietras K (2010) Genetic and pharmacological targeting of activin receptor-like kinase 1 impairs tumor growth and angiogenesis. J Exp Med 207:85–100
221. Carmeliet P, Dor Y, Herbert JM, Fukumura D, Brusselmans K, Dewerchin M, Neeman M, Bono F, Abramovitch R, Maxwell P, Koch CJ, Ratcliffe P, Moons L, Jain RK, Collen D, Keshert E (1998) Role of HIF-1alpha in hypoxia-mediated apoptosis, cell proliferation and tumour angiogenesis. Nature 394:485–490
222. Ehrlich P (1909) Über den jetzigen Stand der Karzinomforschung. Ned Tijdschr Geneeskd 5:273–290
223. Mapara MY, Sykes M (2004) Tolerance and cancer: mechanisms of tumor evasion and strategies for breaking tolerance. J Clin Oncol 22:1136–1151
224. Parmiani G, De Filippo A, Novellino L, Castelli C (2007) Unique human tumor antigens: immunobiology and use in clinical trials. J Immunol 178:1975–1979
225. Lee PP, Yee C, Savage PA, Fong L, Brockstedt D, Weber JS, Johnson D, Swetter S, Thompson J, Greenberg PD, Roederer M, Davis MM (1999) Characterization of circulating T cells specific for tumor-associated antigens in melanoma patients. Nat Med 5:677–685
226. Stockert E, Jager E, Chen YT, Scanlan MJ, Gout I, Karbach J, Arand M, Knuth A, Old LJ (1998) A survey of the humoral immune response of cancer patients to a panel of human tumor antigens. J Exp Med 187:1349–1354

227. Ben-Neriah Y, Daley GQ, Mes-Masson AM, Witte ON, Baltimore D (1986) The chronic myelogenous leukemia-specific P210 protein is the product of the bcr/abl hybrid gene. Science 233:212–214
228. Kawakami Y, Dang N, Wang X, Tupesis J, Robbins PF, Wang RF, Wunderlich JR, Yannelli JR, Rosenberg SA (2000) Recognition of shared melanoma antigens in association with major HLA-A alleles by tumor infiltrating T lymphocytes from 123 patients with melanoma. J Immunother 23:17–27
229. Khong HT, Wang QJ, Rosenberg SA (2004) Identification of multiple antigens recognized by tumor-infiltrating lymphocytes from a single patient: tumor escape by antigen loss and loss of MHC expression. J Immunother 27:184–190
230. Hogquist KA, Baldwin TA, Jameson SC (2005) Central tolerance: learning self-control in the thymus. Nat Rev Immunol 5:772–782
231. Ramsdell F, Fowlkes BJ (1990) Clonal deletion versus clonal anergy: the role of the thymus in inducing self tolerance. Science 248:1342–1348
232. Kappler JW, Roehm N, Marrack P (1987) T cell tolerance by clonal elimination in the thymus. Cell 49:273–280
233. Ramsdell F, Lantz T, Fowlkes BJ (1989) A nondeletional mechanism of thymic self tolerance. Science 246:1038–1041
234. Farrar JD, Katz KH, Windsor J, Thrush G, Scheuermann RH, Uhr JW, Street NE (1999) Cancer dormancy. VII. A regulatory role for CD8+ T cells and IFN-gamma in establishing and maintaining the tumor-dormant state. J Immunol 162:2842–2849
235. Nishimura MI, Avichezer D, Custer MC, Lee CS, Chen C, Parkhurst MR, Diamond RA, Robbins PF, Schwartzentruber DJ, Rosenberg SA (1999) MHC class I-restricted recognition of a melanoma antigen by a human CD4+ tumor infiltrating lymphocyte. Cancer Res 59:6230–6238
236. Ruiter DJ, Mattijssen V, Broecker EB, Ferrone S (1991) MHC antigens in human melanomas. Semin Cancer Biol 2:35–45
237. Aptsiauri N, Cabrera T, Mendez R, Garcia-Lora A, Ruiz-Cabello F, Garrido F (2007) Role of altered expression of HLA class I molecules in cancer progression. Adv Exp Med Biol 601:123–131
238. Bubenik J (2003) Tumour MHC class I downregulation and immunotherapy (Review). Oncol Rep 10:2005–2008
239. Cabrera T, Lara E, Romero JM, Maleno I, Real LM, Ruiz-Cabello F, Valero P, Camacho FM, Garrido F (2007) HLA class I expression in metastatic melanoma correlates with tumor development during autologous vaccination. Cancer Immunol Immunother 56:709–717
240. Carretero R, Romero JM, Ruiz-Cabello F, Maleno I, Rodriguez F, Camacho FM, Real LM, Garrido F, Cabrera T (2008) Analysis of HLA class I expression in progressing and regressing metastatic melanoma lesions after immunotherapy. Immunogenetics 60:439–447
241. van Houdt IS, Sluijter BJ, Moesbergen LM, Vos WM, de Gruijl TD, Molenkamp BG, van den Eertwegh AJ, Hooijberg E, van Leeuwen PA, Meijer CJ, Oudejans JJ (2008) Favorable outcome in clinically stage II melanoma patients is associated with the presence of activated tumor infiltrating T-lymphocytes and preserved MHC class I antigen expression. Int J Cancer 123:609–615
242. Mannie MD, Prevost KD, Marinakis CA (1995) Prostaglandin E2 promotes the induction of anergy during T helper cell recognition of myelin basic protein. Cell Immunol 160:132–138
243. Wahl SM, Swisher J, McCartney-Francis N, Chen W (2004) TGF-beta: the perpetrator of immune suppression by regulatory T cells and suicidal T cells. J Leukoc Biol 76:15–24
244. Liu CM, Okayasu T, Goldman P, Suzuki Y, Suzuki K, Wheelock EF (1986) Immune regulation of the L5178Y murine tumor-dormant state. I. In vivo and in vitro effects of prostaglandin E2 and indomethacin on tumor cell growth. J Exp Med 164:1259–1273
245. Li XC, Rothstein DM, Sayegh MH (2009) Costimulatory pathways in transplantation: challenges and new developments. Immunol Rev 229:271–293
246. Jenkins MK, Taylor PS, Norton SD, Urdahl KB (1991) CD28 delivers a costimulatory signal involved in antigen-specific IL-2 production by human T cells. J Immunol 147:2461–2466

247. Freeman GJ, Long AJ, Iwai Y, Bourque K, Chernova T, Nishimura H, Fitz LJ, Malenkovich N, Okazaki T, Byrne MC, Horton HF, Fouser L, Carter L, Ling V, Bowman MR, Carreno BM, Collins M, Wood CR, Honjo T (2000) Engagement of the PD-1 immunoinhibitory receptor by a novel B7 family member leads to negative regulation of lymphocyte activation. J Exp Med 192:1027–1034
248. Kroczek RA, Mages HW, Hutloff A (2004) Emerging paradigms of T-cell co-stimulation. Curr Opin Immunol 16:321–327
249. Dong H, Strome SE, Salomao DR, Tamura H, Hirano F, Flies DB, Roche PC, Lu J, Zhu G, Tamada K, Lennon VA, Celis E, Chen L (2002) Tumor-associated B7-H1 promotes T-cell apoptosis: a potential mechanism of immune evasion. Nat Med 8:793–800
250. Gao Q, Wang XY, Qiu SJ, Yamato I, Sho M, Nakajima Y, Zhou J, Li BZ, Shi YH, Xiao YS, Xu Y, Fan J (2009) Overexpression of PD-L1 significantly associates with tumor aggressiveness and postoperative recurrence in human hepatocellular carcinoma. Clin Cancer Res 15:971–979
251. Hamanishi J, Mandai M, Iwasaki M, Okazaki T, Tanaka Y, Yamaguchi K, Higuchi T, Yagi H, Takakura K, Minato N, Honjo T, Fujii S (2007) Programmed cell death 1 ligand 1 and tumor-infiltrating CD8+ T lymphocytes are prognostic factors of human ovarian cancer. Proc Natl Acad Sci USA 104:3360–3365
252. Hirano F, Kaneko K, Tamura H, Dong H, Wang S, Ichikawa M, Rietz C, Flies DB, Lau JS, Zhu G, Tamada K, Chen L (2005) Blockade of B7-H1 and PD-1 by monoclonal antibodies potentiates cancer therapeutic immunity. Cancer Res 65:1089–1096
253. Nomi T, Sho M, Akahori T, Hamada K, Kubo A, Kanehiro H, Nakamura S, Enomoto K, Yagita H, Azuma M, Nakajima Y (2007) Clinical significance and therapeutic potential of the programmed death-1 ligand/programmed death-1 pathway in human pancreatic cancer. Clin Cancer Res 13:2151–2157
254. Hori S, Nomura T, Sakaguchi S (2003) Control of regulatory T cell development by the transcription factor Foxp3. Science 299:1057–1061
255. Nishizuka Y, Sakakura T (1969) Thymus and reproduction: sex-linked dysgenesia of the gonad after neonatal thymectomy in mice. Science 166:753–755
256. Sakaguchi S, Sakaguchi N, Asano M, Itoh M, Toda M (1995) Immunologic self-tolerance maintained by activated T cells expressing IL-2 receptor alpha-chains (CD25). Breakdown of a single mechanism of self-tolerance causes various autoimmune diseases. J Immunol 155:1151–1164
257. Ahmadzadeh M, Felipe-Silva A, Heemskerk B, Powell DJ Jr, Wunderlich JR, Merino MJ, Rosenberg SA (2008) FOXP3 expression accurately defines the population of intratumoral regulatory T cells that selectively accumulate in metastatic melanoma lesions. Blood 112:4953–4960
258. Clark RA, Huang SJ, Murphy GF, Mollet IG, Hijnen D, Muthukuru M, Schanbacher CF, Edwards V, Miller DM, Kim JE, Lambert J, Kupper TS (2008) Human squamous cell carcinomas evade the immune response by down-regulation of vascular E-selectin and recruitment of regulatory T cells. J Exp Med 205:2221–2234
259. Curiel TJ, Coukos G, Zou L, Alvarez X, Cheng P, Mottram P, Evdemon-Hogan M, Conejo-Garcia JR, Zhang L, Burow M, Zhu Y, Wei S, Kryczek I, Daniel B, Gordon A, Myers L, Lackner A, Disis ML, Knutson KL, Chen L, Zou W (2004) Specific recruitment of regulatory T cells in ovarian carcinoma fosters immune privilege and predicts reduced survival. Nat Med 10:942–949
260. Liyanage UK, Moore TT, Joo HG, Tanaka Y, Herrmann V, Doherty G, Drebin JA, Strasberg SM, Eberlein TJ, Goedegebuure PS, Linehan DC (2002) Prevalence of regulatory T cells is increased in peripheral blood and tumor microenvironment of patients with pancreas or breast adenocarcinoma. J Immunol 169:2756–2761
261. Woo EY, Yeh H, Chu CS, Schlienger K, Carroll RG, Riley JL, Kaiser LR, June CH (2002) Cutting edge: regulatory T cells from lung cancer patients directly inhibit autologous T cell proliferation. J Immunol 168:4272–4276

Chapter 9
Tumor Dormancy: Long-Term Survival in a Hostile Environment

Bruno Quesnel

Abstract Tumor dormancy occurs when cancer cells are present but the tumor does not grow. Following treatment, patients may enter complete remission in which persistent cells represent the minimal residual disease (MRD). Experimental models and clinical data suggest that the absolute quantity of this MRD is extremely low. Very few cancer cells can persist for years or decades under these hostile conditions that include continuous exposure to maintenance treatment, autologous anti-tumor immune response, and a nonpermissive microenvironment. Dormant tumor cells may survive despite these destruction factors if they adapt and develop strategies to escape from cell death. Escape may result in a state of equilibrium between MRD and the patient. Equilibrium between the immune response and tumor cells can result in long-term tumor dormancy; however, after variable lengths of time, tumor dormancy ends, and the disease progresses. Experimental models have shown that dormant tumor cells may over-express B7-H1 and B7.1 and inhibit cytotoxic T-cell-mediated lysis. This resistance could be therapeutically targeted using drugs like MEK inhibitors that modulate pathways involved in B7-H1 expression. Dormant tumor cells may also develop nonspecific resistance mechanisms to cell death, such as deregulation of JAK/STAT and mTORC2/AKT pathways or autocrine and paracrine production of cytokines. This deregulation leads to cross-resistance between the immune response and cytotoxic drugs, indicating that the long-term selection that occurs in vivo during tumor dormancy may ultimately result in resistant relapse. Long-term selection of cancer cells in vitro using tyrosine kinase inhibitors selects cells that harbor the same resistance mechanisms as dormant tumor cells. Elucidating the mechanisms underlying the equilibrium that allows for the persistence of dormant tumor cells presents a novel strategy for targeted drug treatment in the context of maintenance therapy.

B. Quesnel, MD, PhD (✉)
Service des Maladies du Sang, Centre Hospitalier et Universitaire de Lille,
Rue Polonovski, 59037 Lille, France
e-mail: brunoquesnel@hotmail.com

H. Enderling et al. (eds.), *Systems Biology of Tumor Dormancy*,
Advances in Experimental Medicine and Biology 734,
DOI 10.1007/978-1-4614-1445-2_9, © Springer Science+Business Media New York 2013

Keywords Tumor dormancy • B7-H1 • PD-L1 • Interferon • TLR • Stem cell • NK • Cytotoxic T-cells • NK • Imatinib • AKT • MEK • GILZ • JAK • STAT • Resistance • Apoptosis • Persistence

Introduction

There is now evidence that patients can harbor cancer cells for extended periods and, in some cases indefinitely, without relapsing. This phenomenon of long-term, persistent cancer cells that do not grow is called tumor dormancy and may occur following treatment in the form of MRD. There is now growing evidence that this same phenomenon occurs early in tumor development, indicating that tumor growth is not continuous and may pass through a long period of subclinical disease [1].

It is important to note that dormant tumor cells may persist as quiescent and furtive cells, with little interaction with their microenvironment, or they may develop active survival mechanisms. The first hypothesis is intuitive in that tumor cells originate from the individual and develop in the host for years or decades. If residual cancer cells share some properties of physiological stem cells such as low-cycling, niche location, and sensitivity to inhibitory signals from the microenvironment, they could hide and persist for years as normal cells [2]. The presence of residual cancer cells, however, is not necessarily neutral to the host. Malignant transformation can result in significant alterations compared with the normal cellular phenotype. Cancer cells are no longer a part of the individual and must survive despite a less adapted environment and in some cases a continuous exposure to cytotoxic drugs. This latter situation suggests that a substantial proportion of cellular death among residual cancer cells is compensated for by active replication. Tumor dormancy would then result from equilibrium rather than from cellular quiescence. Such long-term equilibrium may be due to several conditions. The host could control the tumor outgrowth, the tumor cells could resist the host defenses, or the microenvironment may not present an optimal environment for tumor growth (e.g., limited angiogenesis, lack of appropriate extracellular matrix to stimulate adhesion molecules) [3–7]. Replicative and quiescent cells could also coexist if only replicative cells are being destroyed by the host defenses. All of these conditions, which it should be noted are not mutually exclusive, have been demonstrated using experimental models [8, 9].

Determining the Number of Cells That Exist at Equilibrium

If equilibrium can be established between host and dormant tumor cells, it is important to determine the lowest number of cells that are necessary to allow for long-term tumor dormancy. If this number is low, then cells are likely to have developed efficient survival strategies. Alternatively, some cancers may enter into tumor dormancy with a relatively high tumor mass [10]. Patients with low-grade non-Hodgkin's lymphoma can carry clinically significant tumor mass for years prior to tumor progression. Rare cases of metastatic melanoma exhibit spontaneous tumor regressions that may last for extended periods. However, in most cancers presenting with long-term complete remission, residual disease remains undetectable or detection requires extremely sensitive techniques. In patients in complete remission from solid tumors, frequencies of circulating tumor cells (CTCs) and disseminated tumor cells (DTCs) are generally approximately one per million [11, 12]. In hematological malignancies, long-term persisting cells are mostly below the limit of molecular detection, which is approximately 10^{-5}–10^{-6}. For example, to detect residual AML1-ETO (Acute Myeloid Leukemia 1—Eight Twenty One) positive cells in the bone marrow of leukemia patients in long-term complete remission, serial long-term colony assays and extensive PCR testing for the fusion transcript are required [13]. Chronic myeloid leukemia (CML) patients who have undergone imatinib therapy for years frequently enter into molecular complete remission defined by undetectable BCR/ABL fusion transcript; however, detection of BCR/ABL at the DNA level shows that nearly every patient still carries leukemia cells. These data suggest that the number of residual cells is indeed very low; however, the definition of "low" remains undetermined. Experimental models can present an indication of this parameter. The DA1-3b model is a BCR-ABL mouse model of leukemia dormancy (Fig. 9.1) [14–16]. Mice are first vaccinated with DA1-3b cells expressing CD40L or IL-12 and are then challenged with live DA1-3b cells (Fig. 9.1) [14, 15, 17–20]. Unvaccinated animals usually rapidly develop leukemia, while a fraction of vaccinated animals do not develop disease. Mice in complete remission after 1 year exhibit few dormant tumor cells, usually no more than 1,000 and sometimes no more than 100. When these cells are isolated and injected into naive mice, they can still induce acute myeloid leukemia (AML). Tumor dormancy may then result from a very small population of residual cells that persists in balance with the immune system [17]. These results may illustrate the function of the dormant tumor cell mass in humans, and if this is the case, then the model may accurately reflect the relationship between patients and MRD. In our opinion, it is crucial to determine the number of dormant tumor cells. A large tumor mass should already contain numerous heterogeneous clones that ensure statistically that some may exhibit long-term survival even in the presence of efficient cytotoxicity. A small population of cells, however, must quickly develop powerful and active mechanisms of survival to avoid elimination, as possibilities of clonal selection are limited.

Equilibrium with the Immune System

If a small number of residual cancer cells can persist for years, then it is of interest to determine if they can be recognized and controlled by the immune system (Fig. 9.2). This subject has been under debate for a number of years. The existence of an equilibrium between the host immune response and dormant tumor cells has been demonstrated using several experimental models [9, 21, 22]. Significant

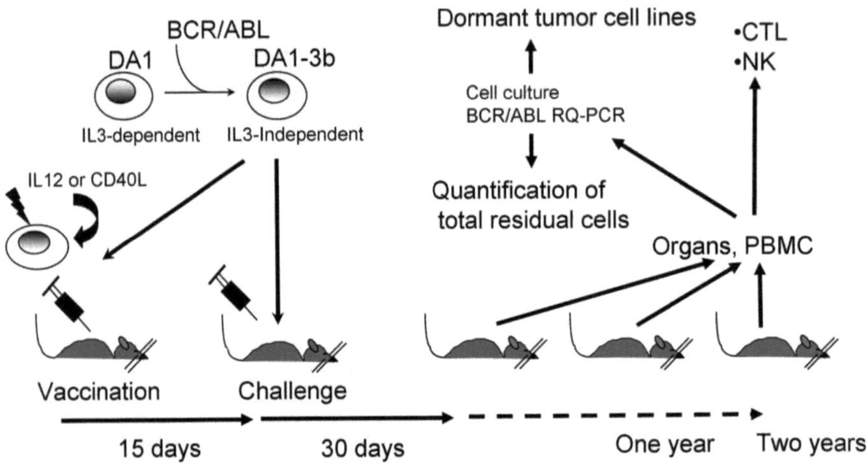

Fig. 9.1 The DA1-3b mouse model of tumor dormancy. C3H/Hej mice are vaccinated with irradiated BCR/ABL$^+$ DA1-3b leukemic cells transduced with IL12 or CD40L. Immunity is challenged by injection of DA1-3b cells and mice that remain in remission are randomly euthanized at selected times during 1 year follow-up. Total quantity of minimal residual disease (MRD) is evaluated by BCR/ABL Quantitative-PCR performed on all organs and dormant tumor cells are isolated to generate dormant tumor cells-derived cell lines. From Quesnel [9], reprinted with permission from John Wiley and Sons

clinical observations have also suggested that this equilibrium occurs in human disease. Residual lymphoma cells have been found in patients treated with anti-idiotype antibodies and who have been in continuous remission for at least 3–8 years [23]. Nearly 85% of CML patients that received allogeneic hematopoietic stem cell transplant remained BCR/ABL PCR-positive after years of follow-up testing, suggesting that most or all of these patients carry residual leukemic cells controlled by a graft-versus-leukemia (GVL) effect [24]. Patients who relapse from CML, AML, and to a lesser degree myelodysplastic syndrome (MDS) after allogenic stem cell transplantation can reenter complete remission (CR) after donor lymphocytes infusion and some have remained disease-free for more than a decade. Cancers exhibiting a donor origin have been reported in organ transplant recipients where in some cases, the donor had been in complete remission for decades, and it is suspected that immunosuppression allowed the tumor to escape from dormancy [25]. Epstein-Barr Virus (EBV) lymphomas can occur following transplantation or in immunocompromised patients. When immunosuppressive treatment is reduced, prognosis improves. This finding also indicates that in humans, an equilibrium may exist between the anti-tumor immune response and dormant tumor cells. Another important clinical observation is that in colon cancer, prognosis may depend on infiltration of early metastases by cells from the immune system, suggesting that equilibrium may occur [26, 27]. However, the progression and late relapses of cancer indicate that this equilibrium can be altered, possibly through the immunoescape of dormant tumor cells.

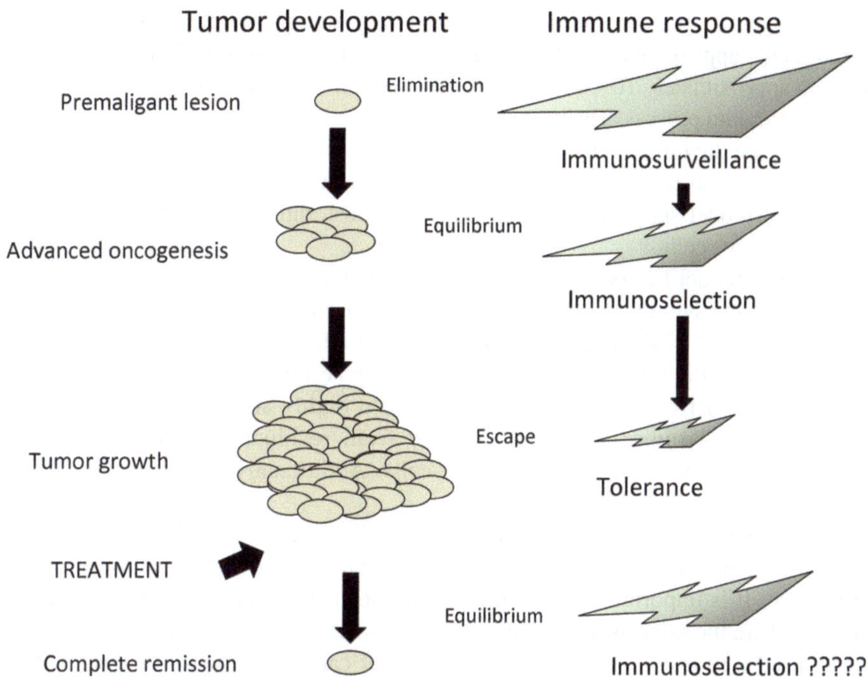

Fig. 9.2 Crosstalk of cancer cells and host immune system. During early steps of carcinogenesis, premalignant clones are continuously eliminated by host immune response. Fully malignant clones that survive enter in equilibrium and tumor dormancy until they subvert the immune response. After therapy, the MRD enter in a new phase of long-term equilibrium between dormant malignant cells that may still evolve to escape immune response, and cytotoxic T and NK cell response that also adapt to kill residual cells

Experimental Models and Long-Term Equilibrium

There are too few studies to firmly establish the mechanisms that lead to long-term equilibrium. Most of the studies that have been performed have been conducted in a therapeutic context that does not necessarily reflect spontaneous immune response. It is extraordinarily difficult to isolate dormant tumor cells from humans, and the timeframe of tumor dormancy in humans has limited the field to animal models where a reasonable experimental duration can be achieved. Current knowledge of equilibrium between dormant tumor cells and the immune response has been almost exclusively derived from experimental models. Although it is possible to maintain cells in quiescence in vitro for a few weeks, relevant longer durations of dormancy can only be obtained in vivo. These models are absolutely required to understand the complex interactions that tumor cells establish with a microenvironment.

Several models of tumor dormancy involve immunization of mice followed by tumor cell challenge. In the BCL1 mouse lymphoma model, tumor dormancy can be induced by immunization with BCL1-derived immunoglobulin (Ig) to generate an anti-idiotype immune response [28–30]. After subsequent injection with BCL1 tumor cells, approximately 70% of mice develop dormant tumors in the spleen that contain approximately 10^6 cells. Mice relapse 2–24 months later at a steady rate, indicating a random process. The host antibodies act as an agonist by hyper-cross-linking surface Ig on the tumor cells, which usually induces cell cycle arrest and apoptosis. $CD8^+$ T-cells and IFN-γ collaborate with humoral immunity to maintain dormancy [31]. This model supports the hypothesis that host anti-tumor immunity controls dormant tumor cells.

Another model has been described by Shirmarer et al. In athymic *nu/nu* mice, Gal-expressing syngeneic tumor cells (ESbL-Gal) can cause tumors [32]. Adoptive transfer of LacZ (Gal)-reactive T-cells into these mice prevents tumor growth, and many Gal-specific $CD8^+$ T-cells persist in the bone marrow and spleen. Memory T-cells derived from bone marrow exhibited a highly significant turnover rate, suggesting that tumor-associated antigen from residual dormant tumor cells maintains Gal-specific $CD8^+$ memory T-cells. Additionally, T-cell depletion was able to remove dormancy.

In the above models, tumor dormancy is established after induction of an efficient immune response. Tumors can also be dormant during early disease prior to treatment. It is important to determine if the mechanisms controlling dormancy at this stage are the same as those controlling dormancy later during remission. Koebel et al. used a mouse model of primary chemical carcinogenesis to investigate this idea [33]. Mice were injected with 3′-methylcholanthrene (MCA), and those that developed progressively growing sarcomas were removed. The remaining mice exhibited small stable masses at the injection site. These masses contained transformed fibroblasts, $CD3^+$ T-cells, $B220^+$ cells, and mononuclear phagocytes. When these fibroblasts were injected into immunodeficient Rag $2^{-/-}$ mice, they formed tumors, confirming that they were fully transformed. The stable masses showed increased apoptosis and decreased proliferation. When CD4, CD8, IL12, and IFN-γ were blocked, the stable masses transformed into growing tumors. Natural killer cells (NK) depletion exhibited no effect. In this model, adaptive immunity restrained tumor growth for long periods of time. These results suggest that an equilibrium between host immunity and malignant cells can also induce dormancy early in disease. They also confirm the central role of T-cell-mediated immunity that has been observed in other models.

The Role of T-Cell Inhibition in the Immuno-Escape of Dormant Tumor Cells

If dormant tumor cells can persist in the presence of an efficient immune response and can induce frequent relapses, they must counteract the immune effectors or resist their cytotoxic effects. One possible mechanism is through the active suppression of T-cells. As mentioned above, in several experimental models, inhibition of T-cell-mediated immunity through T-cell depletion or IFN-γ or IL12 blocking

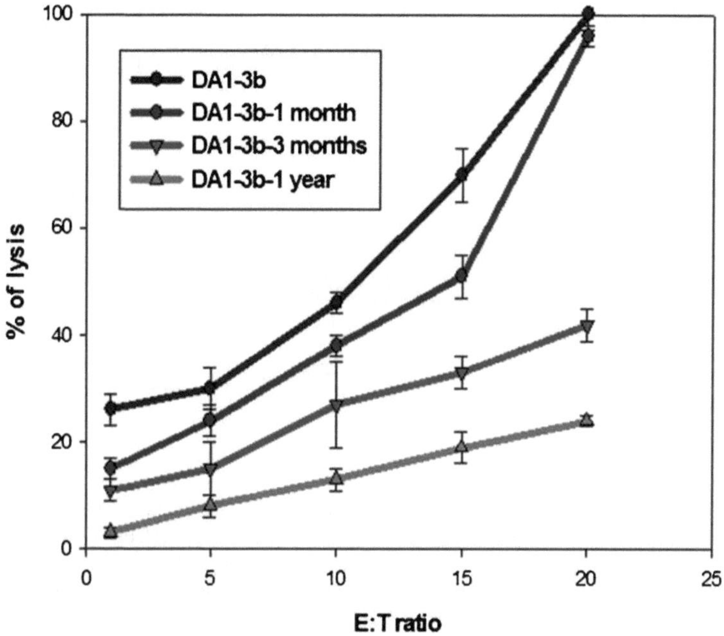

Fig. 9.3 Acquired resistance of dormant tumor cells to CTLs. CTLs were isolated 1 month after challenge from mice vaccinated with irradiated DA1 3b/IL12 cells and tested for their ability to kill leukemic cells that had persisted in other animals for 1 month, 3 months, and 1 year. The more the cells remained dormant, the more they resisted to CTLs

induces escape from dormancy. In the DA1-3b mouse model, DA1-3b-specific CD8⁺ cytotoxic lymphocytes (CTLs) killed DA1-3b cells with the same efficiency, whether isolated 1 month or 1 year after vaccination [14, 15, 17]. However, the dormant tumor cells became less sensitive to lysis over time and lost the ability to stimulate CTLs to secrete IFN-γ and TNF-α. Thus, over time, dormant tumor cells became more resistant to specific CTL-mediated killing (Fig. 9.3). This gradual development of resistance suggests that the tumor does not merely survive passively, but that there is a continuous struggle between the host immune response and the dormant tumor cells. In the DA1-3b model, dormant tumor cells escape the immune response by over-expressing B7-H1 (also known as PD-L1 or CD274), a B7 family member and ligand for PD-1 (programmed death-1, a member of the CD28 family) (Fig. 9.4) [22]. B7-H1 is normally expressed after exposure to inflammatory cytokines, particularly IFN-γ, and is broadly distributed in various tissues. B7-H1 interacts with PD-1 on T-cells and inhibits T-cell activation and CTL-mediated lysis. In other systems, B7-H1 increases T-cell activation. In DA1-3b mice, dormant tumor cells express elevated B7-H1 and B7.1 in proportion to the time that they persisted in the host [17]. In vitro, blocking of B7-H1 or the B7.1/CTLA-4 interaction enhanced CTL-mediated killing of these persistent cells. In vivo, blocking B7-H1, B7.1, or CTLA-4 prolonged the survival of naive mice injected with dormant tumor cells.

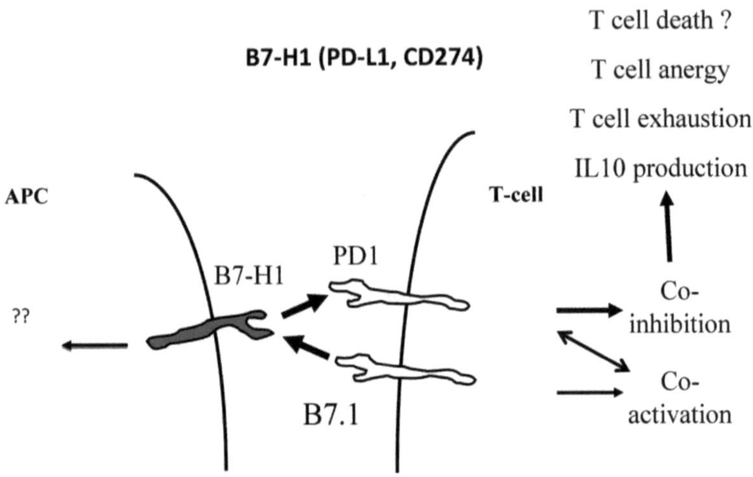

Fig. 9.4 Functions of B7-H1 in immunological synapse. B7-H1 expressed on antigen-presenting cells (APC) may deliver a co-inhibitory signal to T-cells via PD-1. B7-H1/PD-1 interaction results in T-cell anergy, T-cell exhaustion, and possibly T-cell death. B7-H1 and B7.1 may also interact through a different domain than those involved between B7-H1 and PD-1. However, the role of B7-H1/B7.1 interaction remains unclear

Over-expressing B7-H1 may allow tumors to escape the host immune response. Human carcinomas abundantly express B7-H1, and in mice, this expression enhances tumor growth [34, 35]. Blocking B7-H1 enhances cancer vaccine effectiveness. In the DA1-3b model, the progressively increasing B7-H1 expression observed in dormant tumor cells may progressively inhibit CTLs in vivo. Thus, if B7-H1 can aid in tumor cell escape from the host immune system, perhaps inhibition of CTL-mediated killing could help tumor cells to establish dormancy.

Another surprising observation was that B7.1 expression was increased in dormant tumor cells. In vitro blocking of B7.1 was able to enhance $CD8^+$ CTL-mediated killing of MRD-derived cells and enhanced production of IFN-γ and TNF-α, which was mediated by CTLA-4. Additionally, B7.1 can induce peripheral tolerance depending on the basal expression level of this molecule on antigen-presenting cells. In several tumor models, blocking CTLA-4 in vivo led to rejection, and clinical trials using anti-CTLA-4 antibodies have demonstrated activity in some patients. Given these findings, B7.1 may also contribute to the long-term persistence of dormant tumor cells by inhibiting $CD8^+$ CTL-mediated killing. A recent report has confirmed that in colon carcinoma and melanoma models, low surface expression of B7.1 provides an advantage to cancer cells against the immune response [36].

The exact mechanisms of inhibition by B7.1 and B7-H1 are unknown. In some models, they directly affect CTLs. It is possible that in tumor dormancy, they stimulate regulatory T-cells (Treg); however, this theory remains untested. An intriguing recent finding is that B7-H1 binds to B7.1 in addition to PD-1. B7.1 binds less strongly than PD-1, but can still cause immunosuppression [37, 38]. Thus, the immunosuppressive effects of excessive B7-H1 observed in several models may result from B7-H1/B7.1 interaction. T-cells or tumor cells could express either marker, and there are reports of T-cells expressing B7-H1 infiltrating lung or breast cancer. This resembles the reverse interaction between CTLA-4 expressed on Treg-cells and B7.1 on dendritic cells: a reverse signal induces production of immuno-suppressive signals by dendritic cells (i.e., indoleamine-2, 3-dioxygenase [IDO]). Thus, immuno-escape is probably the result of the co-expression of several B7 family molecules inside the tumor, where the unusual microenvironment makes interpretation of in vivo blocking experiments extremely difficult.

An interesting function of B7-H1 is its role in T-cell exhaustion. T-cells derived from patients presenting with chronic viral infections such as HIV or chronic hepatitis show over-expression of PD-1, which is the receptor for B7-H1. T-cell exhaustion has been demonstrated in a wide variety of animal models and in humans suffering from chronic viral, bacterial, and parasitic infections as well as during human cancer. Exhausted T-cells typically lose IL2 production first, after which they become no longer able to produce IFN-γ and degranulate; finally, they are deleted if chronic antigen exposure remains. There is a clear correlation between antigen load and exhaustion. Blockade of PD-1 can reverse T-cell exhaustion. Additionally, several other inhibitory molecules are coexpressed with PD-1 on exhausted T-cells. Blocking of PD1 and Tim3, the receptor of galectin 9, acts synergistically to suppress T-cell exhaustion. It is of interest to determine if such mechanisms occur during tumor dormancy. Dormant tumor cells persist for years, similar to chronic infections. However, as mentioned above, the total number of cells is extremely low. It is currently unknown if this small antigen load is sufficient to exhaust T-cells. It is tempting to think that both dormant tumor cells and T-cells modify the phenotype of each other during their prolonged cohabitation, and this hypothesis warrants exploration.

Recent data have also shown that blast cells derived from acute leukemia patients relapsing following allogenic stem cell transplantation over-express B7-H1. Additionally, allogenic T-cells express PD-1, which suggests that escape from the graft vs. leukemia effect induced by an allogenic response may result from B7-H1/PD-1 interaction and that allogenic T-cells harbor makers of T-cell exhaustion. These data must be confirmed, but this a clear demonstration of a break in equilibrium between immunity and cancer cells.

A direct antiapoptotic effect of B7-H1 in tumor cells, independently from the immune system, has also been reported. The mechanism of this prosurvival role of B7-H1 cells remains completely unknown and suggests that dormant tumor cells that express B7-H1 may possess broad protection; this hypothesis certainly deserves further exploration.

Other mechanisms of immuno-escape have been observed or suspected. In the ESbL-Gal model, dormant tumor cells isolated from bone marrow and established as cell lines showed a decrease in the expression of adhesion molecules such as ICAM-1, suggesting malfunction of the immunological synapse [32].

Another possible mechanism of escaping T-cells is silencing tumor antigen expression. A reduction in tumor antigen presentation to antigen-presenting cells (APCs) through the loss of antigen expression or alteration of MHC has been frequently described in tumors escaping from a natural immune response or from a response induced by tumor vaccines. For instance, in a long-term survivor of melanoma, there was initial loss of antigen from tumor cells where the immune response shifted to target escape variants [39]. This finding shows both immune editing and immune adaptation and resulted in equilibrium. Thus, dormant tumor cells may be selected by immune pressure. Decreased expression of surface idiotype (Id) has also been reported in a small number of mice with BCL1 tumor dormancy, where dormancy was induced by anti-Id-mediated cross-linking of surface IgM [40, 41]. However, only a few mice whose disease relapsed showed a loss of surface Id, indicating that other mechanisms must exist. If tumor cells in vivo do not express tumor antigens, then it will be extremely difficult to evaluate the precise role of antigen selection in the escape from dormancy.

Targeting Immunoevasion

An important question is whether it is possible to block B7-H1 to facilitate eradication of dormant tumor cells by T-cells. Clinical grade antibodies against the B7-H1/PD-1 interaction are currently under development. For instance, a CT-011 humanized monoclonal antibody was tested in a Phase I clinical trial in 17 patients diagnosed with advanced hematological malignancies [42]. Thirty-three patients (100%) responded, with one complete remission, and the tolerance seemed good. To test the efficacy of this antibody against dormant tumor cells would require a maintenance schedule of several years. Another possibility could be to target the other B7 molecule that appears to be involved in tumor dormancy (B7.1). As mentioned above, B7.1 was over-expressed in the DA1-3b dormant tumor cells. B7.1 suppresses T-cell immunity through its binding to the inhibitory receptor CTLA4. Clinical grade antibodies against CTLA4 already exist. Ipilimumab has demonstrated its efficacy in advance metastatic melanoma [43]. We currently do not know if B7.1 is expressed in dormant tumor cells and if it can interact and suppress T-cells by influencing CTLA4 action or through its direct binding with B7-H1. In any case, new antibodies targeting this interaction would be more effective than anti-CTLA4 in clearing residual cancer cells.

Another option to target immuno-escape mechanisms in residual cells is to use pathway inhibitors. B7-H1 is over-expressed in malignant plasma cells from multiple myeloma [44, 45]. Moreover, this expression can be enhanced by IFN-γ, or toll-like-receptor (TLR) ligands. Interferon receptor and TLR mediate B7-H1 expression through a common pathway involving MyD88, TRAF6, MEK1/2, and STAT1 [44]. These results lead to a hypothesis that infections may contribute to immune escape if pathogens activate TLRs and induce B7-H1 expression. Blocking MEK1/2 or MyD88 resensitizes malignant plasma cells to CTLs. These data have been confirmed in AML [46]. Several potent MEK inhibitors are currently in Phase I/II trials and some may abrogate B7-H1 expression, even if they do not directly induce cancer cell death [47, 48]. Other tumor types seem to modulate B7-H1 expression through different pathways. Glioma cells, for instance, express

B7-H1 through an IGF1/PI3K/AKT pathway. A last possibility would be to use gene therapy vectors targeted to B7-H1. Adenoviral vectors with chimeric fibers that use B7-H1/B7.1 interaction ensure efficient transduction of B7-H1 overexpressing cells and are specific to dormant tumor cells derived from the DA1-3b model [49, 50]. It seems likely that targeting this immunoevasion mechanism is feasible using currently available drugs. It remains to be demonstrated whether the necessary long-term exposure with pathways inhibitors during the tumor dormancy period is effective for controlling any residual disease, with no severe side effects.

Chemotherapeutic drugs may induce immunogenic cancer cell death. We also observed that cytarabine, a cytotoxic drug that is commonly used to treat AML, can modulate B7 expression at the surface of leukemia cells by inducing ROS and activating Nf-κB [20, 51]. Thus, even currently used chemotherapy may modulate cancer cell sensitivity to CTLs; however, the use of chemotherapy as long-term maintenance therapy to control tumor dormancy could be problematic due to cumulated toxicity.

Another way to avoid B7-H1- and B7.1-mediated immune suppression is to stimulate NK cells. In the DA1-3b mouse model, an injection of irradiated leukemic cells that had been transduced by the chemokine CXCL10 led to prophylactic immunity with long-term memory. It also cured pre-established leukemia [19]. CXCL10 (also called Interferon Inducible Protein 10 or IP-10) is an essential mediator of the anti-tumor effect of IL-12 and attracts activated T-cells and NK cells. We searched for persistent leukemic cells in every organ of mice that were vaccinated with leukemic cells transduced with CXCL10 and that survived 1 year after challenge. We did not detect any, which indicates that the anti-leukemic immunity both improves survival rates and eradicates dormant tumor cells. Immunization by CXCL10-transfected leukemia cells recruits NK cells to the lesion and activates them. Dormant leukemia cells resist T cell-mediated cytotoxicity but remain sensitive to NK cell lysis, suggesting that NK cells may be the major effectors of cancer vaccines. Logically, dormant tumor cells should develop immune mechanisms to escape NK cells, but evidence is still lacking.

Stroma Immuno-escape

Destruction of stromal cells within tumors by CTLs is essential for eradicating large, well-established solid tumors. Stromal cells present antigen released from cancer cells to CTLs, resulting in the killing of cancer cells, through direct cytotoxicity and a bystander effect that results in destruction of the stroma. Among stromal cells, myeloid-derived suppressor cells (MDSCs), which include immature myeloid cells and macrophages, contribute to CTL inhibition, inhibit dendritic cell maturation, promote Treg generation, favor tumor growth through inflammation, and favor resistance to anti-angiogenic therapies [52, 53]. Zhang et al. showed in an experimental mouse model that specific targeting of stromal cells without killing cancer cells leads to an equilibrium between residual cancer cells and hosts [54]. Additionally, CD8 depletion led to escape from tumor dormancy. There was a continuous turnover of T-cells and cancer cells, confirming a state of equilibrium and demonstrating that dormancy did not result from cellular quiescence. The MDSCs from tumor stroma inhibited CTL-mediated killing by nitric oxide and arginase and promoted angiogenesis. Thus, tumor dormancy can be established through equilibrium not only between dormant tumor cells and CTLs but also between stromal cells and CTLs through the killing of MDSCs. This equilibrium can be broken by immunosuppressive enzymes such as arginase and IDO. Interestingly, in parallel with this immunosuppressive effect, metabolites from these enzymes may have tumor promoting effects. IDO and a related enzyme, TDO, catabolize tryptophan in kynurenine [55]. Kynurenine is an endogenous ligand for aryl hydrocarbon receptor that promotes cancer cell survival and motility [56]. Other interactions are likely between cells of the microenvironment and cells of the immune system, such as IFN-γ-mediated angiostasis that plays a role in tumor rejection by $CD8^+$ T-cells [57]. These immunoregulatory roles for tumor cell stroma compose only a part of the interactions. The extracellular matrix regulates tumor cell proliferation, but its stiffness promotes escape and dormancy [58]. The network comprising dormant tumor cells, stroma, and immune cells appears extremely complex and will certainly been difficult to model in vitro.

Resistance to Apoptosis: Escape to Death

Dormant tumor cells must survive for long time periods despite numerous hostile factors such as immune response, continuous drug treatment, or a nonpermissive microenvironment. Additionally, as the number of residual cancer cells is extremely low, even a strong proliferation rate would not be sufficient to compensate cell death. To persist, these cells have to develop resistance mechanisms to cell death. Evidences of such resistances in humans are indirect. Cancer cells from relapses are often more resistant to various treatments than cancer cells at diagnosis. We do not know, however, if resistance to cell death is acquired during dormancy or relapse. Experimental models lend support to the first hypothesis.

In the DA1-3b model, dormant cells showed a progressive decrease of SOCS1 gene expression, as methylation of this gene deregulated the JAK/STAT pathway [18]. Dormant cells resisted apoptosis induced by specific CTLs, but resistance was decreased when SOCS1 expression was restored, either by de-methylation or by gene transfer. An interesting finding is that IFN-γ reinforces the resistance of tumor cells. When a T-cell recognizes a tumor antigen, it produces IFN-γ, which aids in tumor cell killing through various mechanisms including the induction of MHC expression at the cell surface. However, if regulation of JAK/STAT is lifted, IFN-γ may instead activate survival mechanisms. Thus, hyperactivation of the JAK/STAT pathway allows tumor cells to resist CTL-mediated killing. Deregulation of the JAK/STAT pathway has been reported in relapsed AML, suggesting that it may have contributed to the survival of residual cells.

Dormant tumor cells in the DA-3b model resisted apoptosis following irradiation, cytarabine, or imatinib mesylate; however, gene transfer of SOCS1 reduced this resistance [18]. This cross-resistance was induced by IL3 over-production by dormant tumor cells and was reversed by an anti-IL3 antibody. Tumor cells that persist for long periods may deregulate their JAK/STAT pathways and develop cross-resistance to various treatments through an autocrine loop. Other cytokines may be implicated. Over-expression of IL6 also protected, at least partially, dormant tumor cells from apoptosis (see Fig. 9.2). Antibody-arrays indicated that cytokine profiles were modified, suggesting that dormant tumor cells can modify their microenvironment (unpublished data). We sought to determine if these mechanisms could be therapeutically targeted. A Jak2 inhibitor was able to resensitize dormant tumor cells to cell death. Several Jak1/2 inhibitors are currently in clinical trials for the treatment of myeloproliferative diseases. Their use as consolidation therapy to eradicate residual disease in other cancers should be explored.

Similar mechanisms may be observed when resistance to cell death is induced in vitro. Patients presenting with CML can become resistant to the Abl kinase inhibitor imatinib through different mechanisms including BCR-ABL mutations. These mutations cause amino-acid substitution in the kinase domain of BCR-ABL and disrupt the binding of imatinib to the tyrosine kinase. To model the evolution of resistance, we exposed the mouse DA1-3b BCR-ABL$^+$ leukemic cell line to imatinib for several months and obtained resistant cells carrying BCR/ABL mutations [59]. In co-culture, mutated cells were able to spread resistance to non-mutated cells by over-expressing IL3, activating MEK/ERK and JAK2/STAT5 pathways, and down-regulating Bim. Even the presence of less than 10% of mutated cells was sufficient to protect the non-mutated cells. Blocking JAK2 inhibited the protective effect in co-culture. Mutated cells were also sensitive to JAK2 inhibition. Another group has reported a similar mechanism in which the over-expression of GM-CSF by TKI-resistant cells also protects sensitive cells through the JAK2/STAT5 pathway [60]. Thus, models of drug resistance mimic what is observed in dormant tumor cells that acquire resistance to cell death in vivo, which suggests a possible escape mechanism of dormant tumor cells where continuous drug treatment may select drug-resistant cells that are also resistant to autologous immune response.

Another example of identical mechanisms of resistance to death in dormant- and drug-resistant cells is the silencing of glucocorticoid-induced leucine zipper protein (GILZ). GILZ is an essential mediator of glucocorticoid action. GILZ inhibits two of the main pathways involved in oncogenesis, the NFκB and Raf/Ras/ERK pathways. We found that some imatinib-resistant BCR-ABL$^+$ cells generated by long-term culture with this tyrosine kinase inhibitor demonstrated reduced expression of GILZ [61]. We also observed an even more pronounced silencing of GILZ in dormant tumor cells in the DA1-3b model. Increasing GILZ expression by transfection or by glucocorticoids overcame imatinib resistance and suppressed BCR-ABL$^+$ tumor growth through mTORC2 inhibition. GILZ interacts with mTORC2 and inhibits P-AKT (Ser473) to activate FoxO3a-mediated transcription of the pro-apoptotic protein Bim, thus providing another example that dormant tumor cells and drug-resistant cells harbor the same characteristics.

Other mechanisms of resistance to apoptosis that allow for tumor escape have also been reported. In the BCL1 lymphoma model, cross-linking of membrane IgM by anti-idiotypic antibodies induces apoptosis [62]. However, cross-linking also arrests the cell cycle through the over-expression of $p21^{waf1}$, which may protect cells from apoptosis [63]. Thus, the interaction of dormant tumor cells with the immune response may have bivalent effects. Most animals, however, escaped from dormancy through alteration in signaling from cross-linked surface IgM, including loss of Lyn kinase activity [64]. These models suggest that cell signaling, in protecting the dormant cell population from the immune response, is an essential component of tumor dormancy.

These cross-resistance mechanisms may select more aggressive tumor subclones that would cause relapses to be more difficult to treat than the initial disease and also make them resistant to immune responses, whether spontaneous or therapeutically induced. Targeting a cross-resistance mechanism could have synergistic therapeutic effects by helping chemotherapy or targeted drugs to kill the tumor cells and the immune system to clear residual cells. Cytokine-protective effects may also favor clonal heterogeneity in dormant tumor cells by cross-protecting cells with different genotype, despite continuous selective pressure from drugs or from the immune response. A number of these cells would persist within the protective microenvironment but not outside.

If dormancy is induced by escape from the immune response and resistance to cell death, what prevents the expansion of the resistant population of tumor cells? What restricts the number of tumor cells as an MRD? In absence of definitive scientific evidence, this question remains mostly unanswered. It is difficult to imagine a precise and totally stable long-term balance between residual cells and factors of cell death. Any oscillation of death or replication rate of tumor cells would rapidly result in tumor outgrowth. As mentioned above, microenvironment constraints may limit this expansion through various factors like space limitation, availability of supportive cytokines, nutriments, or oxygen [65]. This hypothesis implies that residual cancer cells reside only in specific niches where they compete with physiological cells for occupancy. The stem cell niche would be a likely candidate. Recent data show that metastatic cancer cells may instruct the stem cell niche to adapt the microenvironment in favor of tumor cells [66, 67]. This specific anatomical entity would limit dormant tumor cell growth rate oscillations and reduce the probability of stochastic tumor outgrowth.

Cancer Stem Cells and Immuno-Escape

A possible explanation for tumor dormancy is a hierarchy inside tumors, with cancer stem cells that self renew by asymmetric division and that can reproduce the entire tumor. This concept is relatively well established for acute leukemias and CML [68–71], and there are now indications that such cells may exist in solid tumors [72]. Definitive evidence of tumor stem cells remains to be discovered, and the markers used to purify these cells remain controversial [73]; however, it is clear there is heterogeneity of cancer cells within tumors. Thus, we can hypothesize that dormant tumor cells are merely stem cells that ensure long-term persistence of the malignancy. To avoid elimination, these cells presumably develop immuno-escape mechanisms, and as there are very few of these cells, such mechanisms must be very effective. There are few data available concerning such mechanisms. A report on CML has shown that allogeneic T-cells target leukemic cells by recognizing minor histocompatibility antigens (mHags) [74]. These allogeneic cells, however, cannot target leukemic stem cells because stem cells do not express mHags. Recently, it has been shown that the hematopoietic stem cell niche is an immunoprivileged site because a large number of Tregs occupy the niche [75]. Although it remains to be demonstrated that such protective effect occurs in other niche and in a neoplastic context, this result indicates that the most primitive cancer stem cell would be difficult to eliminate through immunotherapy.

Clonal Heterogeneity of Dormant Tumor Cells as a Moving Target

If tumor stem cells do or do not exist, another type of heterogeneity also may contribute to tumor dormancy. A growing body of evidence indicates that cancer relapses occur from minor ancestral subclones already present at diagnosis. Systematic comparisons between blasts derived from acute lymphoblastic leukemia (ALL) at diagnosis and relapse have shown that less than 10% of relapses that occur form an exactly identical clone as those that produced the bulk of the disease at diagnosis [76]. More recently, CGH-array profiles of blast cells collected at diagnosis of ALL were compared with the same cells but isolated from xenografts after implantation in immunodeficient mice and to blasts from the same patients at relapse [77]. There was a genomic similarity between xenograft blast cells from diagnosis and blast cells collected at relapse, which indicates that the leukemia cells at diagnosis are composed of different subclones. Among them, those that are able to graft into mice will cause relapse in patients. Such clonal heterogeneity has also been observed in AML and other malignancies and is likely to be found in a growing number of cancers with the use of deep-sequencing technologies [78, 79]. A key question is if these minor subclones behave like the major clone in the context of the microenvironment during residual disease. Their ability to induce relapse and to graft in immunodeficient mice suggests that they may be more aggressive. Genomic analysis of ALL and AML is in favor of this hypothesis, showing additional genetic lesions when compared to the major clone. The comparison of relapse-specific vs. primary tumor mutations in AML revealed an increase in transversions, probably caused by cytotoxic chemotherapy, suggesting that clonal evolution is influenced by the drugs that the patients receive to establish and maintain remissions [78]. Thus, they are also likely resistant to the immune response (Fig. 9.5). An adequate immunotherapy regimen during the tumor dormancy period would be required to target these aggressive clones. Extensive description of the different cell population that composes the tumor at diagnosis will be required.

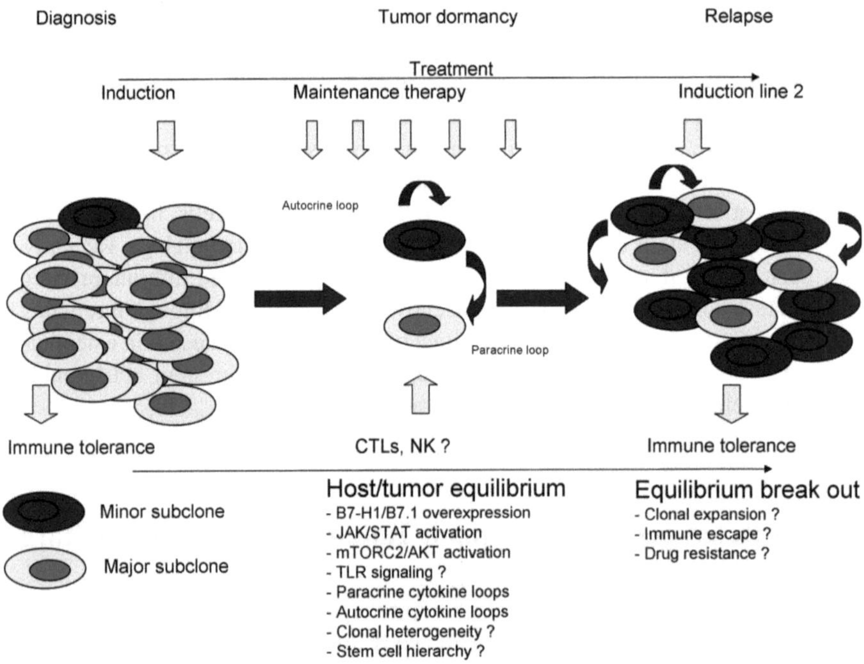

Fig. 9.5 Hypothetical survival mechanisms of dormant tumor cells. Tumor mass at diagnosis is composed of several subclones. The minor drug-resistant subclone may survive as an MRD during the tumor dormancy period where cells are continuously exposed to drugs given as maintenance therapy, and immune response mediated by CTLs and NK cells. Paracrine cytokine loops produced by the minor subclone protect residual cells from the major subclone. Dormant tumor cells develop survival mechanisms that lead to drug and immune resistance

Conclusions

There is now substantial evidence that tumor dormancy results from equilibrium between host and residual cancer. This delicate equilibrium may be affected by immune escape mechanisms resulting from active suppression of adaptive immunity and intrinsic resistance to apoptosis and drugs. Targeting immuno-escape and drug resistance to cell death could allow MRD to be eliminated. This strategy could also prolong tumor dormancy indefinitely. Such clinical results have already been obtained with drugs that can be delivered for long periods because of their favorable tolerance. It is of note that 2 years maintenance therapy with rituximab antibody in follicular lymphoma can prolong progression-free survival [80]. Some patients with CML have undergone imatinib therapy for nearly a decade and are still in complete remission. Thus, the conversion of cancer to a chronic disease maintained as tumor dormancy is possible and could be a valuable result for patients.

Acknowledgment Supported by the Institut National du Cancer (IMMUDORM Project) and Canceropole Nord-Ouest, the Ligue Nationale Contre le Cancer (Equipe labellisée et Comité du septentrion), the Association de Recherche sur le Cancer, the Fondation de France, and the Institut de Recherche sur le Cancer de Lille.

References

1. Weinberg RA (2008) The many faces of tumor dormancy. APMIS 116(7–8):548–551
2. Clevers H (2011) The cancer stem cell: premises, promises and challenges. Nat Med 17(3): 313–319
3. Naumov GN, Akslen LA, Folkman J (2006) Role of angiogenesis in human tumor dormancy: animal models of the angiogenic switch. Cell Cycle 5(16):1779–1787
4. Naumov GN, Bender E, Zurakowski D, Kang SY, Sampson D, Flynn E et al (2006) A model of human tumor dormancy: an angiogenic switch from the nonangiogenic phenotype. J Natl Cancer Inst 98(5):316–325
5. Almog N, Henke V, Flores L, Hlatky L, Kung AL, Wright RD et al (2006) Prolonged dormancy of human liposarcoma is associated with impaired tumor angiogenesis. FASEB J 20(7):947–949
6. Naumov GN, MacDonald IC, Weinmeister PM, Kerkvliet N, Nadkarni KV, Wilson SM et al (2002) Persistence of solitary mammary carcinoma cells in a secondary site: a possible contributor to dormancy. Cancer Res 62(7):2162–2168
7. Aguirre-Ghiso JA, Estrada Y, Liu D, Ossowski L (2003) ERK(MAPK) activity as a determinant of tumor growth and dormancy; regulation by p38(SAPK). Cancer Res 63(7):1684–1695
8. Quesnel B (2008) Dormant tumor cells as a therapeutic target? Cancer Lett 267(1):10–17
9. Quesnel B (2008) Tumor dormancy and immunoescape. APMIS 116(7–8):685–694
10. Decaudin D, Brousse N, Brice P, Haioun C, Bourhis JH, Morel P et al (2000) Efficacy of autologous stem cell transplantation in mantle cell lymphoma: a 3-year follow-up study. Bone Marrow Transplant 25(3):251–256
11. Meng S, Tripathy D, Frenkel EP, Shete S, Naftalis EZ, Huth JF et al (2004) Circulating tumor cells in patients with breast cancer dormancy. Clin Cancer Res 10(24):8152–8162
12. Black PC, Mize GJ, Karlin P, Greenberg DL, Hawley SJ, True LD et al (2007) Overexpression of protease-activated receptors-1,-2, and-4 (PAR-1, -2, and -4) in prostate cancer. Prostate 67(7):743–756
13. Miyamoto T, Weissman IL, Akashi K (2000) AML1/ETO-expressing nonleukemic stem cells in acute myelogenous leukemia with 8;21 chromosomal translocation. Proc Natl Acad Sci USA97(13):7521–7526
14. Saudemont A, Buffenoir G, Denys A, Desreumaux P, Jouy N, Hetuin D et al (2002) Gene transfer of CD154 and IL12 cDNA induces an anti-leukemic immunity in a murine model of acute leukemia. Leukemia 16(9):1637–1644
15. Vereecque R, Buffenoir G, Preudhomme C, Hetuin D, Bauters F, Fenaux P et al (2000) Gene transfer of GM-CSF, CD80 and CD154 cDNA enhances survival in a murine model of acute leukemia with persistence of a minimal residual disease. Gene Ther 7(15):1312–1316
16. Vereecque R, Buffenoir G, Gonzalez R, Preudhomme C, Fenaux P, Quesnel B (1999) A new murine aggressive leukemic model. Leuk Res 23(4):415–416
17. Saudemont A, Quesnel B (2004) In a model of tumor dormancy, long-term persistent leukemic cells have increased B7-H1 and B7.1 expression and resist CTL-mediated lysis. Blood 104(7):2124–2133
18. Saudemont A, Hamrouni A, Marchetti P, Liu J, Jouy N, Hetuin D et al (2007) Dormant tumor cells develop cross-resistance to apoptosis induced by CTLs or imatinib mesylate via methylation of suppressor of cytokine signaling 1. Cancer Res 67(9):4491–4498

19. Saudemont A, Jouy N, Hetuin D, Quesnel B (2005) NK cells that are activated by CXCL10 can kill dormant tumor cells that resist CTL-mediated lysis and can express B7-H1 that stimulates T cells. Blood 105(6):2428–2435
20. Vereecque R, Saudemont A, Quesnel B (2004) Cytosine arabinoside induces costimulatory molecule expression in acute myeloid leukemia cells. Leukemia 18(7):1223–1230
21. Quesnel B (2006) Cancer vaccines and tumor dormancy: a long-term struggle between host antitumor immunity and persistent cancer cells? Expert Rev Vaccines 5(6):773–781
22. Boissel N, Leroy H, Brethon B, Philippe N, de Botton S, Auvrignon A et al (2006) Incidence and prognostic impact of c-Kit, FLT3, and Ras gene mutations in core binding factor acute myeloid leukemia (CBF-AML). Leukemia 20(6):965–970
23. Davis TA, Maloney DG, Czerwinski DK, Liles TM, Levy R (1998) Anti-idiotype antibodies can induce long-term complete remissions in non-Hodgkin's lymphoma without eradicating the malignant clone. Blood 92(4):1184–1190
24. Chomel JC, Brizard F, Veinstein A, Rivet J, Sadoun A, Kitzis A et al (2000) Persistence of BCR-ABL genomic rearrangement in chronic myeloid leukemia patients in complete and sustained cytogenetic remission after interferon-alpha therapy or allogeneic bone marrow transplantation. Blood 95(2):404–408
25. MacKie RM, Reid R, Junor B (2003) Fatal melanoma transferred in a donated kidney 16 years after melanoma surgery. N Engl J Med 348(6):567–568
26. Galon J, Fridman WH, Pages F (2007) The adaptive immunologic microenvironment in colorectal cancer: a novel perspective. Cancer Res 67(5):1883–1886
27. Pages F, Berger A, Camus M, Sanchez-Cabo F, Costes A, Molidor R et al (2005) Effector memory T cells, early metastasis, and survival in colorectal cancer. N Engl J Med 353(25): 2654–2666
28. Engert A, Diehl V, Schnell R, Radszuhn A, Hatwig MT, Drillich S et al (1997) A phase-I study of an anti-CD25 ricin A-chain immunotoxin (RFT5-SMPT-dgA) in patients with refractory Hodgkin's lymphoma. Blood 89(2):403–410
29. Yefenof E, Picker LJ, Scheuermann RH, Tucker TF, Vitetta ES, Uhr JW (1993) Cancer dormancy: isolation and characterization of dormant lymphoma cells. Proc Natl Acad Sci USA90(5):1829–1833
30. Uhr JW, Tucker T, May RD, Siu H, Vitetta ES (1991) Cancer dormancy: studies of the murine BCL1 lymphoma. Cancer Res 51(18 suppl):5045s–5053s
31. Farrar JD, Katz KH, Windsor J, Thrush G, Scheuermann RH, Uhr JW et al (1999) Cancer dormancy. VII. A regulatory role for CD8+ T cells and IFN-gamma in establishing and maintaining the tumor-dormant state. J Immunol 162(5):2842–2849
32. Mahnke YD, Schwendemann J, Beckhove P, Schirrmacher V (2005) Maintenance of long-term tumour-specific T-cell memory by residual dormant tumour cells. Immunology 115(3):325–336
33. Koebel CM, Vermi W, Swann JB, Zerafa N, Rodig SJ, Old LJ et al (2007) Adaptive immunity maintains occult cancer in an equilibrium state. Nature 450(7171):903–907
34. Dong H, Strome SE, Salomao DR, Tamura H, Hirano F, Flies DB et al (2002) Tumor-associated B7-H1 promotes T-cell apoptosis: a potential mechanism of immune evasion. Nat Med 8(8):793–800
35. Iwai Y, Ishida M, Tanaka Y, Okazaki T, Honjo T, Minato N (2002) Involvement of PD-L1 on tumor cells in the escape from host immune system and tumor immunotherapy by PD-L1 blockade. Proc Natl Acad Sci USA99(19):12293–12297
36. Tirapu I, Huarte E, Guiducci C, Arina A, Zaratiegui M, Murillo O et al (2006) Low surface expression of B7-1 (CD80) is an immunoescape mechanism of colon carcinoma. Cancer Res 66(4):2442–2450
37. Keir ME, Butte MJ, Freeman GJ, Sharpe AH (2008) PD-1 and its ligands in tolerance and immunity. Annu Rev Immunol 26:677–704
38. Butte MJ, Keir ME, Phamduy TB, Sharpe AH, Freeman GJ (2007) Programmed death-1 ligand 1 interacts specifically with the B7-1 costimulatory molecule to inhibit T cell responses. Immunity 27(1):111–122
39. Yamshchikov GV, Mullins DW, Chang CC, Ogino T, Thompson L, Presley J et al (2005) Sequential immune escape and shifting of T cell responses in a long-term survivor of melanoma. J Immunol 174(11):6863–6871

40. Marches R, Scheuermann R, Uhr J (2006) Cancer dormancy: from mice to man. Cell Cycle 5(16):1772–1778
41. Uhr JW, Marches R (2001) Dormancy in a model of murine B cell lymphoma. Semin Cancer Biol 11(4):277–283
42. Berger R, Rotem-Yehudar R, Slama G, Landes S, Kneller A, Leiba M et al (2008) Phase I safety and pharmacokinetic study of CT-011, a humanized antibody interacting with PD-1, in patients with advanced hematologic malignancies. Clin Cancer Res 14(10):3044–3051
43. Robert C, Thomas L, Bondarenko I, O'Day S, Jeffrey W, Garbe C et al (2011) Ipilimumab plus dacarbazine for previously untreated metastatic melanoma. N Engl J Med 364(26):2517–2526
44. Liu J, Hamrouni A, Wolowiec D, Coiteux V, Kuliczkowski K, Hetuin D et al (2007) Plasma cells from multiple myeloma patients express B7-H1 (PD-L1) and increase expression after stimulation with IFN-{gamma} and TLR ligands via a MyD88-, TRAF6-, and MEK-dependent pathway. Blood 110(1):296–304
45. Kuranda K, Berthon C, Dupont C, Wolowiec D, Leleu X, Polakowska R et al (2010) A subpopulation of malignant CD34+ CD138+ B7-H1+ plasma cells is present in multiple myeloma patients. Exp Hematol 38(2):124–131
46. Berthon C, Driss V, Liu J, Kuranda K, Leleu X, Jouy N et al (2010) In acute myeloid leukemia, B7-H1 (PD-L1) protection of blasts from cytotoxic T cells is induced by TLR ligands and interferon-gamma and can be reversed using MEK inhibitors. Cancer Immunol Immunother 59(12):1839–1849
47. Quesnel B (2007) MEK inhibitor: the MM magic bullet? Blood 110:1402–1403
48. Tai YT, Fulciniti M, Hideshima T, Song W, Leiba M, Li XF et al (2007) Targeting MEK induces myeloma-cell cytotoxicity and inhibits osteoclastogenesis. Blood 110(5):1656–1663
49. Grellier E, Lecolle K, Rogee S, Couturier C, D'Halluin JC, Hong SS et al (2011) A fiber-modified adenoviral vector interacts with immunoevasion molecules of the B7 family at the surface of murine leukemia cells derived from dormant tumors. Mol Cancer 10:105
50. Rogee S, Grellier E, Bernard C, Jouy N, Loyens A, Beauvillain JC et al (2010) Influence of chimeric human-bovine fibers on adenoviral uptake by liver cells and the antiviral immune response. Gene Ther 17(7):880–891
51. Vereecque R, Buffenoir G, Gonzalez R, Cambier N, Hetuin D, Bauters F et al (2000) gamma-ray irradiation induces B7.1 expression in myeloid leukaemic cells. Br J Haematol 108(4):825–831
52. Rabinovich GA, Gabrilovich D, Sotomayor EM (2007) Immunosuppressive strategies that are mediated by tumor cells. Annu Rev Immunol 25:267–296
53. Nagaraj S, Gabrilovich DI (2008) Tumor escape mechanism governed by myeloid-derived suppressor cells. Cancer Res 68(8):2561–2563
54. Zhang B, Zhang Y, Bowerman NA, Schietinger A, Fu YX, Kranz DM et al (2008) Equilibrium between host and cancer caused by effector T cells killing tumor stroma. Cancer Res 68(5):1563–1571
55. Corm S, Berthon C, Imbenotte M, Biggio V, Lhermitte M, Dupont C et al (2009) Indoleamine 2,3-dioxygenase activity of acute myeloid leukemia cells can be measured from patients' sera by HPLC and is inducible by IFN-gamma. Leuk Res 33(3):490–494
56. Opitz CA, Litzenburger UM, Sahm F, Ott M, Tritschler I, Trump S et al (2011) An endogenous tumour-promoting ligand of the human aryl hydrocarbon receptor. Nature 478(7368):197–203
57. Qin Z, Schwartzkopff J, Pradera F, Kammertoens T, Seliger B, Pircher H et al (2003) A critical requirement of interferon gamma-mediated angiostasis for tumor rejection by CD8+ T cells. Cancer Res 63(14):4095–4100
58. Schrader J, Gordon-Walker TT, Aucott RL, van Deemter M, Quaas A, Walsh S et al (2011) Matrix stiffness modulates proliferation, chemotherapeutic response, and dormancy in hepatocellular carcinoma cells. Hepatology 53(4):1192–1205
59. Liu J, Joha S, Idziorek T, Corm S, Hetuin D, Philippe N et al (2008) BCR-ABL mutants spread resistance to non-mutated cells through a paracrine mechanism. Leukemia 22(4):791–799
60. Wang Y, Cai D, Brendel C, Barett C, Erben P, Manley PW et al (2007) Adaptive secretion of granulocyte-macrophage colony-stimulating factor (GM-CSF) mediates imatinib and nilotinib resistance in BCR/ABL+ progenitors via JAK-2/STAT-5 pathway activation. Blood 109(5):2147–2155
61. Joha S, Nugues AL, Hetuin D, Berthon C, Dezitter X, Dauphin V et al (2012) GILZ inhibits the mTORC2/AKT pathway in BCR-ABL(+) cells. Oncogene 31(11):1419–1430

62. Racila E, Hsueh R, Marches R, Tucker TF, Krammer PH, Scheuermann RH et al (1996) Tumor dormancy and cell signaling: anti-mu-induced apoptosis in human B-lymphoma cells is not caused by an APO-1-APO-1 ligand interaction. Proc Natl Acad Sci USA93(5):2165–2168
63. Marches R, Scheuermann RH, Uhr JW (1998) Cancer dormancy: role of cyclin-dependent kinase inhibitors in induction of cell cycle arrest mediated via membrane IgM. Cancer Res 58(4):691–697
64. Scheuermann RH, Racila E, Tucker T, Yefenof E, Street NE, Vitetta ES et al (1994) Lyn tyrosine kinase signals cell cycle arrest but not apoptosis in B-lineage lymphoma cells. Proc Natl Acad Sci USA91(9):4048–4052
65. Enderling H, Anderson AR, Chaplain MA, Beheshti A, Hlatky L, Hahnfeldt P (2009) Paradoxical dependencies of tumor dormancy and progression on basic cell kinetics. Cancer Res 69(22):8814–8821
66. Joseph J, Shiozawa Y, Jung Y, Kim JK, Pedersen E, Mishra A et al (2012) Disseminated prostate cancer cells can instruct hematopoietic stem and progenitor cells to regulate bone phenotype. Mol Cancer Res 10(3):282–292
67. Malanchi I, Santamaria-Martinez A, Susanto E, Peng H, Lehr HA, Delaloye JF et al (2012) Interactions between cancer stem cells and their niche govern metastatic colonization. Nature 481(7379):85–89
68. Eppert K, Takenaka K, Lechman ER, Waldron L, Nilsson B, van Galen P et al (2011) Stem cell gene expression programs influence clinical outcome in human leukemia. Nat Med 17(9):1086–1093
69. Chu S, McDonald T, Lin A, Chakraborty S, Huang Q, Snyder DS et al (2011) Persistence of leukemia stem cells in chronic myelogenous leukemia patients in prolonged remission with imatinib treatment. Blood 118(20):5565–5572
70. Corbin AS, Agarwal A, Loriaux M, Cortes J, Deininger MW, Druker BJ (2011) Human chronic myeloid leukemia stem cells are insensitive to imatinib despite inhibition of BCR-ABL activity. J Clin Invest 121(1):396–409
71. Perl A, Carroll M (2011) BCR-ABL kinase is dead; long live the CML stem cell. J Clin Invest 121(1):22–25
72. Li Y, Laterra J (2012) Cancer stem cells: distinct entities or dynamically regulated phenotypes? Cancer Res 72(3):576–580
73. Kuranda K, Berthon C, Lepretre F, Polakowska R, Jouy N, Quesnel B (2011) Expression of CD34 in hematopoietic cancer cell lines reflects tightly regulated stem/progenitor-like state. J Cell Biochem 112(5):1277–1285
74. von dem Borne PA, van Luxemburg-Heijs SA, Heemskerk MH, Jedema I, Mulder A, Willemze R et al (2006) Molecular persistence of chronic myeloid leukemia caused by donor T cells specific for lineage-restricted maturation antigens not recognizing immature progenitor-cells. Leukemia 20(6):1040–1046
75. Fujisaki J, Wu J, Carlson AL, Silberstein L, Putheti P, Larocca R et al (2011) In vivo imaging of Treg cells providing immune privilege to the haematopoietic stem-cell niche. Nature 474(7350):216–219
76. Mullighan CG, Phillips LA, Su X, Ma J, Miller CB, Shurtleff SA et al (2008) Genomic analysis of the clonal origins of relapsed acute lymphoblastic leukemia. Science 322(5906):1377–1380
77. Clappier E, Gerby B, Sigaux F, Delord M, Touzri F, Hernandez L et al (2011) Clonal selection in xenografted human T cell acute lymphoblastic leukemia recapitulates gain of malignancy at relapse. J Exp Med 208(4):653–661
78. Ding L, Ley TJ, Larson DE, Miller CA, Koboldt DC, Welch JS et al (2012) Clonal evolution in relapsed acute myeloid leukaemia revealed by whole-genome sequencing. Nature 481(7382):506–510
79. Greaves M, Maley CC (2012) Clonal evolution in cancer. Nature 481(7381):306–313
80. Salles G, Seymour JF, Offner F, Lopez-Guillermo A, Belada D, Xerri L et al (2011) Rituximab maintenance for 2 years in patients with high tumour burden follicular lymphoma responding to rituximab plus chemotherapy (PRIMA): a phase 3, randomised controlled trial. Lancet 377(9759):42–51

Chapter 10
A Review of Mathematical Models of Cancer–Immune Interactions in the Context of Tumor Dormancy

Kathleen P. Wilkie

Abstract The role of the immune system in tumor dormancy is now well established. In an immune-induced dormant state, potentially lethal cancer cells persist in a state where growth is restricted, to little or no increase, by the host's immune response. To describe this state in the context of cancer progression and immune response, basic temporal (spatially homogeneous) quantitative predator–prey constructs are discussed, along with some current and proposed augmentations that incorporate potentially significant biological phenomena such as the cancer cell transition to a quiescent state or the time delay in T-cell activation. Advances in cancer-immune modeling that describe complex interactions underlying the ability of the immune system to both promote and inhibit tumor growth are emphasized. Finally, the review concludes by discussing future mathematical challenges and their biological significance.

Keywords Theoretical Cancer Immunology • Tumor Dormancy • Mathematical Modeling • Cancer-Immune Interactions • Ordinary Differential Equations (ODE)

Introduction

The presence of cancer elicits a host immune response, the dynamics of which have been the subject of many current mathematical analyses. Recent technologies have yielded data and discoveries across diverse areas of cancer research which require quantitative analysis and point to the increasing importance of mathematical

---

K.P. Wilkie, PhD (✉)
Center of Cancer Systems Biology, Steward Research & Specialty Projects Corp.,
St. Elizabeth's Medical Center, Tufts University School of Medicine, 736 Cambridge St.,
Boston, MA 02135, USA
e-mail: kpwilkie@alumni.uwaterloo.ca

H. Enderling et al. (eds.), *Systems Biology of Tumor Dormancy*, Advances in Experimental Medicine and Biology 734, DOI 10.1007/978-1-4614-1445-2_10,
© Springer Science+Business Media New York 2013

modeling in this arena. Theoretical modeling can help to elucidate and interpret experimental findings and to identify the most significant biological mechanisms involved in the complex spectrum of processes driving tumor initiation, progression, and metastasis. The goal of theoretical modeling is to develop simple models that capture the essence of the varied and complex interactions involved in all stages of carcinogenesis, allowing the fundamental nature of the process to be elucidated. This knowledge is paramount to understanding cancer development, estimating cancer risk, and designing effective treatment strategies.

Instances of tumor dormancy can arise from mechanisms that are intrinsic to the tumor cell population itself, or from mechanisms that involve the interaction of tumor cells with other cell types, such as angiogenic or immune-mediated dormancy. Cellular dormancy describes cancer cells in a state of senescence or quiescence, where cellular proliferation is essentially halted. Angiogenic dormancy describes a tumor mass whose growth is essentially limited by blood and nutrient supply, which can occur either pre-vascularization (see Chap. 1) or post-vascularization (see Chap. 2). Immune-mediated dormancy describes the state where tumor growth is modulated by immune interactions that induce cancer cell death, or the transition of the cancer cell to a state of senescence or quiescence, such that net tumor growth is halted. Angiogenesis and immunomodulation are important targets for therapies since the target cells are non-transformed and therefore more genetically stable and less likely to evolve a resistance to the therapy [1]. Consequently, this review focuses on mathematical models of tumor–immune interactions aimed at improving the understanding of tumor growth and the induction of tumor dormancy. A brief biological motivation is given before proceeding into a discussion of the mathematical models used in the current literature, with particular focus on predictions and conclusions related to tumor dormancy.

Cancer and the Immune Response

The human immune system is comprised of an innate immune component and an adaptive immune component. Briefly, the innate immune system recognizes and responds to pathogens in a generic way. It recruits immune cells through cytokines, provides an immediate defense mechanism through the initiation of inflammation, activates the complement cascade (that identifies bacteria and activates T cells), removes foreign substances, and activates the adaptive immune response [2]. Innate immune cells include, among others, natural killer (NK) cells, mast cells, macrophages, neutrophils, dendritic cells, and platelets. The adaptive immune system recognizes and remembers specific pathogens, increases in efficacy each time a pathogen is encountered, requires activation before the cytotoxic cells become effective, generates a tailored immune response, develops immunological memory, and requires time to develop an effective response [2]. Adaptive immune cell types include B and T lymphocytes such as plasma cells, cytotoxic $CD8^+$ T cells, and helper $CD4^+$ T cells.

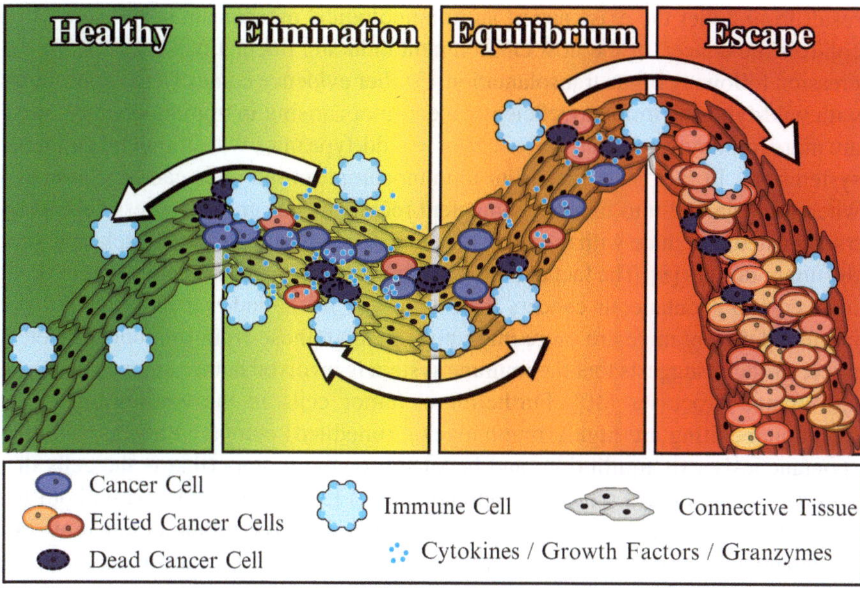

Fig. 10.1 The dynamic process of cancer immunoediting involves elimination, equilibrium, and escape

The immune response to tumor presence typically involves many different cell types and activities. The degeneration of a normal cell into a malignant cell is accompanied by the development of tumor-specific molecules on the cell membrane called tumor antigens [3, 4]. These antigens stimulate an immune response in the host. As a result, B cells are recruited to the tumor site to produce antibodies which are molecules that specifically bind to the tumor antigens either neutralizing the cell or labeling it as a target. Recruited cytotoxic T lymphocytes, natural killer cells, and macrophages can destroy tumor cells through antibody–antigen bonding (T cells), release of granzymes (NK and T cells), or phagocytosis (macrophages) [5]. Neutrophils, tumor-associated macrophages, $CD4^+$ T cells, and other immune cell types are also recruited, which infiltrate the tumor site, induce inflammation, and modify the tumor microenvironment [6, 7].

As the immune system interacts with cancer cells, it may sculpt the cell phenotype, developing a less immunogenic variant that facilitates tumor growth and immune evasion. This process is known as immunoediting and it consists of three stages: elimination, equilibrium, and escape [3]. In the elimination stage, cancer cells are successfully recognized and eliminated by the immune system. Cancer cells that are not completely eliminated progress to the equilibrium stage, where the immune system may sculpt the tumor phenotype by allowing the survival of less immunogenic cells [8]. In the third stage, cancer cells escape immune control and rapidly expand. See Fig. 10.1 for a schematic of the immunoediting process.

Perhaps one of the most convincing pieces of evidence suggesting immune-mediated tumor dormancy is the increased incidence of cancers in immune-suppressed recipients following organ transplants [3, 8, 9]. These clinical observations may result from either the existence of an immunological mechanism in the donor that controls the cancer, or the host environment inducing fluctuations in oncogene expression following organ transplantation. Further evidence comes from experiments with wild-type and immunodeficient mice: tumors arising in immunodeficient mice are more immunogenic than those arising in wild-type mice with competent immune systems [8]. In fact, sarcomas from immunocompromised mice can be eliminated when transplanted into immunocompetent recipients but grow progressively when transplanted into mice with the same immune deficiency and genetic background as the tumor donors [8]. The fact that immunodeficient mice develop more carcinogen-induced and spontaneous cancers than wild-type mice, and that tumor cells from immunodeficient mice are more immunogenic than those from immunocompetent mice, strongly suggests that the immune system is involved in the tumor elimination and escape processes [10]. Furthermore, tumor cells in the equilibrium phase of immunoediting are highly immunogenic (unedited) compared to the cells that spontaneously exit equilibrium and become growing tumors (that is the cells that progress to the escape phase of immunoediting) [3]. Interestingly, while both the innate and adaptive immune responses are required for tumor elimination [4], the adaptive immune response plays the most significant role in maintaining the dormant state [10].

Tumor dormancy has been observed in many different types of cancer and is often found in tissue samples from individuals who died from other non-cancer-related causes [11–13]. The equilibrium phase of tumor development is a subclass of immune-mediated tumor dormancy wherein the immune response balances tumor growth and is responsible for the maintenance of a consistent tumor size. Dormancy can occur during tumor development and after successful remission, as tumor cell populations can persist undetected following primary tumor excision. Bone marrow and lymph nodes have been shown to be particular sites where potentially lethal tumor cells are controlled in a dormant state specifically by the immune system [14]. For example, in an animal model of B cell lymphoma, dormant tumors were found to persist in the spleen (a hematopoietic organ and member of the lymphatic system) at around one million cells, with the majority of cells existing in a quiescent state [15, 16]. Some of the cells were proliferating, but growth was kept in check by an equal rate of cell death.

Dormant avascular tumors were described by Folkman [17], in conjunction with his seminal work on tumor angiogenesis. As the immune system is responsible for wound healing, a process that includes the production of pro-angiogenic growth factors, the immune system may play a role in the vascularization of dormant avascular tumors, leading to tumor regrowth. In fact, the similarities between the wound healing process and cancer progression prompted Dvorak to describe tumors as wounds that do not heal [18]. Escape from immune-mediated dormancy may be due to a reduced expression of tumor-associated antigens and other mechanisms for tumor evasion of the immune system [19]. Stress, trauma, and old age are other mechanisms that can potentially disrupt the balance maintaining a dormant state.

Within the immune response to tumor presence, the many different types of immune cells, and the cytokines and chemokines that they produce for communication and recruitment purposes, create a complex integrated network of interactions

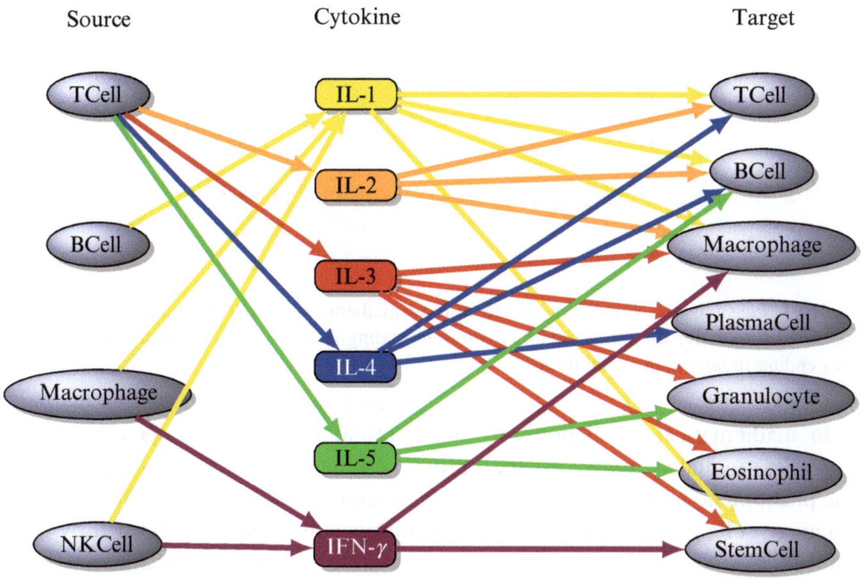

Fig. 10.2 A small sample of the known cytokine cross-communication that occurs between various immune cells. Note that here stem cells refer to hematopoietic stem cells. Adapted from [20]

(see Fig. 10.2). Describing each of these interactions explicitly in a single model produces a system that, though numerically soluble, is too complicated to be useful, due to the vast numbers of equations and model parameters. The large parameter space not only hinders the understanding of the parameter dependence but also obscures biological interpretation. Thus, if simpler models of the immune response can be created which capture the essence of the fundamental underlying processes, they will prove to be more useful in practice. This review focuses on spatially homogeneous, ordinary differential equation-based models that provide a simple analytical framework within which these complex biological interactions can be explored.

Tumor Dormancy and Mathematical Modeling

A period of tumor dormancy is defined as an interval of time over which a tumor does not grow or shrink significantly. In mathematical terms, this can either be the case where the overall growth or regression rate is very small (approximately equal to zero as in an equilibrium solution), or the case where small oscillations occur in the tumor size over time, but on average the size remains approximately constant (as in a limit cycle). Theoretical models of tumor–immune interactions (as shown in Fig. 10.3) can be used to investigate the mechanisms behind tumor dormancy when immune cell activities and treatments are considered.

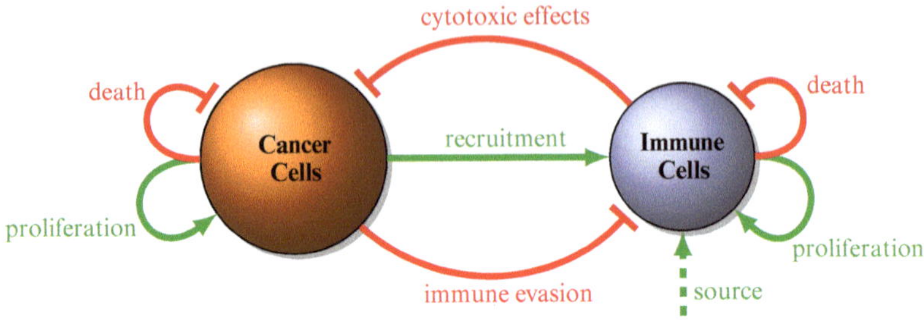

Fig. 10.3 Possible interactions considered in most mathematical models of tumor growth in the presence of an active immune system. *Green lines* ending in *arrows* indicate stimulation and *red lines* ending in *bars* indicate inhibition

In mathematical simulations of cancer–immune interactions, possible tumor outcomes include: tumor escape, tumor elimination, and tumor dormancy. The phenomenon known as *sneaking through* refers to the ability of a small number of cancer cells (usually injected) to form tumors similar to those formed by large numbers of cancer cells, and where medium numbers of cancer cells are rejected by the hosts immune response [21]. The small number of cancer cells may initially exist beneath the detection threshold of the immune system, allowing them to progress and form a viable tumor. This threshold is called the *immune barrier*. On one side of the immune barrier *sneaking through* is possible, and on the other side cancer cells are detected and possibly eliminated by the immune response.

Ordinary differential equation models are the focus of this review as they provide a simple framework within which the complex dynamics of the cancer–immune interactions can be studied. Other mathematical methods include stochastic differential equation-based models [22], models based on the kinetic theory for active particles [23–25], spatiotemporal models [26, 27], and cellular automaton models [27, 28].

Existing contributions in the literature for mathematical models of solid tumor growth and tumor–immune interactions are many and varied. A comprehensive review of this literature is a daunting, task. As a result, literature reviews tend to focus their efforts on specific aspects of tumor growth and immune interactions. Araujo and McElwain [29] provide an historical review of solid tumor modeling with connections to experimental observations. A review of the varied mathematical techniques used in tumor–immune interaction models, aimed at clinicians and experimental researchers, is given by Woelke et al. [30]. Roose et al. [27] present a review of avascular tumor growth including both continuum or partial differential equation-based models and discrete or cellular automata-based models. Bellomo et al. [31] review discrete tumor–immune cellular competition models using lattice Boltzmann methods based on classical kinetic theory. And a review of ordinary differential equation-based models for cancer–immune interactions was recently provided by Eftimie et al. [32]. This chapter discusses some of the main approaches to cancer–immune interaction models using ordinary differential equations with an emphasis on a few models specific to tumor dormancy.

Equations for Tumor Growth

The simplest models used to describe tumor growth dynamics in the absence of an immune response involve only one equation. Both avascular and vascular tumors can be modeled in this manner.

Consider a population of cancer cells of size $C(t)$ at time t described by Fig. 10.4. If these cells proliferate at a rate proportional to their population size, say $gC(t)$, and die at a rate proportional to their population size, say $dC(t)$, then the overall change of the population size can be described by the differential equation

$$\frac{\mathrm{d}C}{\mathrm{d}t} = gC(t) - dC(t) = \alpha C(t), \tag{10.1}$$

where $\alpha = g - d$. This equation describes the process of exponential growth if $\alpha > 0$ (or $g > d$) and exponential decay if $\alpha < 0$ (or $g < d$). Such an equation can describe the growth dynamics of a small population of cells that grows rapidly due to an abundance of nutrients and a lack of contact inhibition. This equation, however, fails to predict growth for large vascular tumors, which are observed to grow at an increasingly slower rate [33].

To construct a more general framework, as done by d'Onofrio [34], assume that the growth and death rates are not constant, but instead are functions of the cell population size. Then, a generalized equation for the growth of a tumor can be written as

$$\frac{\mathrm{d}C}{\mathrm{d}t} = \alpha f(C)C(t), \tag{10.2}$$

where $f(C)$ is usually a positive function over the domain of interest, representing tumor growth. Note that if $f = 1$ this model simplifies to the exponential model, Eq. (10.1).

Self-limited Growth

Empirical evidence suggests that tumors grow rapidly from a single cell, or small collection of cells, to a size detectable by current imaging or palpation techniques, and then continue to grow at a slowly decreasing rate [33]. This type of growth is described as self-limited because the growth rate slowly decreases to zero, resulting theoretically in a maximum tumor size. From the generalized model

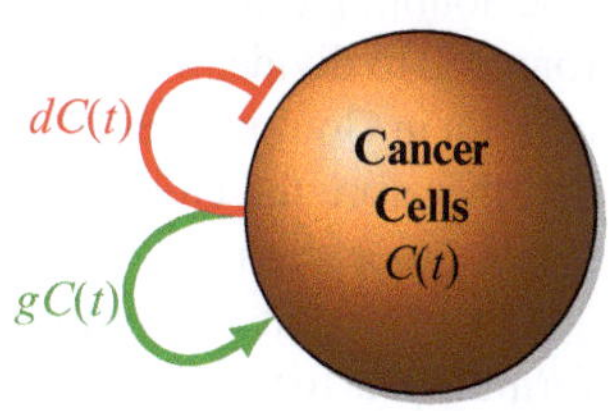

Fig. 10.4 A cancer cell population of size $C(t)$ proliferates at a rate $gC(t)$ and cell death occurs at a rate $dC(t)$, resulting in a net population growth rate of $\alpha C(t)$ where $\alpha = g - d$

introduced above, Eq. (10.2), a decreasing growth rate implies that $f(C)$ is a decreasing function, or $f'(C) \leq 0$. Biological explanations for such self-limited growth dynamics can include the increasing burden on nutrient supply and cell inhibition as a result of cell contact or crowding.

An early equation used to describe the self-limited growth characteristic of tumors is the Gompertz equation. Laird [35] used this equation to fit the growth kinetics of a variety of primary and transplanted tumors. The differential equation describing Gompertzian tumor growth is

$$\frac{\mathrm{d}C}{\mathrm{d}t} = \alpha \ln\left(\frac{K}{C}\right) C, \tag{10.3}$$

where K is the maximum size, or carrying capacity, of the cancer cell population. In the generalized framework, this is the special case where $f(C) = \ln\left(\frac{K}{C}\right)$.

Other popular choices to describe self-limited tumor growth are logistic growth,

$$\frac{\mathrm{d}C}{\mathrm{d}t} = \alpha \left(1 - \frac{C}{K}\right) C, \tag{10.4}$$

with $f(C) = \left(1 - \frac{C}{K}\right)$, or generalized logistic growth,

$$\frac{\mathrm{d}C}{\mathrm{d}t} = \frac{\alpha}{\upsilon} \left(1 - \left(\frac{C}{K}\right)^{\upsilon}\right) C, \tag{10.5}$$

with $f(C) = \frac{1}{\upsilon}\left(1 - \left(\frac{C}{K}\right)^{\upsilon}\right)$. Generalized logistic growth incorporates the Gompertz equation in the limit as $\upsilon \to 0$. Similar to Gompertzian growth, logistic growth limits the tumor growth rate through a self-regulated feedback mechanism from the environment, namely the carrying capacity K. Usually K is assumed to be a constant, but it may be more accurate and informative to take it as a function of time, increasing in response to tumor-produced stimulatory and inhibitory growth factors [36].

Gompertzian growth exhibits an exponential retardation of growth. This feature, not incorporated in logistic growth, makes it a useful model for describing normal growth kinetics over a large growth range. Logistic growth, however, is now generally preferred over Gompertzian growth due to the infinite growth rate predicted by Gompertzian growth at very small tumor sizes ($\lim_{C\to 0} f(C) = +\infty$ for Gompertzian growth compared to $\lim_{C\to 0} f(C) = 1$ for logistic growth) [34]. The doubling time of a population of cells cannot be smaller than the time required for a cell to divide, which is approximately equal to 1 day. Thus, the unboundedness of $f(C)$ in Gompertzian growth is not consistent with this biological constraint [37], and therefore the Gompertz equation should be used with caution when modeling small tumor sizes.

Other models for tumor growth have been proposed, mostly using a power law formulation for $f(C)$ in our general framework, Eq. (10.2). Examples include the

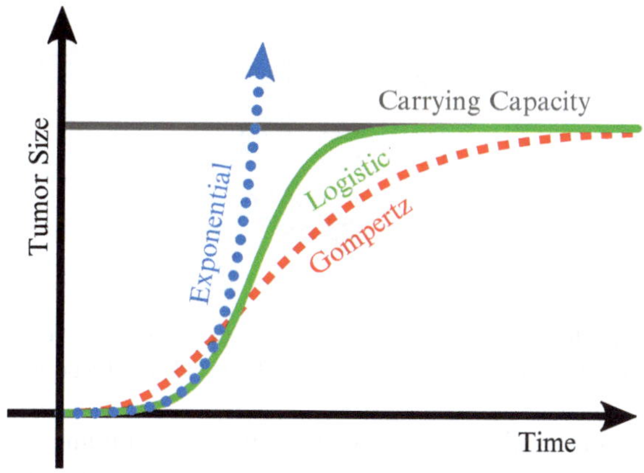

Fig. 10.5 The common choices for tumor growth models include exponential growth (*blue dotted line*), Gompertzian growth (*red dashed line*), and logistic growth (*green solid line*). The self-limited growth predicted by the Gompertz and logistic curves both have a finite carrying capacity

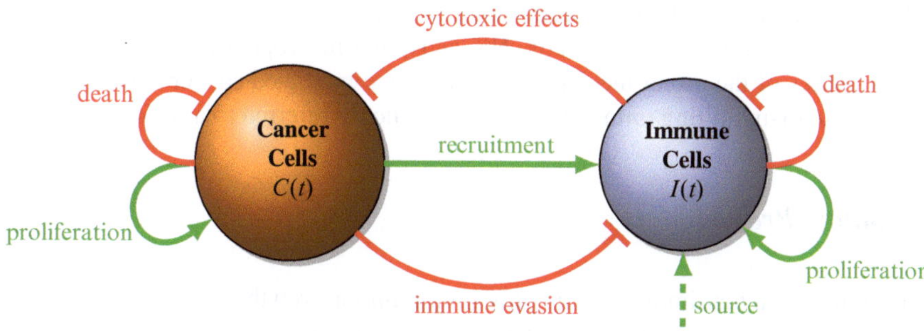

Fig. 10.6 A cancer cell population of size $C(t)$ and an immune cell population of size $I(t)$ interact with each other through stimulatory and inhibitory actions

von Bertanlaffy model [38], the Hart–Schochat–Agur model [39], and the Guiot et al. model [40]. The three main models of tumor growth, exponential, Gompertz, and logistic, are shown in Fig. 10.5.

Models of Tumor–Immune Interactions

The concept of cancer immunosurveillance suggests that the immune system can recognize and, in some cases eliminate, the precursors of cancer. Experimental evidence (see Zitvogel et al. [41] for a review) suggests that immunosurveillance has a significant role in cancer development and in anticancer therapy. Mathematical models can be used to elucidate the mechanisms involved in the immune system's response to cancer presence.

To incorporate the actions of a host's immune response to cancer into a model, a differential equation describing the immune cell population, $I(t)$, is required in addition to the equation describing the cancer cell population, $C(t)$. These two equations form a system of differential equations for the model's state space described by $(C(t), I(t))$: see Fig. 10.6 for a model schematic.

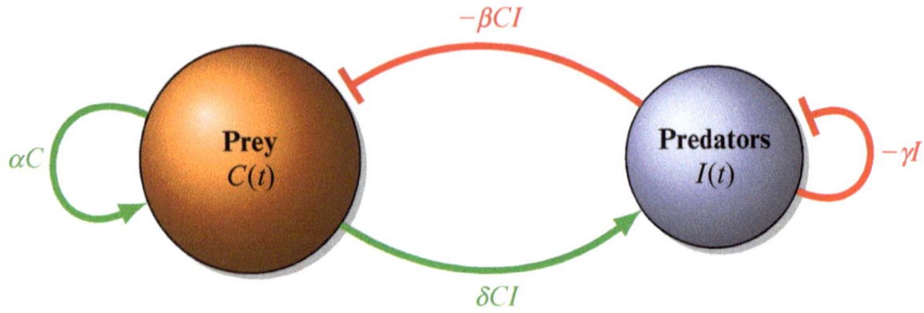

Fig. 10.7 The classic Lotka–Volterra predator–prey model. The prey population, $C(t)$, and predator population, $I(t)$, interact with each other through stimulatory and inhibitory actions

Generally, in mathematical models, immune cells are assumed to only have inhibitory effects on cancer cells. That is, immune cells are considered to only interact with cancer cells in a direct tumor-suppressing manner, such as cell death induced by cytotoxic $CD8^+$ T cells, NK cells, and macrophages. See [41, Fig. 2] for examples of the direct mechanisms of immune-mediated cancer cell lysis. The various tumor-inhibiting immune cells are commonly referred to as effector cells.

Predator–Prey Models

The assumption that immune cells only inhibit tumor growth leads to the common perception of immune cells as predatory cells with the cancer cells as their prey. In mathematics, such models are commonly referred to as Lotka–Volterra models. The classic Lotka–Volterra equations are

$$\frac{\mathrm{d}C}{\mathrm{d}t} = \alpha C - \beta CI, \tag{10.6a}$$

$$\frac{\mathrm{d}I}{\mathrm{d}t} = \delta CI - \gamma I, \tag{10.6b}$$

where $C(t)$ are the prey and $I(t)$ are the predators. Each term can be interpreted as follows: αC is the growth rate of the prey with ample nutritional supply, βCI is the death rate of the prey induced by the predators, δCI is the predator growth rate which is dependent on the prey population, and γI is the natural death rate of the predators. See Fig. 10.7 for a diagram of this model.

These equations typically have periodic solutions with the predator oscillations delayed compared to the prey oscillations, see Fig. 10.8. The equilibrium points of the system can be found by setting $\frac{\mathrm{d}C}{\mathrm{d}t} = \frac{\mathrm{d}I}{\mathrm{d}t} = 0$ which gives two equilibria, $(C,I) = (0,0)$ and $(C,I) = (\frac{\gamma}{\delta}, \frac{\alpha}{\beta})$. The first equilibrium point is the trivial solution and represents the cancer-free state in the cancer–immune system. The second equilibrium point represents an immune-induced dormant state. Periodic solutions will oscillate around this second equilibrium point in closed cycles determined by the initial conditions and the invariant $z = \alpha \ln I(t) - \beta I(t) - \delta C(t) + \gamma \ln C(t)$.

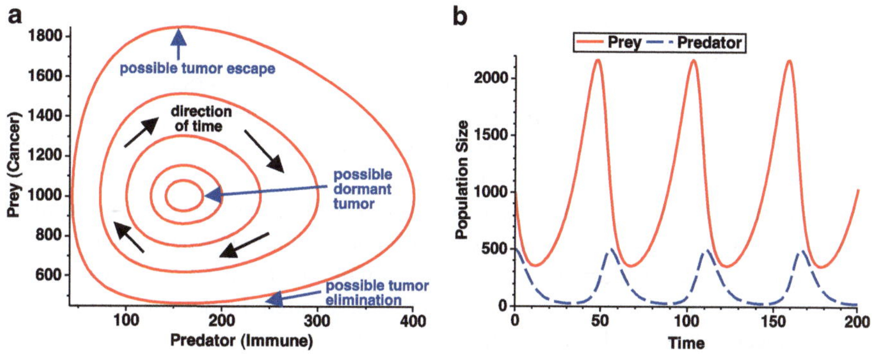

Fig. 10.8 An example of the Lotka–Volterra predator–prey model showing closed cycles around an equilibrium point. A phase portrait is shown in (**a**). Each *curve* represents the state of the system (predator, prey), which evolves over time. The direction of increasing time is indicated by the *black arrows*. A time series plot for both the predator population (*dashed blue curve*) and prey population (*solid red curve*) is shown in (**b**). Model parameters are $\alpha = 0.08$, $\beta = 0.0005$, $\delta = 0.0002$, and $\gamma = 0.2$

That is, each cycle is a level curve of the surface defined by z, and the dormant state is the peak of this surface. For the tumor–immune system, this model does not adequately predict tumor escape or elimination, as most solutions are cycles. Very large amplitude cycles may be interpreted as tumor escape if the cancer population becomes large enough that it may be considered lethal. Or, they may be interpreted as tumor elimination if the cancer population becomes small enough that it is effectively eliminated by the immune system. Small amplitude cycles may be considered dormant tumors since the tumor size does not significantly change over time. See Fig. 10.8 for an example of a Lotka–Volterra phase portrait.

The model proposed by Bell in 1973 [42] simplifies to a form similar to the Lotka–Volterra model. Bell initially derived the model for antigen (the prey or cancer cell) and antibody (the predator or immune cell) concentrations. In the notation used here, we write

$$\frac{\mathrm{d}C}{\mathrm{d}t} = \alpha C - \beta \frac{CI}{1+C+I} \tag{10.7a}$$

$$\frac{\mathrm{d}I}{\mathrm{d}t} = \delta \frac{CI}{1+C+I}(1-\nu I) - \gamma I. \tag{10.7b}$$

If $\nu = 0$, for unlimited antibody production, these equations take on a similar form to the predator–prey system, Eq. (10.6). In the Lotka–Volterra model, the prey growth rate is given by $\frac{1}{C}\frac{\mathrm{d}C}{\mathrm{d}t} = \alpha - \beta I$ which becomes arbitrarily large in magnitude for large I. Similarly, $\frac{1}{I}\frac{\mathrm{d}I}{\mathrm{d}t}$ becomes arbitrarily large for large C. This implies that the rate of prey decrease (or predator increase) becomes arbitrarily large as the concentration of predator (or prey) becomes large. In Bell's model, however, there are finite limits on the population growth rates, which may be more physical. For example, as I becomes arbitrarily large, $\frac{1}{C}\frac{dC}{dt} \to \alpha - \beta$, which imposes a bound on the growth rate.

In contrast to the Lotka–Volterra equations, which have periodic solutions for positive initial conditions (except for equilibrium points), Bell's model (with $\nu = 0$) only has periodic solutions when $\beta = \delta$. The cancer (or antigen) will proliferate uncontrolled if $\beta\delta - \beta\gamma - \delta\alpha \leq 0$. If $\beta\delta - \beta\gamma - \delta\alpha > 0$, the cancer and immune populations will tend to oscillate, and either converge on a steady-state (or dormant tumor) when $\beta < \delta$, or diverge to large amplitude oscillations when $\beta > \delta$ [42]. In this final case, the large amplitude oscillations will either cause host death through an uncontrolled tumor mass or tumor elimination through the destruction of a very small cancer population by the immune response.

Incorporating More Complex Interactions

Typical interactions between cancer and immune cells are more complex than Lotka–Volterra type models allow. For example, the models do not account for a direct source of immune cells through recruitment from the blood and bone marrow. They also do not account for immune cell death due to interactions with cancer cells. This section discusses extensions of the predator–prey type models that incorporate more complex interactions of the cancer–immune system and implications of these new models for tumor dormancy.

A simple model that incorporates more mechanisms for cancer–immune interactions was proposed by Sotolongo-Costa et al. [43]. In their model, the cancer population is assumed to grow exponentially, αC, and cancer cell death is assumed to occur at a rate proportional to the interaction of cancer cells with immune cells (or cytotoxic lymphocyte effectors), $-\beta CI$. This is equivalent to the prey equation of the Lotka–Volterra model, Eq. (10.6a). The immune population is assumed to grow at a rate proportional to the interaction of cancer cells with immune cells, δCI, and to have a constant supply rate, σ. Immune cell death is assumed to occur at a natural rate proportional to the population size, $-\gamma I$, and in response to the growing tumor, $-\kappa C$. This last term may include T cell inactivation or other immunosuppressive mechanisms originating from the tumor. Together, these assumptions form the Sotolongo-Costa et al. model given below

$$\frac{dC}{dt} = \alpha C - \beta CI \tag{10.8a}$$

$$\frac{dI}{dt} = \delta CI - \gamma I - \kappa C + \sigma. \tag{10.8b}$$

Rescaling effectively sets $\beta = \delta = 1$ and $\alpha = \frac{1}{\gamma}$. With respect to the rescaled model, Sotolongo-Costa et al. [43] showed that if $\sigma < \kappa\gamma$, the tumor escapes immune control, but if $\kappa\gamma < \sigma < 1$, the system approaches a dormant cancer state through damped oscillations. In other words, if the source of immune cells

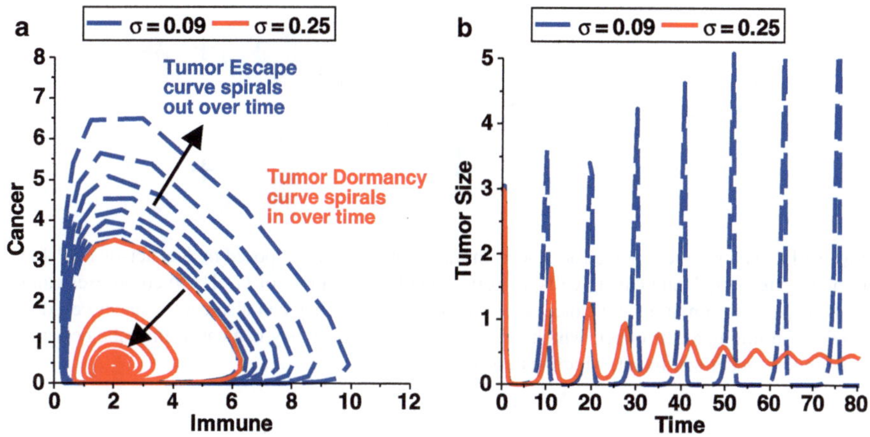

Fig. 10.9 Simulation of the rescaled Sotolongo-Costa et al. model [43] given by Eq. (10.8). A phase portrait is shown in (**a**) and a time plot is shown in (**b**) for two values of the immune source parameter σ. The *blue dashed curves* represent a tumor escaping immune control (with parameter value $\sigma = 0.09 < \kappa\gamma$), and the *red solid curves* represent a tumor approaching an immune-induced dormant state (with parameter value $\sigma = 0.25 > \kappa\gamma$). Other parameter values used are $\gamma = 0.5$, $\kappa = 0.2$, $C(0) = 3$ and $I(0) = 1$

is too small, the tumor will grow unbounded, but if the source is sufficient but not unlimited, the tumor will be controlled to a dormant state. If $\sigma > 1$, then the immune source is strong and the tumor is eliminated. See Fig. 10.9 for simulations of this model demonstrating tumor escape and tumor dormancy.

Analysis of this system in the presence of periodic immunotherapy treatment is also presented by Sotolongo-Costa et al. [43]. The model predicts short-term oscillations of tumor size and long-term tumor relapse. Parameters for the frequency and dose amplitude of the immunotherapy treatment are found to be significant in controlling tumor outcome. In cases where dormant tumors are predicted, the tumor size can be reduced by increasing the amplitude or dose of the immunotherapy treatment.

Taking a different approach, Kuznetsov et al. [44] consider the interactions between immune effector cells and cancer cells to follow the kinetic scheme shown in Fig. 10.10. Activated cytotoxic T cells and NK cells are denoted by I, cancer cells are denoted by C, and I^* and C^* denote dead or inactivated immune and cancer cells, respectively.

By imposing the condition that immune cells and cancer cells must first form a complex before cytolysis of either cell can occur, Kuznetsov et al. [44] derive the following model:

$$\frac{\mathrm{d}C}{\mathrm{d}t} = \alpha C\left(1 - \frac{C}{K}\right) - \beta CI \tag{10.9a}$$

$$\frac{\mathrm{d}I}{\mathrm{d}t} = \delta\frac{CI}{\nu + C} - \gamma(1 + \xi C)I + \sigma. \tag{10.9b}$$

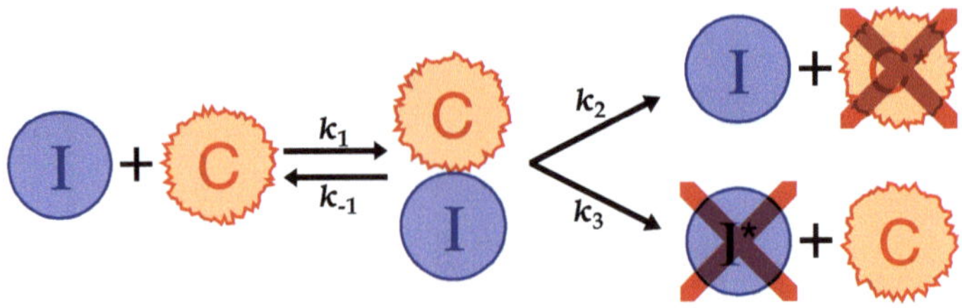

Fig. 10.10 The kinetic scheme for cancer–immune interactions proposed by Kuznetsov et al. [44]. Immune effectors, I, and cancer cells, C, form complexes at a rate of k_1. These complexes can be broken apart into immune and cancer cells at a rate of k_{-1}. They can also cause cancer cell death leaving the immune cell intact, which occurs at the rate k_2, or they can cause immune cell death leaving the cancer cell intact, which occurs at the rate k_3

This model can have zero, one, or three equilibrium points but no closed orbits or limit cycles. Tumor dormancy is predicted through damped oscillations, or an inward spiral in the phase portrait. The phenomenon of *sneaking through* is demonstrated nicely, as an increase in initial immune presence can change the prediction of a dormant tumor to one that appears to become dormant but that eventually escapes. Parameters identified as significant include γ, the immune cell death rate, and σ, the baseline immune source rate. From their analysis, Kuznetsov et al. suggest that there may be a connection between the phenomena of immunostimulation of tumor growth, *sneaking through* of a tumor, and the state of tumor dormancy, even though the mechanisms leading to these phenomena are unknown and most likely diverse.

The ratio of $\frac{k_3}{k_2}$, or $\frac{\gamma\xi}{\beta}$ in the model parameters, is identified as a critical ratio particularly in the phenomena of *sneaking through* and immunostimulation. This ratio relates the rate at which immune cells are inactivated by cancer cells to the rate at which cancer cells are inactivated by immune cells. A small ratio implies that $k_3 \ll k_2$ which implies that tumors do not effectively inactivate immune cells. The model proposed by Kuznetsov et al., Eq. (10.9), only predicts immunostimulation and *sneaking through* when this ratio is above a certain threshold, which suggests that a small immune inactivation rate is required for these phenomena to occur. This intriguing result suggests that some form of tumor-mediated immune inhibition is necessary for the process of immune-mediated tumor stimulation.

Building on the model of Kuznetsov et al. [44] from Fig. 10.10, Joshi et al. [45] also require the immune cells to form complexes with the cancer cells before cytolysis is induced in either but not both cells of the complex. They investigate the role of antigen presentation and co-stimulatory signaling pathways on cancer dormancy and recurrence. From their model results, they identify a few model parameters that modulate the dormancy period but that may also result in larger tumor burdens upon recurrence.

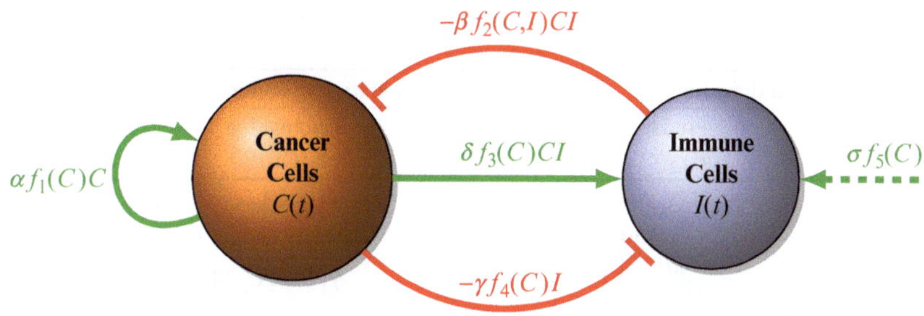

Fig. 10.11 A generalized model of cancer–immune interactions given by Eq. (10.10). The cancer population, $C(t)$, and immune population, $I(t)$, interact with each other through stimulatory and inhibitory actions. Each action of the immune response may depend on the cancer population

Cancer–Immune Model Generalization

Many of the mathematical models in the literature can be expressed as special cases of the general model proposed by d'Onofrio [34], which we write as

$$\frac{\mathrm{d}C}{\mathrm{d}t} = \alpha f_1(C)C - \beta f_2(C,I)CI \tag{10.10a}$$

$$\frac{\mathrm{d}I}{\mathrm{d}t} = \delta f_3(C)CI - \gamma f_4(C)I + \sigma f_5(C). \tag{10.10b}$$

Here, $\alpha f_1(C)C$ is the growth rate of the cancer population and $\beta f_2(C,I)CI$ is the predation rate mediated by the immune response. For the immune population, $\delta f_3(C)CI$ is the proliferation and recruitment rate, $\gamma f_4(C)I$ is the death and inactivation rate, and $\sigma f_5(C)$ is the source or supply rate, all of which may depend on the tumor size. A diagram of this model is shown in Fig. 10.11.

Common functional forms used in Eq. (10.10) are given in Table 10.1. As discussed in the section "Equations for Tumor Growth," tumor growth rates are commonly assumed to follow exponential growth, Gompertzian growth, logistic growth, or generalized logistic growth. Generalizations of Gompertzian growth [37, 46] are less commonly used. The immune predation rate of cancer cells is commonly assumed to be proportional to the interaction of the two populations, as used by Stepanova [47] and others [45,48], or to saturate to a maximum possible rate for very large tumor size, C. These saturating predation rates are based on the Michaelis–Menten form of $\frac{\beta C}{\eta + C}$. In a similar manner, DeLisi and Rescigno [49] assume that the tumor geometry provides protection from immune predation and prescribe a predation rate that depends on the immune response, $f_2(C,I) = \frac{C^{-1/3}}{1+I}$. Thus, as the immune response grows large, the predation rate approaches $-\beta C^{2/3}$.

Table 10.1 Common functional forms for the generalized model Eq. (10.10) of tumor–immune interactions

Cancer population dynamics		Immune population dynamics		
Growth	Predation	Recruitment	Inactivation	Influx
$\alpha f_1(C)$	$\beta f_2(C,I)$	$\delta f_3(C)$	$\gamma f_4(C)$	$\sigma f_5(C)$
α [L–V]	β [L–V]	δ [L–V]	γ [L–V]	0 [L–V]
$\alpha \ln\left(\frac{K}{C}\right)$ [35]	$\frac{\beta}{\eta+C}$ [50]	$\frac{\delta}{\nu+C}$ [44]	$\gamma(1+\xi C^2)$ [47]	σ [47]
$\alpha\left(1-\frac{C}{K}\right)$	$\frac{\beta I^\lambda}{\eta C^\lambda + I^\lambda}\frac{1}{I}$ [51]	$\frac{\delta C}{\nu+C^2}$ [51]	$\gamma(1+\xi C)$ [44]	$\sigma(1+\rho C)$ [50]
$\frac{\alpha}{\upsilon}\left(1-\left(\frac{C}{K}\right)^\upsilon\right)$ [52]				$\sigma(1-\rho C)$ [43]

Note that α, β, δ, γ, σ, η, ν, ξ, ρ, and K are positive real numbers and that υ is any real number. The Lotka–Volterra predator–prey model is denoted by [L–V]

The rate of immune effector proliferation and recruitment is commonly assumed to be proportional to the tumor size or to saturate to a maximum rate for very large tumor size. Immune death and inactivation rates typically include a death rate proportional to the immune response and an inactivation rate proportional to the sizes of both the immune response and the tumor. The immune influx or source rate may be constant, proportional to the tumor size, or it may saturate for very large tumor size.

For example, the model proposed by Kirschner and Panetta [50] assumes logistic tumor growth and $\beta f_2(C,I) = \frac{\beta}{\eta+C}$, $\gamma f_4(C) = \gamma$, and $\sigma f_5(C) = \sigma(1+\rho C)$. They introduce the concentration of cytokine IL-2 as a third model variable that mediates the activation and recruitment of immune cells. Thus, $\delta f_3(C) = \frac{\delta \text{IL-2}}{\nu+\text{IL-2}}\frac{1}{C}$, which saturates to a maximum rate of δ when the concentration of IL-2 becomes large. Their results suggest that the interactions between cancer cells, immune cells, and the cytokine IL-2 can explain both short-term oscillations in tumor size and long-term tumor relapses.

Arciero et al. [53] build on this model and consider not only a cytokine that stimulates the immune system (IL-2) but also a cytokine that suppresses the immune system (TGF-β). This second cytokine also stimulates tumor growth and is the target of an siRNA treatment. Their findings suggest that the tumor antigenicity parameter controls the dormant state. Small antigenicity allows a single unstable steady state for passive tumors characterized by sustained oscillations with periods of large tumor burden and longer periods of dormancy. As the antigenicity increases, however, the amplitude and period of the limit cycles decrease, eventually leading to a stable dormant state approached through damped oscillations. The imposed siRNA treatment is assumed to inhibit TGF-β production and thus remove the immune escape mechanism used by aggressive tumors. Their model suggests that this treatment may lead to controlled oscillatory tumor behavior but that a dormant state will not persist.

In 2001, de Pillis and Radunskaya [48] proposed a model that considers the interactions between tumor, immune, and normal cells. Immune recruitment is modeled with $f_3(C) = \frac{1}{\nu+C}$, $f_4(C) = 1+\xi C$, and $f_5(C) = 1$. Later, de Pillis et al. [51] separate the cytotoxic actions of NK cells and T cells. For NK cells they prescribe $f_3(C) = \frac{C}{\nu + C^2}$, but for T cells they prescribe $f_3(C,I) = \frac{D^2}{\nu + D^2}\frac{1}{C}$, where $D = \frac{dI^{\lambda}C}{hC^{\lambda}+I^{\lambda}}$. The cancer cell predation rates are modeled by βCI for NK cells, and by D for T cells. They showed that the fractional cell kill term, D, provides a better fit to experimental data for T cell-mediated cell lysis. Their simulations use different sets of parameter values at different times. This caused some confusion [54, 55] but may be more physical as the immune system evolves over time. In 2010, d'Onofrio [56] also separated the innate and adaptive immunity by prescribing the innate predation rate as βC and the adaptive predation rate as $\frac{\beta C^2}{\eta + C^2}$.

In 2004, de Vladar and González [57] modified the Stepanova model [47] by replacing exponential growth with Gompertzian growth,

$$\frac{\mathrm{d}C}{\mathrm{d}t} = -\alpha C \log \frac{C}{K} - \beta CI \tag{10.11a}$$

$$\frac{\mathrm{d}I}{\mathrm{d}t} = \delta CI - \gamma(1 + \xi C^2)I + \sigma. \tag{10.11b}$$

For immune-induced tumor dormancy, the model predicts a corresponding equilibrium point which can be reached if the immune response is sufficiently fast. This system has two bifurcation points. If $\frac{\alpha}{\beta}$ is small only one equilibrium point exists representing a small dormant tumor. If $\frac{\alpha}{\beta}$ increases, two new equilibrium points arise (for a total of three), one representing a large, presumably lethal, tumor, and one representing a dormant tumor of a size in between these other two. Finally, as $\frac{\alpha}{\beta}$ increases further, the smaller two equilibrium points are annihilated leaving only the large, lethal, equilibrium point. The smallest of these equilibria corresponds to the immunosurveillance hypothesis, and thus, this model suggests that immunosurveillance depends on both the tumor growth rate and the predation rate.

When modeling very small tumors or the immune surveillance hypothesis, however, Gompertzian growth should be used with caution. The cancer free equilibrium point (where $C = 0$) is unstable for Gompertzian growth since the derivative of $f_1(C)C$ tends to $+\infty$ as C becomes small. This implies that very small tumors grow faster than is biologically possible and that the immune system would never be able to totally eliminate these tumors [34, 57]. In comparison, for logistic growth, both $f_1(C)C$ and its derivative are bounded as C becomes small, implying that the cancer-free equilibrium should be locally asymptotically stable [34].

One way in which a tumor escapes immune surveillance is through mutation towards an immune resistant clone. The development of such a cancer subpopulation was considered by d'Onofrio [58]. He suggests that a dormancy phase may exist prior to tumor escape if the time lag before the first mutation is long enough that an equilibrium state can be reached between the immune response and the wild-type cancer cells. Cattani and Ciancio [59] take this idea further by allowing the immune efficacy and recruitment to increase in time as the immune system learns to recognize the cancer and then decrease in time as the cancer learns to evade the immune system. Experimental evidence suggests that the tolerance of cytotoxic T cells to cancer cells may be removed by depleting the endogenous dendritic cells or by injecting antigen-loaded mature dendritic cells directly into the tumor site [60]. With direct injections, the tolerized T cells reactivate and remain that way while the dendritic cells persist in the tumor. Such treatments would alter the dynamics of tumor–immune interactions and immune-induced tumor dormancy, but have yet to be investigated with mathematical modeling.

Dormancy in Mathematical Models

This section focuses on a few mathematical models presenting different approaches to model the interactions between cancer and immune cells, specifically focusing on tumor dormancy.

Proliferation Versus Quiescence and Intercellular Signaling

Michelson et al. [61], although not directly modeling immune dynamics, present a two-compartment model for tumor growth with control mechanisms based on intercellular signaling. Regardless of whether a signal is autocrine (self-receiving), paracrine (other-receiving from the local environment), or endocrine (other-receiving from a systemic source), growth factors and chemical signals control and shape tumor growth. Immune cells are one example of a paracrine source of tumor-stimulating or tumor-inhibiting signals. In [61], they present the following model for proliferating (P) and quiescent (Q) cancer cell growth

$$\frac{\mathrm{d}P}{\mathrm{d}t} = \alpha P\left(1 - \frac{P}{K} - cQ\right) - mP \tag{10.12a}$$

$$\frac{\mathrm{d}Q}{\mathrm{d}t} = mP. \tag{10.12b}$$

Growth of the proliferating population is assumed to be exponential, αP, and to be controlled by two factors: competition among the proliferating cells for nutrients and growth factors or inhibitory autocrine growth signals, $-\alpha\frac{P^2}{K}$, and competition between proliferating and quiescent cells or inhibitory paracrine growth factors, $-\alpha cPQ$. They interpret this last term as a negative growth signal produced by a necrotic core, but it could also be produced by an immune response.

In their analysis, they focus on the effects of growth factor signals on the three model parameters that control the proliferating population, α, K, and c. They conclude that the growth rate α has no appreciable effect on tumor dormancy. The carrying capacity K, and its relationship to the transition to quiescence rate m, on the other hand, create interesting growth dynamics in terms of the resulting tumor size and the cell-type composition. For example, as K increases, the proliferating population grows, but the quiescent population grows faster, resulting in larger tumors with smaller fractions of proliferating cells.

Tumor dormancy may occur through suppression signals, possibly emerging from the primary tumor itself, or from an immune response stimulated by the primary tumor. In such cases, Michelson et al. [61] suggest that the suppression signal may force $m > \alpha$ resulting in a purely quiescent tumor. If the suppression signal induces a state of zero proliferation, $\alpha = 0$, and if $m > 0$, a purely quiescent tumor results. If $\alpha = m = 0$, a dormant tumor of viable but nonproliferating cells results. Removal of the primary tumor may terminate the suppression signal and restart proliferation in this viable dormant tumor, resulting in recurrence.

As long as the quiescence transition rate is positive, $m > 0$, any dormant tumor will result in a purely quiescent population. From this model, the quiescent population (Q) can only become constant if $P = 0$ or if $m = 0$. Thus, all dormant tumors predicted by this model are purely quiescent. Note that they do not consider the case where $m = 0$ in which case the proliferating population is only limited by the capacity K, which may correspond to a lethal tumor size. If some inhibitory signaling causes the transition rate m to turn off, $m = 0$, once some population size of quiescent cells is reached, $Q = Q^*$, then a dormant tumor could be attained with both proliferating and quiescent populations at the point $(P, Q) = (K(1 - cQ^*), Q^*)$ in the positive quadrant. This total tumor size, however, would be controlled by the carrying capacity, K, and thus may be lethal to the host.

A local wound-healing event, mediated by the immune response, can result in systemic signals interpreted by the tumor–stroma microenvironment to increase the carrying capacity K. They propose that m and K could be functions of growth factors and stimulatory or inhibitory signals, although no functional forms are provided. One possible form for the carrying capacity as a function of such signals has been proposed elsewhere [36].

Proliferating, Quiescent, and Immune Cell Interactions

In 2005, Page and Uhr [1] presented a series of models to describe tumor dormancy in the presence of an immune response, focusing on the role of quiescence in the tumor population. The first, and simplest, model they present considers the densities of both proliferating, P, and quiescent, Q, cancer cells, as well as the concentration of antibody, I. The model equations are

$$\frac{\mathrm{d}P}{\mathrm{d}t} = rP - \alpha_1 IP - \alpha_2 IP - mP \qquad (10.13a)$$

$$\frac{\mathrm{d}Q}{\mathrm{d}t} = \alpha_1 IP - \alpha_3 IQ - \lambda Q + mP \qquad (10.13b)$$

$$\frac{\mathrm{d}I}{\mathrm{d}t} = \gamma(P + Q) - dI. \qquad (10.13c)$$

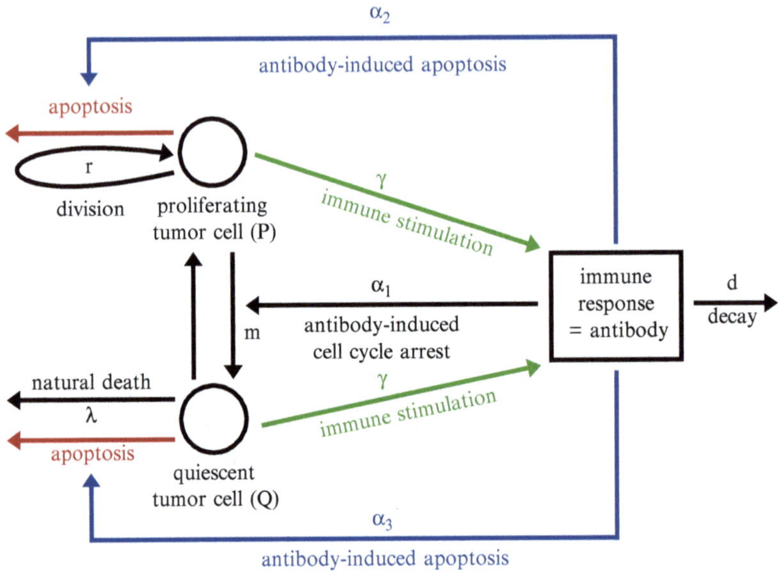

Fig. 10.12 The diagram of a model for tumor dormancy proposed by Page and Uhr [1] considering proliferating, quiescent, and immune cells, described by Eq. (10.13)

Proliferating cancer cells are assumed to grow at the rate rP, to transition naturally into the quiescent state at the rate mP, to transition into the quiescent state in the presence of antibody at the rate $\alpha_1 IP$, and to undergo antibody-induced apoptosis at the rate $\alpha_2 IP$. Quiescent cells are assumed to grow through the transition of proliferating cells to the quiescent state, $\alpha_1 IP$ and mP, to undergo antibody-induced apoptosis at the rate of $\alpha_3 IQ$, and to have a natural death rate of λQ. Immune produced antibodies are assumed to be produced at a rate proportional to the cancer presence $\gamma(P+Q)$ and to decay at the rate dI. A diagram of this model is shown in Fig. 10.12.

In this model, the cancer-free state is always unstable provided that $r > m$, which is assumed as otherwise tumor growth would not occur. There is also a nonzero steady state, which is stable provided that d is sufficiently large or that α_3 is large. The dependence of this dormant state on the antibody degradation rate, d, is intriguing, but not investigated further by Page and Uhr.

Simulations of the model, Eq. (10.13), are shown in Fig. 10.13. Both sets of parameters used in this figure demonstrate tumor dormancy: one tumor converges with decaying oscillations to a dormant state, while the other tumor converges to a limit cycle where the size and composition of the tumor change periodically. In reality, when the tumor size becomes very small, it is most likely cleared by the immune system instead of regrowing as the model predicts.

Page and Uhr modify their model to demonstrate the effect of immunization on tumor control. First, they assume logistic growth for the proliferating cancer

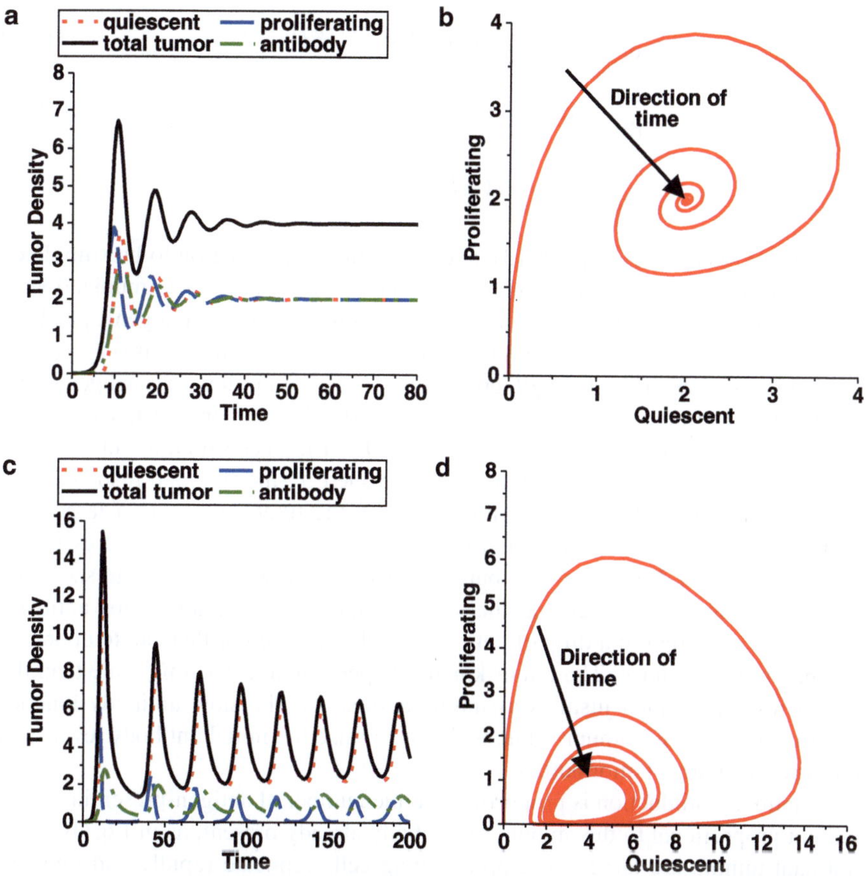

Fig. 10.13 Numerical simulations of the system defined by Eq. (10.13). In (**a**) and (**b**), the tumor undergoes decaying oscillations and eventually converges to a dormant steady state. Parameter values are $P(0) = 0.001$, $Q(0) = 0$, $I(0) = 0.001$, $\gamma = 0.5$, $d = 1$, $r = 1$, $\alpha_1 = 0.5$, $\alpha_2 = 0$, $\alpha_3 = 0$, $\lambda = 1$, and $m = 0$. In (**c**) and (**d**), the tumor burden oscillates much more strongly, reaching a dormant tumor that continually oscillates within a limited size range. Parameter values that differ from those used in (**a**) include $\gamma = 0.1$, $d = 0.4$, $\alpha_1 = 1$, $\alpha_3 = 0.1$, and $\lambda = 0$. Time series are shown in (**a**) and (**c**), and quiescent-proliferating phase portraits are shown in (**b**) and (**d**)

cells, so that the proliferating tumor size is limited by the environment even in the absence of an immune response (antibody). Second, they assume that the antibody production rate through tumor–B cell interactions, $\varepsilon\gamma(P+Q)I$, and the decay rate, εdI, are slow processes (ε is small). Third, they assume that the antibody can be produced faster following immunization (idiotype i produces antibody at a faster rate γiI). This immunization is an idiotype and adjuvant that makes the cancer cells more immunogenic. The proliferating cells, quiescent cells, and immune response (antibody) are now modeled by

$$\frac{dP}{dt} = rP\left(1 - \frac{P}{K}\right) - \alpha_1 IP - \alpha_2 IP - mP \tag{10.14a}$$

$$\frac{dQ}{dt} = \alpha_1 IP - \alpha_3 IQ - \lambda Q + mP \tag{10.14b}$$

$$\frac{dI}{dt} = \gamma i I + \varepsilon I\big(\gamma(P+Q) - d\big). \tag{10.14c}$$

The previous model, Eq. (10.13), allowed antibody production to be stimulated by tumor presence. In reality, this may not happen without prior immunization as idiotype presentation on cancer cells may not be sufficiently immunogenic [1]. If an immunization is given prior to tumor cell injection, then the idiotype is only present transiently and the effect of the γiI term in Eq. (10.14c) simplifies to a changed initial condition $I(0)$. In the simulations shown in Fig. 10.14, $i = 0$, and a value of $I(0) = 0.1$ indicates prior immunization while $I(0) = 0.001$ indicates no immunization.

With the immunization, the tumor is controlled by the immune response to a small dormant size. Without the immunization, the tumor grows to a large, and presumably lethal, size. When high initial antibody concentrations are assumed, the tumor burden drops very low but then grows rapidly to a moderate size. The regrowth is due to the fact that the small tumor size essentially stops antibody production, resulting in a drop of antibody levels and a tumor that can temporarily escape immune control. The tumor kinetics happen on a faster time scale than the B cell response which justifies assuming a small ε. The slow antibody kinetics illustrate the role of immunization, which is to raise the initial antibody levels and thus prevent a large tumor burden.

If antibody production is proportional to the number of antigen-presenting cells, but not proportional to the amount of antibody already present, as in Eq. (10.13c), dormant tumors are large if the proliferating cells replicate rapidly and undergo apoptosis and cell cycle arrest at a slow rate. Replication rate, however, doesn't affect dormant tumor size when antibody concentration is produced at a rate proportional to the amount of antibody already present as well as the number of antigen presenting cells, as in Eq. (10.14c). In this case, the dormant tumor size is determined by the antibody kinetics. This suggests that since greater numbers of cancer cells stimulate more antibody production, which increases the death rate of cancer cells, antibody signaling can cause cell death to balance cell proliferation [1].

In the model described by Eq. (10.14), dynamic clearance of the tumor is likely if the initial concentration of antibody satisfies $I(0) \gg \frac{r}{\alpha_2}$. If $I(0) < \frac{r}{\alpha_2}$, then oscillations are likely to occur, but should converge to a dormant state. Host death will occur if the amplitude of the tumor size oscillations are sufficiently large. Many of the dormant states predicted by this model are approached through damped oscillations. The tumor may then escape from this dormant state through any of the following mechanisms. The antibody may disappear due to depletion of antibody producing B cells or cancer cell evasion of the immune response. Cancer cells may cease to bind the antibody and thus grow tolerant of the immune attack.

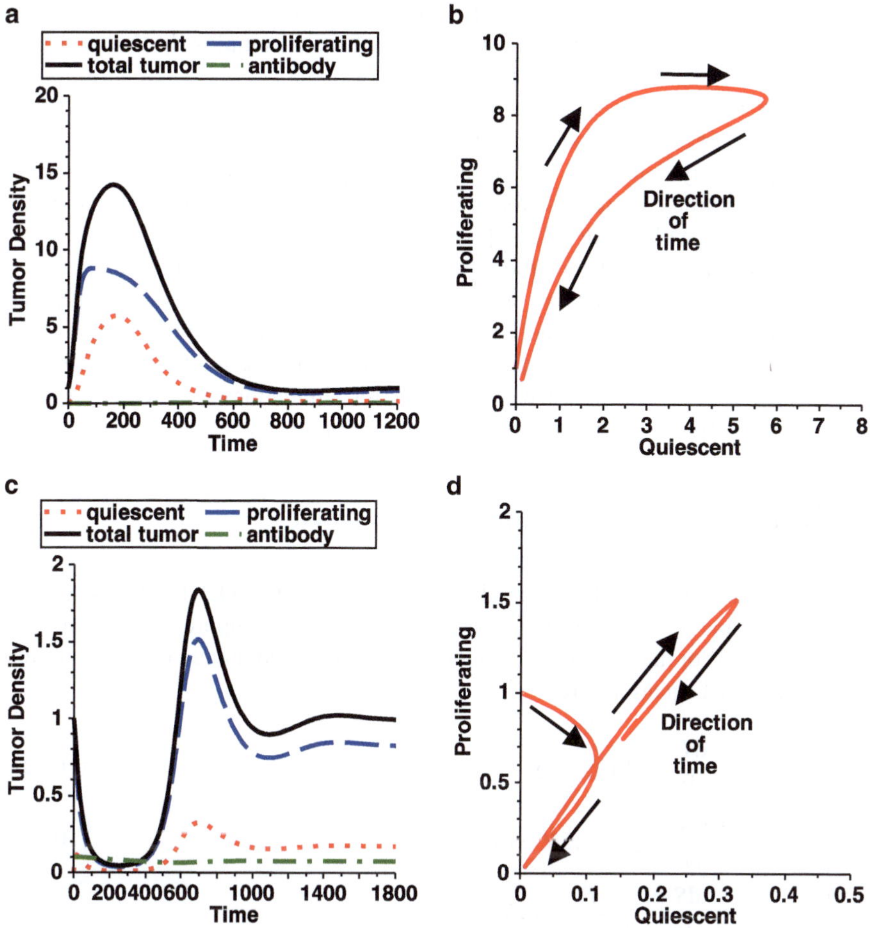

Fig. 10.14 Numerical simulations of the system defined by Eq. (10.14). In (**a**), with no immunization ($I(0) = 0.001$), the tumor grows rapidly to the maximum presumed lethal size. In long time, however, if the host survives, the antibody slowly builds and eventually controls the tumor to a dormant size, as seen in the phase portrait (quiescent and proliferating cell composition) in (**b**). In (**c**), the immunization ($I(0) = 0.1$) increases the initial concentration of antibody and the immune system is thus able to control the tumor to a dormant state, as seen in the phase portrait in (**d**). Parameter values are $r = 0.1$, $\gamma = \alpha_2 = d = 1.0$, $\alpha_1 = 0.1$, $m = 0.01$, $\lambda = 0.01$, $\alpha_3 = 1.0$, $\varepsilon = 0.001$, $K = 10.0$, $P(0) = 1.0$, and $Q(0) = 0$, as in [1]

And downstream signaling pathways of the antibody–cancer cell receptors may be disrupted, essentially stopping the antibody-triggered signals to apoptose or arrest the cell cycle. Such mechanisms, however, are beyond the scope of these models.

The models of Page and Uhr [1] specifically consider the antibody response to tumor growth rather than the cytotoxic T cell response. Recall that in the model of

Kuznetsov et al. [44], Eq. (10.9), tumor cells could kill immune cells if and only if the tumor cell was not killed in the cancer–immune complex. In the models of Page and Uhr [1], antibody is either consumed (model not discussed here), or not consumed, in the process of initiating cancer cell death. The models discussed here, where antibodies are not consumed, may be more realistic because antibody levels may be primarily controlled by the number of specific B cells (if the kinetics of antibody secretion and degradation are assumed to be fast) and B cells may never directly interact with tumor cells [1].

Another difference between these two modeling approaches is that Kuznetsov et al. assume that immune cells are stimulated by the tumor in a manner that saturates. Page and Uhr, on the other hand, assume that this stimulation is proportional to the product of the two population densities. The saturation assumption may be more physical, but it also may not be significant for simulations that focus on small dormant tumors, in which case simplicity may be preferred.

Delay in Activation of the Immune Response

In 1988, Kuznetsov [62] presented a multicomponent model for tumor growth considering the actions of tumor cells (T), NK cells (N), suppressor cells such as T lymphocytes or macrophages (S), memory cells (P), and mature cytotoxic T cells (C). The model equations are

$$\frac{\mathrm{d}T}{\mathrm{d}t} = T\left(c_1 - c_2C - c_3N\right) \tag{10.15a}$$

$$\frac{\mathrm{d}N}{\mathrm{d}t} = j_N - c_4N - c_5NT \tag{10.15b}$$

$$\frac{\mathrm{d}S}{\mathrm{d}t} = \frac{c_6T}{1+\frac{T}{c_7}} - c_8S \tag{10.15c}$$

$$\frac{\mathrm{d}P}{\mathrm{d}t} = \frac{c_9CT}{c_{10}+S} - c_{11}P - c_{12}PT \tag{10.15d}$$

$$\frac{\mathrm{d}C}{\mathrm{d}t} = c_{13}H(t-\tau)P(t-\tau) + \frac{j_C}{1+\frac{S}{c_{14}}} - c_{15}CT - c_{16}C, \tag{10.15e}$$

where H is the Heaviside function that turns on the memory cell production of cytotoxic T cells after time τ, j_N and j_C are the constant source rates of NK cells and cytotoxic T cells, and c_i for $i = 1,\ldots,16$ are the parameters for the interactions of the different cell types.

The model is simplified by assuming that the innate immune response occurs quickly compared to the tumor doubling time. Thus N and S are fast variables and reach an approximate equilibrium state. The memory cells, P, are also assumed to reach a quasi-steady state. Further simplifications occur under the assumption that the tumor induces suppressor cell production ($T \ll c_7$) and that suppressors affect the generation of cytotoxic T cells ($S \ll c_{14}$). These two assumptions omit the inhibition (or saturating effect) of suppressors by tumor cells and of cytotoxic T cells by suppressors. These assumptions simplify the above model to the two-equation system below

$$\frac{dT}{dt} = \left(c_1 - c_2 C - \frac{c_3 j_N}{c_4 + c_5 T}\right) T \tag{10.16a}$$

$$\frac{dC}{dt} = \frac{c_{13} c_9 H(t-\tau) C(t-\tau) T(t-\tau)}{\left(c_{10} + \frac{c_6}{c_8} T(t-\tau)\right)\left(c_{11} + c_{12} T(t-\tau)\right)} + j_C - c_{15} C T - c_{16} C. \tag{10.16b}$$

This model describes an exponentially growing tumor, T, with inhibition from cytotoxic T cells, C, proportional to CT and inhibition from the innate immune system proportional to $\frac{T}{g+T}$ which saturates for large T. Before activation of cytotoxic T cells, $t < \tau$, analysis shows that there are no stable stationary points where $T > 0$, thus no dormant state exists. After cytotoxic T cell activation, however, dormant tumors can exist.

Kuznetsov suggests that tumor growth from a single transformed cell is impossible if the growth rate is smaller than the sum of the base concentrations of effector cells multiplied by their corresponding effectiveness coefficients; that is, if

$$c_1 < \frac{c_2}{c_{16}} j_C + \frac{c_3}{c_4} j_N \tag{10.17}$$

a tumor cannot form from a single cell. Furthermore, tumor growth and its prolonged existence are impossible if the initial concentration of tumor cells is below the threshold, T_{cr}, where

$$T_{cr} = \frac{c_4 \left(c_2 \frac{J_C}{c_{16}} - c_3 \frac{J_N}{c_4} - c_1\right)}{c_5 \left(c_1 - c_2 \frac{J_C}{c_{16}}\right)}. \tag{10.18}$$

These conditions are referred to as an *immune barrier*. The base immunity actively clears tumor cells, Eq. (10.17), before they can accumulate sufficient mass to overcome the threshold, Eq. (10.18). This barrier does not allow for the existence of very small tumors (i.e., microscopic dormant tumors), and can exist even in the absence of cytotoxic T cells. Furthermore, the barrier is independent of the activation mechanisms for these T cells. Kuznetsov assumes that the effectiveness of NK cells, T cells, and macrophages are all approximately equivalent. Considering this and the fact that the base level of NK cells and macrophages are about two or three orders of magnitude larger than cytotoxic T cells, Kuznetsov concludes that this barrier is mainly enforced by the innate immunity.

From the experimental data reported by Diefenbach et al. [63], the assumption of equal sensitivity amongst cytotoxic cells may not be valid. They reported the percentage of specific tumor cell lysis for various effector to target ratios. For primed T cells, a $CD8^+$ effector to target ratio of 3 produced a maximum percent lysis of 30, whereas for NK recruited effectors, an effector to target ratio of 100 produced a maximum percent lysis of 45. This data suggests that the sensitivity of tumor cells to cytotoxic T cells is higher than the sensitivity to NK cells. Therefore, even though the base level of T cells is orders of magnitude smaller than that of the innate immunity, the increased effectiveness of T cells may account for this discrepancy.

If the tumor growth rate is large enough that the initial tumor size is greater than the immune barrier, Eq. (10.18), then tumor elimination, escape, and dormancy may occur as simulated in Fig. 10.15. Dormancy may appear as a stable node or as a limit cycle in this model. For simulation purposes, the system is nondimensionalized using $\tilde{t} = c_1 t$, $x = \frac{c_2}{c_1}C$, and $y = \frac{c_{12}}{c_{11}}T$, leading to the simpler form below (note that time is renamed as $t = \tilde{t}$):

$$\frac{\mathrm{d}y}{\mathrm{d}t} = y\left(1 - x - \frac{\mu}{1 + \nu y}\right) \tag{10.19a}$$

$$\frac{\mathrm{d}x}{\mathrm{d}t} = \frac{\alpha H(t-\tau)x(t-\tau)y(t-\tau)}{\left(1 + y(t-\tau)\right)\left(1 + \delta y(t-\tau)\right)} + j - \beta xy - \gamma x,^{1} \tag{10.19b}$$

where $\mu = \frac{c_3 j_N}{c_1 c_4}$, $\nu = \frac{c_5 c_{11}}{c_4 c_{12}}$, $\tilde{\tau} = c_1 \tau$ (but again we drop the tilde), $\alpha = \frac{c_{13} c_9}{c_1 c_{10} c_{12}}$, $\delta = \frac{c_6 c_{11}}{c_8 c_{10} c_{12}}$, $j = \frac{c_2 j_C}{c_1^2}$, $\beta = \frac{c_{11} c_{15}}{c_1 c_{12}}$, and $\gamma = \frac{c_{16}}{c_1}$.

Varying the delay in cytotoxic T cell activation shows that small delays destabilize a dormant tumor possibly leading to tumor elimination, the phenomenon of *sneaking through*, or tumor escape. This suggests that the delay in activation is one factor determining how the immune system both stimulates and inhibits tumor growth. For example, when a spontaneous tumor is first recognized by the immune system, it may be only weakly antigenic and the delay in activation may be small, leading to a perturbation from the dormant state. If the tumor grows larger and more transformed, with the ability to alter its microenvironment, however, the delay in activation may change which could have either an inhibitory effect (tumor is eliminated) or a stimulatory effect (tumor sneaks through and escapes or escapes directly) on tumor growth.

Kuznetsov also examined the relationship between μ, a measure of the innate immune system's strength, and this time delay in cytotoxic T cell activation. He showed that as the natural antitumor resistance increases, the same time delay can be either stimulatory or inhibitory to tumor growth. He concluded that immune stimulation of dormant tumors is due to the existence of two different types of killer cells, cytotoxic T cells and NK cells or macrophages. These different cell types have different initial concentrations, generation velocities, and may have different times to induce cell cytolysis on target tumor cells. These kinetics may lead to

[1]Note that the original article has a positive (+) sign and we have a negative (−) sign on the γx term. This change represents what the author believes to be the correct nondimensionalization of the proposed system.

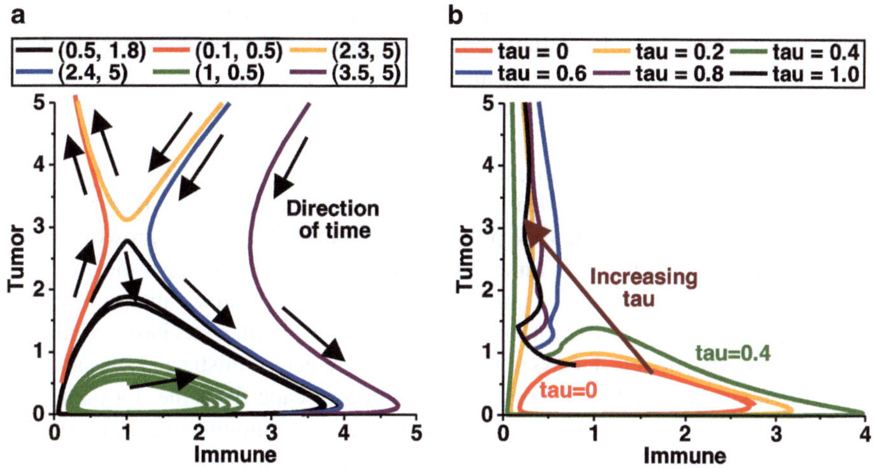

Fig. 10.15 Numerical simulation of Kuznetsov's system defined by Eq. (10.19). In (**a**), there is no time delay ($\tau = 0$), and a stable limit cycle (or dormant tumor) exists. The different *curves* represent different initial densities of tumor and immune cells. In (**b**), as the time delay increases, the tumor is disrupted from its dormant cycle, and it is either eliminated or it escapes immune control. The *curve* with $\tau = 0.4$ is either eliminated by the immune response or it escapes from immune control through the *sneaking through* phenomenon. Parameter values are $\alpha = 5.0$, $\beta = 1.125$, $\gamma = 0.5$, $j = 0.02$, $\mu = 0.4$, $v = 129.6$, and $\delta = 0$, as used in [62]

immunoediting: tumor cells are pruned by NK cells and macrophages, and in the case of low antigenicity, or low cytotoxic T cell activity, increased tumor growth results.

Discussion

This review discussed mathematical models of tumor growth in the presence of an immune response, focusing specifically on mathematical models for the immune induction of tumor dormancy. General findings suggest that in order for a dormant tumor to exist, there must be feedback from the tumor to the immune response. Thus, cancer cells must induce an increase in the immune response, and the effectiveness of the immune predation of cancer cells must increase with the growing immune response. The dormant state may also be sensitive to the immune predation rate, the immune decay rate, and the time delay in immune cell activation.

A limitation of all mathematical models of biological phenomena is accurate parameter estimation. Levels of immune presence in different types and different samples of cancer vary drastically, as do the tumor antigenicities. As a result, the immune response varies significantly between individual cancer manifestations. Mathematical models must be robust to these variations, in order to capture the essential dynamics that describe the physiology of tumor growth and tumor dormancy. The integration of biological data into mathematical models for validation and refinement of the functional forms used in the models pose the next challenges in this field. The resulting models will provide improved theoretical frameworks within which mechanisms of tumor growth in an immunocompetent host can be clarified.

If a tumor is dormant for an extended period of time (perhaps through cellular or angiogenic mechanisms), then it is possible that immune interactions, through wound-healing and inflammatory processes for example, produce growth factors and signals that modify the tumor microenvironment to increase the carrying capacity, modify the dynamics, and disrupt the delicate balance maintaining the dormant state. Such perturbations may be best analyzed with a stochastic modeling approach; all the models discussed in this review are deterministic and thus average over these stochastic effects. Prevention of tumor escape from the dormant state may be attainable once an improved understanding of the immune response to tumor growth is achieved. This requires an improved understanding of homeostasis, and of how the normal homeostatic mechanisms that control the immune system contribute to maintain or disrupt tumor dormancy [1].

Oscillatory solutions are observed in several of the mathematical models reviewed here. Due to the feedback between the cancer and immune populations, this cyclic behavior is mathematically predicted to occur in both populations. Clinical reports exist citing observations of lymphocyte oscillations [64, 65] and direct observations of cancer cell oscillations in experimental systems [66]. Macroscopic cyclic behavior of tumors is not commonly observed in vivo, but some reports of oscillatory behavior do exist for both animals [67] and humans [68, 69]. For example, cyclic oscillations in the number of leukocytes has been reported in chronic human myeloid leukemia [69], and periodic patterns of spontaneous relapse and remission were observed in non-Hodgkin's lymphoma [68]. Interestingly, the time scale of the oscillatory model predictions of Kuznetsov et al., Eq. (10.9), is about 3–4 months, which approximately matches the time for recurrent clinical manifestations of certain human leukemias [44].

If these oscillations exist to an extent in some, or possibly all, manifestations of cancer, then it is natural to wonder what their biological effects may be. For example, after each rise and fall of the immune response, it seems reasonable that the predation efficacy may change due to immune learning or tumor evasion. Furthermore, these oscillations may enhance or delay the tumor immunoediting process as each oscillation provides a period of culling and a period of regrowth. Therefore, these oscillations, if they are in fact a real biological phenomenon, should be investigated as a possible mechanism for immunoediting and immune evasion by tumors.

The models reviewed here that consider a subpopulation of quiescent tumor cells do not allow the quiescent cells to transition back to the proliferating population. Under this assumption, these cells may be better defined as senescent. Quiescence is a possible mechanism used by tumors to evade therapy-induced death. Thus, both quiescent and senescent subpopulations may be interesting to consider in a mathematical framework. The role of the immune system in senescence was recently

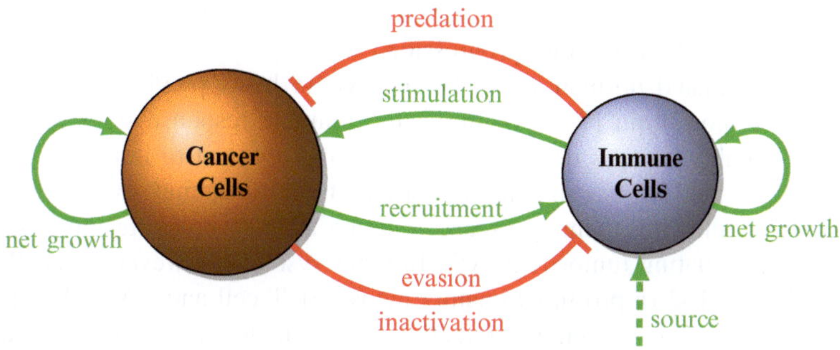

Fig. 10.16 Diagram depicting a new paradigm for cancer–immune interactions, including immune stimulation of tumor growth and tumor evasion of the immune response

found to be significant. Rakhra et al. [70] showed that $CD4^+$ T cells are required for the induction of senescence, the shutdown of angiogenesis, and chemokine expression resulting in sustained tumor regression upon oncogene inactivation in mouse models. They emphasize that intact host immunity is important in evaluating targeted therapies. Typical mathematical models only consider the actions of cytolytic $CD8^+$ T cells, NK cells, or macrophages. The work of Rakhra et al. suggests that in fact, many other immune cell types may play significant roles in tumor growth modulation. Thus, a complete and intact immune response should be emphasized, not only for experimental models, but also for mathematical models, of cancer growth.

The most promising directions for immunotherapies are to maximize the chance of complete tumor clearance as well as to minimize the size of dormant tumors and the chance of escape from dormancy [1]. Contradictory outcomes after immunotherapies occur because the stimulated immune system may directly stimulate tumor growth [71,72]. Mechanisms for immune-mediated tumor growth stimulation include inflammation, angiogenesis, enhanced motility, and clearance of space. Even small numbers of cytotoxic T cells have been reported to stimulate tumor cell growth in vivo [73–75].

Immune stimulation of tumor growth is becoming widely accepted. In 2011, Hanahan and Weinberg named tumor-promoting inflammation as an emerging hallmark of cancer [76]. Despite this, immune stimulation of tumor growth has yet to be considered in a mathematical modeling framework. Much work, therefore, remains to be done, as mathematical models cannot analyze the full effect of an immune response to cancer or predict outcomes of immunotherapies while the models preclude the possibility of immune stimulation. With this in mind, Fig. 10.16 gives a more complete diagram of tumor–immune interactions.

Some of the proposed tumor-promoting actions of chronic inflammation may result from interfering with the adaptive immune response's capacity to maintain the dormant state [10]. Cytokines produced by tumor and inflammatory immune cells can either promote tumor development or exert antitumor effects. Inflammatory mediators including TNF-α, IL-6, and IL-17 lead to eradication of antitumor immunity and accelerated tumor progression [77].

Immune-mediated dormancy can be induced in hosts previously immunized against the tumor strain [78]. Assuming the immune response stays active, the tumor presumably escapes dormancy by becoming resistant to immune-mediated cytolysis. Quesnel [78] has shown that dormant tumor cells may over-express B7-H1 and B7.1 allowing them to escape immune attacks, breaking the dormancy balance, and initiating tumor regrowth. In other systems, however, when B7.1 is combined with IL-2 (a pro-inflammatory cytokine), T cell and NK cell responses are stimulated leading to tumor elimination [79]. Schatton and Frank [80] suggest that the immunomodulatory function of stem cells may be responsible for immune evasion in melanoma, while Noh et al. [81] suggest immune evasion may be due to the activation of Akt, a protein kinase involved in cell proliferation, migration, and apoptosis. Some modeling efforts [58, 59, 82] have already been made to investigate the immune evasion ability of tumors, which is significant for tumor dormancy, but more work remains to be done. The many molecular and cellular mechanisms suggested to underly immune evasion could be analyzed with mathematical models to help determine the most significant contributors to the phenomenon.

Lastly, as the body ages the probability of cellular transformation increases and, most likely, immune surveillance decreases [44]. In mathematical models, this would translate into increased initial cancer population sizes at the time of immune detection and slower immune response kinetics, perhaps through a decreased recruitment potential, or a decreased supply velocity. This presents another area where mathematical modeling could help to improve the understanding of the role of aging in tumor presentation and tumor escape from dormancy.

There is a pressing need to develop biologically directed, mechanistic models that provide real biological insights into the critical parameters controlling the tumor–immune system dynamics. Such tools may help to develop novel therapeutics that can work in conjunction with a host's own immune response to control and shape malignancies into less aggressive dormant lesions. This new frontier in cancer treatment would include ways to revert cancers back to their dormant states by inducing tumor cell senescence, improving vascular normalcy, and increasing the efficacy of the immune response to tumor cells. Hopefully, this review will help to stimulate more theoretical research to compliment and extend experimental research into the basic mechanisms underlying immune-mediated tumor dormancy and their possible therapeutic implications.

Acknowledgments The author wishes to thank Dr. P. Hahnfeldt and Dr. L. Hlatky for their careful editing and support, and Dr. M. La Croix for his support and for the excellent graphical illustrations. This project was supported by the National Cancer Institute under Award Number U54CA149233 (to L. Hlatky) and by the Office of Science (BER), U.S. Department of Energy, under Award Number DE-SC0001434 (to P. Hahnfeldt). The content is solely the responsibility of the author and does not necessarily represent the official views of the National Cancer Institute or the National Institutes of Health.

References

1. Page KM, Uhr JW (2005) Mathematical models of cancer dormancy. Leuk Lymphoma 46(3):313–327
2. Janeway CA Jr, Travers P, Walport M, Shlomchik MJ (2005) Immunobiology the immune system in health and disease. Garland Science Publishing, New York
3. Dunn GP, Old LJ, Schreiber RD (2004) The three Es of cancer immunoediting. Annu Rev Immunol 22:329–360
4. Vesely MD, Kershaw MH, Schreiber RD, Smyth MJ (2011) Natural innate and adaptive immunity to cancer. Annu Rev Immunol 29:235–271
5. Lord EM, Nardella G (1980) The multicellular tumor spheroid model. II. Characterization of the primary allograft response in unsensitized mice. Transplantation 29(2):119–124
6. Cortez-Retamozo V, Etzrodt M, Newton A, Rauch PJ, Chudnovskiy A, Berger C et al (2012) Origins of tumor-associated macrophages and neutrophils. PNAS 109(7):2491–2496
7. Grivennikov SI, Greten FR, Karin M (2010) Immunity, inflammation, and cancer. Cell 140:883–899
8. Teng MWL, Swann JB, Koebel CM, Schreiber RD, Smyth MJ (2008) Immune-mediated dormancy: An equilibrium with cancer. J Leukoc Biol 84(4):988–993
9. MacKie RM, Reid R, Junor B (2003) Fatal Melanoma transferred in a donated kidney 16 years after melanoma surgery. N Engl J Med 348(6):567–568
10. Koebel CM, Vermi W, Swann JB, Zerafa N, Rodig SJ, Old LJ et al (2007) Adaptive immunity maintains occult cancer in an equilibrium state. Nature 450:903–907
11. Breslow N, Chan CW, Dhom G, Drury RA, Franks LM, Gellei B et al (1977) Latent carcinoma of prostate at autopsy in seven areas. Int J Canc 20(5):680–688
12. Black WC, Welch HG (1993) Advances in diagnostic imaging and overestimations of disease prevalence and the benefits of therapy. N Engl J Med 328(17):1237–1243
13. Welch HG, Black WC (2010) Overdiagnosis in cancer. J Natl Canc Inst 102(9):605–613
14. Müller M, Gounari F, Prifti S, Hacker HJ, Schirrmacher V, Khazaie K (1988) EblacZ tumor dormancy in bone marrow and lymph nodes: Active control of proliferating tumor cells by $CD8^+$ immune T cells. Canc Res 58(23):5439–5446
15. Uhr JW, Scheuermann RH, Street NE, Vitetta ES (1997) Cancer dormancy: Opportunities for new therapeutic approaches. Nat Med 3(5):505–509
16. Weiss L, Morecki S, Vitetta ES, Slavin S (1983) Suppression and elimination of BCL1 leukemia by allogeneic bone marrow transplantation. J Immunol 130(5):2452–2455
17. Folkman J (1985) Tumor angiogenesis. Adv Canc Res 43:175–203
18. Dvorak HF (1986) Tumors: Wounds that do not heal. Similarities between tumor stroma generation and wound healing. N Engl J Med 315(26):1650–1659
19. Matsuzawa A, Takeda Y, Narita M, Ozawa H (1991) Survival of leukemic cells in a dormant state following cyclophosphamide-induced cure of strongly immunogenic mouse leukemia (DL811). Int J Canc 49(2):303–309
20. Adam JA, Bellomo N (1997) A survey of models for tumor-immune system dynamics. Modeling and simulation in science, engineering and technology. Birkhäuser, Basel
21. Gatenby PA, Basten A, Creswick P (1981) "Sneaking Through": A T-cell-dependent phenomenon. Br J Canc 44(5):753–756
22. Lefever R, Horsthemke W (1979) Bistability in fluctuating environments. Implications in tumor immunology. Bull Math Biol 41:469–490
23. Angelis ED, Lods B (2008) On the kinetic theory for active particles: A model for tumor–immune system competition. Math Comput Model 47:196–209
24. Bellomo N, Delitala M (2008) From the mathematical kinetic, and stochastic game theory to modelling mutations, onset, progression and immune competition of cancer cells. Phys Life Rev 5:183–206
25. Brazzoli I, Angelis ED, Jabin PE (2010) A mathematical model of immune competition related to cancer dynamics. Math Meth Appl Sci 33(6):733–750

26. Matzavinos A, Chaplain MAJ, Kuznetsov VA (2004) Mathematical modelling of the spatio–temporal response of cytotoxic T–lymphocytes to a solid tumour. Math Med Biol 21(1):1–34
27. Roose T, Chapman SJ, Maini RK (2007) Mathematical models of avascular tumor growth. SIAM Rev 49(2):179–208
28. Takayanagi T, Kawamura H, Ohuchi A (2006) Cellular automaton model of a tumor tissue consisting of tumor cells, cytotoxic T lymphocytes (CTLs), and cytokine produced by CTLs. IPSJ Trans Math Model Appl 47(1):61–67
29. Araujo RP, McElwain DLS (2004) A history of the study of solid tumour growth: The contribution of mathematical modelling. Bull Math Biol 66(5):1039–1091
30. Woelke AL, Murgueitio MS, Preissner R (2010) Theoretical modeling techniques and their impact on tumor immunology. Clin Dev Immunol 2010:271794
31. Bellomo N, Bellouquid A, Angelis ED (2003) The modelling of the immune competition by generalized kinetic (Boltzmann) models: Review and research perspectives. Math Comp Model 37(1–2):65–86
32. Eftimie R, Bramson JL, Earn DJ (2011) Interactions between the immune system and cancer: A brief review of non-spatial mathematical models. Bull Math Biol 73(1):2–32
33. Spratt JS, Meyer JS, Spratt JA (1996) Rates of growth of human neoplasms: Part II. J Surg Oncol 61(1):68–83
34. d'Onofrio A (2005) A general framework for modeling tumor–immune system competition and immunotherapy: Mathematical analysis and biomedical inferences. Physica D 208(3–4):220–235
35. Laird AK (1964) Dynamics of tumor growth. Br J Canc 18(3):490–502
36. Hahnfeldt P, Panigraphy D, Folkman J, Hlatky L (1999) Tumor development under angiogenic signaling: A dynamical theory of tumor growth, treatment response, and postvascular dormancy. Canc Res 59(19):4770–4775
37. Wheldon TE (1988) Mathematical models in cancer research. A. Hilger Publishing, Bristol
38. Vaidya VG, Alexandro FJJ (1982) Evaluation of some mathematical models for tumor growth. Int J Biomed Comput 13(1):19–36
39. Hart D, Shochat E, Agur Z (1998) The growth law of primary breast cancer as inferred from mammography screening trials data. Br J Canc 78(3):382–387
40. Guiot C, Degiorgis PG, Delsanto PP, Gabriele P, Deisboeck TS (2003) Does tumor growth follow a "universal law"? J Theor Biol 225(2):147–151
41. Zitvogel L, Tesniere A, Kroemer G (2006) Cancer despite immunosurveillance: Immunoselection and immunosubversion. Nat Rev Immunol 6(10):715–727
42. Bell GI (1973) Predator–prey equations simulating an immune response. Math Biosci 16(3–4):291–314
43. Sotolongo-Costa O, Molina LM, Perez DR, Antoranz JC, Reyes MC (2003) Behavior of tumors under nonstationary therapy. Physica D 178(3–4):242–253
44. Kuznetsov VA, Makalkin IA, Taylor MA, Perelson AS (1994) Nonlinear dynamics of immunogenic tumors: Parameter estimation and global bifurcation analysis. Bull Math Biol 56(2):295–321
45. Joshi B, Wang X, Banerjee S, Tian H, Matzavinos A, Chaplain MA (2009) On immunotherapies and cancer vaccination protocols: A mathematical modelling approach. J Theor Biol 259(4):820–827
46. d'Onofrio A, Fasano A, Monechi B (2011) A generalization of Gompertz law compatible with the Gyllenberg-Webb theory for tumour growth. Math Biosci 230(1):45–54
47. Stepanova NV (1980) Course of the immune reaction during the development of a malignant tumor. Biophysics 24:917–923
48. de Pillis LG, Radunskaya A (2001) A mathematical tumor model with immune resistance and drug therapy: An optimal control approach. J Theor Med 3:79–100
49. DeLisi C, Rescigno A (1977) Immune surveillance and neoplasia–1 a minimal mathematical model. Bull Math Biol 39(2):201–221
50. Kirschner D, Panetta JC (1998) Modeling immunotherapy of the tumor–immune interaction. J Math Biol 37(3):235–252

51. de Pillis LG, Radunskaya AE, Wiseman CL (2005) A validated mathematical model of cell–mediated immune response to tumor growth. Canc Res 65(17):7950–7958
52. Sachs RK, Hlatky LR, Hahnfeldt P (2001) Simple ODE models of tumor growth and anti–angiogenic or radiation treatment. Math Comput Model 33(12–13):1297–1305
53. Arciero JC, Jackson TL, Kirschner DE (2004) A mathematical model of tumor–immune evasion and siRNA treatment. Discrete Cont Dyn-B 4(1):39–58
54. Gowal S, Giacomi Md, Boudec JYL (2007) Comment on: A validated mathematical model of cell–mediated immune response to tumor growth. Canc Res 67(17):8419–8420
55. de Pillis LG, Radunskaya AE, Wiseman CL (2007) Comment on: A validated mathematical model of cell–mediated immune response to tumor growth. Canc Res 67(17):8420
56. d'Onofrio A (2010) Bounded-noise-induced transitions in a tumor–immune system interplay. Phys Rev E 81(2):021923:1–7
57. de Vladar HP, González JA (2004) Dynamic response of cancer under the influence of immunological activity and therapy. J Theor Biol 227(3):335–348
58. d'Onofrio A (2007) Tumor evasion from immune control: Strategies of a MISS to become a MASS. Chaos, Solitons and Fractals 31:261–268
59. Cattani C, Ciancio A (2010) Early stage of tumor–immune competition with time dependent parameter and external source. UPB Sci Bull A 72(3):113–128
60. Higham EM, Shen CH, Wittrup KD, Chen J (2010) Cutting edge: Delay and reversal of T cell tolerance by intratumoral injection of antigen-loaded dendritic cells in an autochthonous tumor model. J Immunol 184(11):5954–5948
61. Michelson S, Leith JT (1994) Dormancy, regression, and recurrence: towards a unifying theory of tumor growth control. J Theor Biol 169:327–338
62. Kuznetsov VA (1988) Mathematical modeling of the development of dormant tumors and immune stimulation of their growth. Cybern Syst Anal 23(4):556–564
63. Diefenbach A, Jensen ER, Jamieson AM, Raulet DH (2001) Rae1 and H60 ligands of the NKG2D receptor stimulate tumour immunity. Nature 413(6852):165–171
64. Rodriguez AR, Lutcher CL (1976) Marked cyclic leukocytosis–leukopenia in chronic myelogenous leukemia. Am J Med 60(7):1041–1047
65. Chikkappa G, Chanana AD, Chandra P, Cronkite EP, Thompson KH (1980) Cyclic oscillation of blood neutrophils in a patient with multiple myeloma. Blood 55(1):61–66
66. Posadas EM, Criley SR, Coffey DS (1996) Chaotic oscillations in cultured cells: Rat prostate cancer. Canc Res 56(16):3682–3688
67. Uhr JW, Tucker T, May RD, Siu H, Vitetta ES (1991) Cancer dormancy: Studies of the murine BCL1 lymphoma. Canc Res 51(18 Suppl):5045s–5053s
68. Krikorian JG, Portlock CS, Cooney P, Rosenberg SA (1980) Spontaneous regression of non-Hodgkin's lymphoma: A report of nine cases. Cancer 46(9):2093–2099
69. Mehta BC, Agarwal MB (1980) Cyclic oscillations in leukocyte count in chronic myeloid leukemia. Acta Haematol 63(2): 68–70
70. Rakhra K, Bachireddy P, Zabuawala T, Zeiser R, Xu L, Kopelman A et al (2010) $CD4^{+}$ T cells contribute to the remodeling of the microenvironment required for sustained tumor regression upon oncogene inactivation. Canc Cell 18(5):485–498
71. Prehn RT (1983) The dose-response curve in tumor-immunity. Int J Immunopharmacol 5(4):255–257
72. Sampson D, Peters TG, Lewis JD, Metzig J, Kurtz BE (1977) Dose dependence of immunopotentiation of tumor regression induced by levamisole. Canc Res 37(10):3526–3529
73. Prehn RT (1972) The immune reaction as a stimulator of tumor growth. Science 176(31): 170–171
74. Jeejeebhoy HF (1974) Stimulation of tumor growth by the immune response. Int J Canc 13(4):665–678
75. Umiel T, Trainin N (1974) Immunological enhancement of tumor growth by syngeneic thymus-derived lymphocytes. Transplantation 18(3):244–250
76. Hanahan D, Weinberg RA (2011) Hallmarks of cancer: The next generation. Cell 144(5): 646–674

77. Lin WW, Karin M (2007) A cytokine-mediated link between innate immunity, inflammation, and cancer. J Clin Invest 117(5):1175–1183
78. Quesnel B (2008) Tumor dormancy and immunoescape. APMIS 116(7–8):685–694
79. Chan L, Hardwick NR, Guinn BA, Darling D, Gäken J, Galea-Lauri J et al (2006) An immune edited tumour versus a tumour edited immune system: Prospects for immune therapy of acute myeloid leukaemia. Canc Immunol Immunother 55(8):1017–1024
80. Schatton T, Frank MH (2009) Antitumor immunity and cancer stem cells. Ann NY Acad Sci 1176:154–169
81. Noh KH, Kang TH, Kim JH, Pai SI, Lin KY, Hung CF et al (2009) Activation of Akt as a mechanism for tumor immune evasion. Mol Ther 3:439–447
82. Galante A, Tamada K, Levy D (2012) B7-H1 and a mathematical model for cytotoxic t cell and tumor cell interaction. Bull Math Biol 74:91–102

Part IV
Mathematical Biosciences and Dynamical Systems Modeling

Chapter 11
Regulation of Tumor Dormancy and Role of Microenvironment: A Mathematical Model

Yangjin Kim and Khalid Boushaba

Abstract Herein, a mathematical model of a molecular control system for the regulation of secondary tumors is formulated and analyzed to explore how secondary tumors can be controlled by a primary tumor with/without a surgery and the microenvironment. This control system is composed of fibroblast growth factor-2 (FGF2), urokinase-type plasminogen activator (uPA), plasmin, transforming growth factor-beta (TGFβ), latent TGFβ (LTGFβ), and tumor density. The control of secondary tumors by primary tumors was first modeled by Boushaba, Nilsen-Hamiton and Levine in [46]. The model is based on the idea that the vascularization of a secondary tumor can be suppressed by inhibitors from a larger primary tumor. The emergence of tumors at secondary sites 5–7 cm from a primary site was observed after surgical removal of the primary tumor in silico. The model supports the notion that the fate of secondary tumors after surgery depends on the distance from the primary tumor and the surrounding microenvironment. As such, the primary tumor did not influence the growth of remote secondary tumors, but it could effectively suppress the growth of the secondary tumors if they were too close to the primary tumor, even after it was removed. Thus, the model predicts the emergence of secondary tumors after the excision of the primary tumor when the distance between these tumors is in the "distance window." It also predicts that the growth behaviors of the secondary tumors depend on the local microenvironment. Based on these findings, we propose several treatment options for better clinical outcomes.

Keywords Tumor dormancy • Microenvironment • Mathematical model

Y. Kim, PhD
Department of Mathematics and Statistics, University of Michigan Dearborn,
Dearborn, MI, USA

K. Boushaba, PhD (✉)
The Johns Hopkins Carey Business School,
100 International Drive, Baltimore, MD 21202, USA
e-mail: kboushaba@jhu.edu

H. Enderling et al. (eds.), *Systems Biology of Tumor Dormancy*,
Advances in Experimental Medicine and Biology 734,
DOI 10.1007/978-1-4614-1445-2_11, © Springer Science+Business Media New York 2013

Introduction

Tumor dormancy can be observed in both metastases and local recurrences [1] and can be attributed to slow accumulation of genetic mutations that lead to transformation and immortalization through carcinogenesis [2]. Dormancy of loco-regional lesions refers to a pause in cancer progression after clinical treatment of a primary lesion [3, 4]. Such "cancer-free" periods have been shown to last as long as 20–25 years [5, 6], implying that a pause in cancer progression occurs often in the different forms of dormancy, though the exact timeframe and basic mechanism underlying this pause in different cancers remain unclear. It is well documented that the survival rate in several cancers, including prostate, breast, colorectal, and lung cancers, is relatively high for patients who have tumors with a low probability of metastases, whereas it is quite low for patients with distant metastases [7]. Indeed, about 90 % of all cancer deaths are associated with formation of metastasis [8, 9]. Because local recurrences or metastases are derived from local or remaining disseminated tumor cells after the treatment of a diagnosed cancer, it is important to understand the dormancy of these cells in order to prevent subsequent disease [1].

The fate of disseminated tumor cells may be determined by their ability to adapt to the non-orthotopic tissue microenvironment in which they lodge [10]. A cell's success or failure to resume proliferation in this new microenvironment may depend on a balance between stress signals from the new non-permissive microenvironment and the cell's ability to remodel and/or reproduce a niche or microenvironment that is conducive to expansion [11, 12]. The ability of tumor cells to attain a differentiated state might also result in cellular tumor dormancy. For example, in breast cancer, rapidly proliferating tumor cells lead to a loss of tissue architecture, whereas blocking beta1-integrins and epidermal growth factor (EGF) signaling has been shown to induce the re-differentiation of tumor cells into growth arrest, resulting in the formation of normal acinar structures [13]. Recent studies indicated that microRNAs might play a significant role in controlling the switch between the proliferating and migratory phases of glioblastoma cells [14]. In particular, miR451 is known to play a key role in glioblastoma growth by modulating the balance of active proliferation (high miR451) and invasion (low miR451) via the miR451–CAB39/LKB1–AMPK pathway in response to metabolic stress in the microenvironment [14, 15]. In head and neck carcinoma [16], this switch between growth arrest and proliferation was shown to be controlled by tumor–microenvironment interactions using mutual antagonism between uPAR–FAK–RAS–ERK [17, 18] and CDC42/GTP–p38 pathways [1, 16]. Therefore, strategies of inhibiting tumor cell-microenvironment crosstalk have been suggested as a reasonable way of achieving, inducing, and/or maintaining tumor cell dormancy, by, for example, blocking receptor tyrosine kinases and uPAR [1].

The surgical removal of a malignant tumor from a host is insufficient to prevent the cancer from reoccurring in the host and can lead to rapid cancer recurrence [19–21]. Using animal models, researchers have well documented that the regulation of major factors in metastasis development can be perturbed by partial or whole surgery [22–26]. For example, removal of a primary tumor can result in accelerated growth of another smaller, residual tumor [25], a switch of stable foci to the angiogenic phenotype that enhances metastasis [26], and changes in the growth kinetics of metastatic foci [21]. There is increasing evidence that supports the rapid acceleration of residual neoplastic disease after tumor cytoreduction [27]. Tumor removal is usually followed by additional treatments, such as peri-operative chemotherapy [26, 28–30], immunotherapy [31, 32], biomodulation [33–35], and/or use of anti-endotoxin agents [36, 37]. Upon assessing data of 120 patients who underwent surgery for metastatic lung cancers between 1975 and 1997, Maniwa et al. [29] found that the early recurrence of pulmonary foci after metastasectomy resulted in poor prognosis. Therefore, longer survival might be achieved by developing a strategy for regulating dormant metastasis.

Zetter [38] suggested that growth inhibitors secreted by a primary tumor might prevent the vascularization of secondary metastatic tumors. Inhibitors with a longer half-life can travel farther, whereas growth factors with a short half-life are localized at the primary site due to degradation, binding to the extracellular matrix (ECM), or deactivation. Among many possible other explanations, the immune response has been long recognized for its role in regulating tumor growth [39–41] and could be a major cause of induction of tumor dormancy [19, 42]. Although the immune system kills most tumor cells in a challenge injection [39, 43], some residual cells persist, leading to clinical dormancy by the immune system [40, 44]. It is not clear whether immune-surveillance controls the dormancy of non-virally induced tumors [45], or how the immunological response is correlated with angiogenic dormancy or cellular dormancy [1].

Herein, we report the recent development of mathematical models of tumor dormancy [46, 47] that include interactions between a primary tumor and secondary tumors via a biomedical network of growth factors and inhibitors (Fig. 11.1). We investigate both the distance-dependent behavior of secondary tumors after the surgical excision of a primary tumor and the role of microenvironment in tumor dormancy and finally suggest possible strategies of therapeutic intervention. We present our results in the Results Section and discuss those results and our future work in the Discussion Section. Finally, we describe the mathematical model in the Materials and Methods Section.

Results

Distance-Dependent Modulation of Secondary Tumors

The control mechanism of a primary tumor has to rely on the diffusion process from the primary site to surrounding tissue and it should depend on the distance from the secondary tumor site. For the simulations in Figs. 11.2, 11.3, 11.4, and 11.5, a primary (secondary) tumor is implanted near $x=L$ ($x=0$). The initial mass of the primary tumor

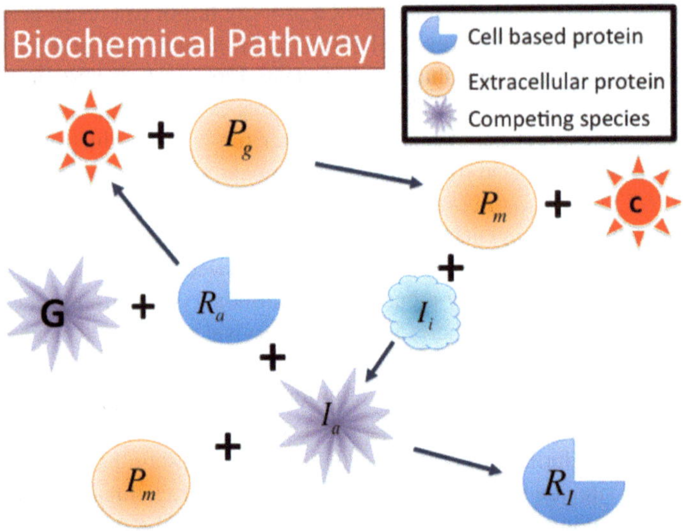

Fig. 11.1 Overview of the biochemical pathway. Active TGFβ (I_a) regulates the active and inactive receptors (R_a and R_i, respectively) on tumor cells. Latent TGFβ (I_i) is activated by plasmin (P_m), which is released from plasminogen (P_g) by uPA (C) from tumor cells. Tumor cells express uPA in response to FGF (G)

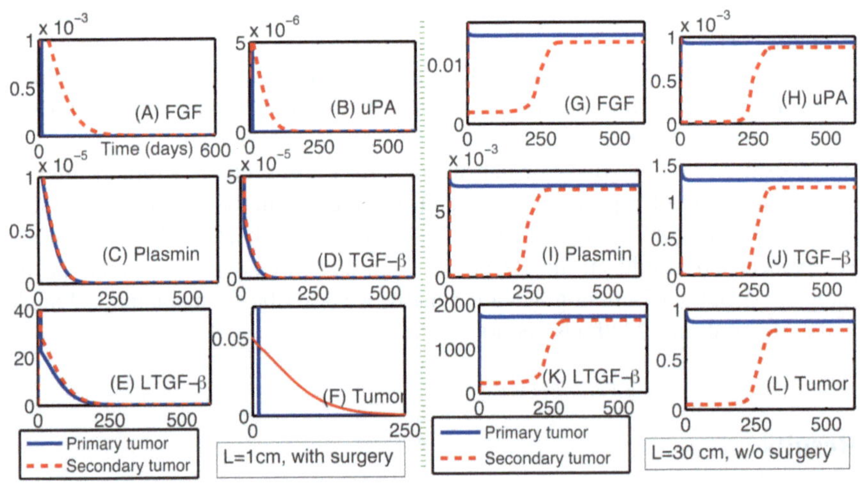

Fig. 11.2 **(a–f)** Time-course of molecular levels (FGF, uPA, plasmin, TGFβ, LTGFβ) and tumor populations at the primary and secondary sites with surgery when the secondary tumor is close to the primary tumor ($L = 1$ cm). **(g–l)** Time-course of all molecular levels and tumor populations without surgery when the secondary tumor is very remote from the primary tumor ($L = 30$ cm)

is larger than that of the secondary tumor. First, we studied the behavior of the secondary tumor in response to surgical resection of a primary tumor for two obvious cases: the secondary tumor near or far away from the primary tumor.

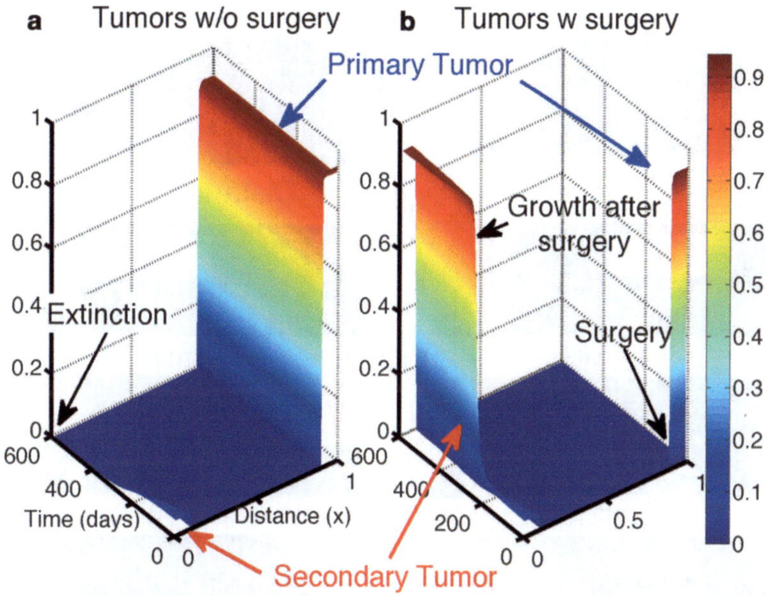

Fig. 11.3 Behavior of a secondary tumor with and without surgery. The secondary tumor begins to grow after surgery (**b**) at $t=10$ days, whereas it becomes extinct without surgery (**a**). $L=6$ cm, where L is the distance between the primary and secondary tumors

Figure 11.2a–f show a time-course of concentrations of growth factors and inhibitors (FGF, uPA, plasmin, TGFβ, LTGFβ) and tumor populations when the secondary tumor is very close to the primary tumor ($L=1$ cm). The primary tumor was removed on day 10. The secondary tumor was effectively controlled by the primary tumor, and even surgical removal of the primary tumor did not induce growth of the secondary tumor (see Fig. 11.2f). On the other hand, the secondary tumor is not affected by the primary tumor when the former is very remote from the latter. For instance, the secondary tumor grew with/without the excision of the primary tumor (see Fig. 11.2l). Figure 11.2g–l show a time-course of all variables without surgery for the remote secondary tumor ($L=30$ cm).

We now study a non-trivial case, i.e., when the secondary tumor exists in the middle range of distance from the primary tumor. As discussed above (cf. Fig. 11.2), the primary tumor has a strong control mechanism of the secondary tumor at nearby sites, but it loses its ability to inhibit a remote secondary tumor. Keeping in mind the "diffusion-dependent" control scheme, we expect to see the onset of nonlinear dynamics of the system as the distance between the primary and secondary tumors is decreased. Figure 11.3 shows different growth behaviors of the secondary tumor in the middle range of the distance ($L=6$ cm) from the primary tumor with/without surgery. While the secondary tumor remained dormant (Fig. 11.3a) without surgery, the secondary tumor started growing at day 10 after the surgery. In the presence of the primary tumor, the secondary tumor died out due to the inhibition mechanism from the primary tumor, and the levels of its growth factors and inhibitors at the

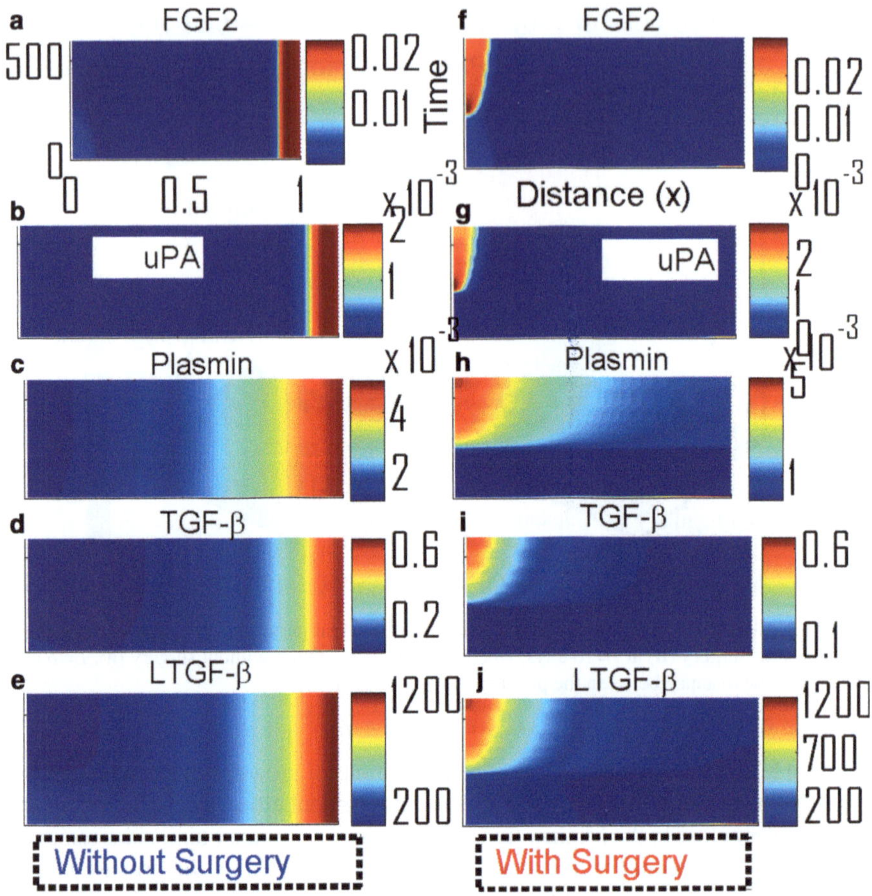

Fig. 11.4 Time evolution of spatial profiles of molecules without (**a–e**) and with (**f–j**) surgery when L is in the distance window ($L = 6$ cm; cf. Fig. 11.3). The secondary tumor ($x = 0$) begins to grow (as shown in Fig. 11.3) after surgery because surgical removal eliminates the inhibition mechanism by the primary tumor. One can observe the localization of molecules near the secondary tumor ($x = 0$). $L = 6$ cm; x-axis = spatial domain [0,1]; y-axis = time

secondary site were decreased. The surgical removal of the primary tumor at day 10 weakened the inhibitory control mechanism of the secondary tumor growth, and the secondary tumor started growing after a delay of roughly 250 days, eventually reaching the size of the original primary tumor. In this case, the levels of the growth factors and inhibitors are increased at the secondary tumor site.

The onset of nonlinear dynamics on growth behaviors of the secondary tumor in the middle range implies nonlinear behaviors of the growth factors and inhibitors in the system exist with and without surgery (cf. Fig. 11.3). Dynamics of these players reveal the underlying mechanism behind the quite opposite growth results—dormancy and growth—of the secondary tumor. Figure 11.4 shows the time evolution of the

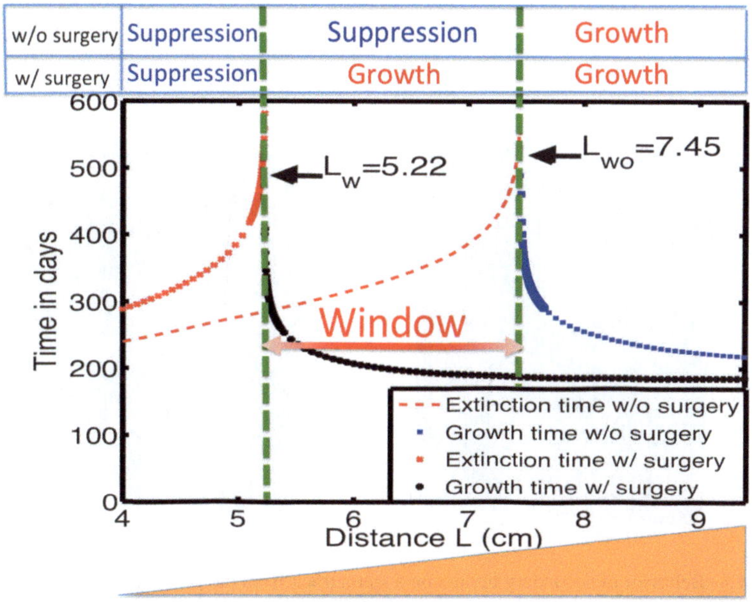

Fig. 11.5 Effect of surgery on growth and extinction of a secondary tumor with various distances between the primary and secondary tumors. There is a window (5.22, 7.45 cm) of the distance (L) from a primary tumor where secondary tumors under tight control of the primary tumor begin to grow only after surgical removal of the primary tumor

profiles of molecules from the simulation in Fig. 11.3 (L=6 cm): FGF2, uPA, plasmin, TGFβ, LTGFβ. Although the profile of all molecules remained constant without surgery, there were many dynamical changes in the profiles of all variables near the secondary tumor site (x=0) after surgery. Without surgery, the inhibitors transported from the primary tumor via the diffusion process were sufficient to suppress the secondary tumor growth. The changes after surgery included a dramatic shift of the high levels of all molecules from the primary tumor site (x=1) to the secondary tumor site (x=0), followed by a localization of these variables near the secondary tumor site. The localization of molecules and growth of the secondary tumor are due to a loss of the ability to sustain the inhibitors at the secondary tumor site.

The modulation of secondary tumors by the primary tumor is dependent on the diffusion of promoters and inhibitors. To study the dependence of secondary tumor growth on the distance to the primary tumor, we simulated secondary tumor growth with/without surgery between 4 and 9 cm away from the primary tumor. Figure 11.5 illustrates the qualitative differences in the secondary tumor growth with/without surgery in response to various distances (L) between the secondary tumors and a primary tumor. The model predicted the existence of a distance window $\Omega_\infty = [5.22\text{cm}, 7.45\text{cm}]$, suggesting the following characteristics of each subdomain: (1) If $L < 5.22$ cm, the growth of the secondary tumor will be suppressed

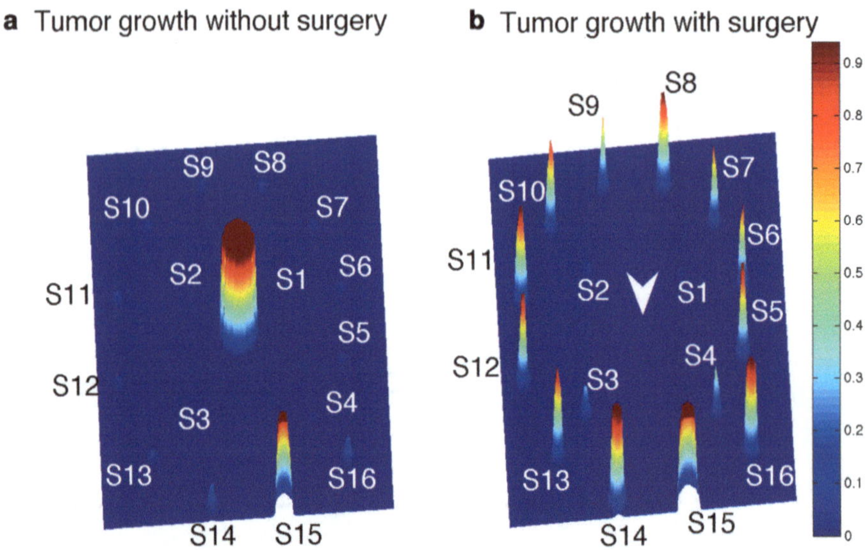

Fig. 11.6 Behavior of secondary tumors on a spiral track from the primary site at the center (0.5, 0.5) of the domain with/without surgery at final time in two-dimensional domain. (**a**) Without removing the primary tumor at the center, the secondary tumors at sites $S_i\,(i = 1,\ldots,13)$ stay in the dormant stage. (**b**) After surgery at $t=10$ days, the secondary tumors in the middle range ($S_i, i = 3,\ldots,13$) begin to grow, whereas those close to the primary tumor (S_1, S_2) remain dormant. The secondary tumors ($S_i, i = 14,\ldots,16$) remote from the primary tumor grow regardless of surgery. S_i=the location of the *i*th secondary tumor $(i = 1,\ldots,16)$

regardless of the time of surgery, even if the time to shrink to less than 1 % of its original size increases as L increases to L_∞ from below; and (2) The growth behavior of the secondary tumor in Ω_∞ will depend on surgery, i.e., the growth arrest without surgery and growth with surgery; and (3) When the secondary tumor is too far away from the primary tumor ($L > L_\infty$), it will grow regardless of surgery.

Role of Microenvironment

The microenvironment plays a significant role in cancer progression and metastasis by enhancing local signaling and modifying the ECM locally [48]. To investigate the role of microenvironment, a primary tumor and 16 secondary tumors $(S_i, i = 1,\ldots,16)$ were implanted at the center (0.5, 0.5) of a two-dimensional domain and a spiral track, respectively, with S_1 being the closest to the primary tumor and S_{16} located at the farthest field from the primary tumor (see Fig. 11.6). Figure 11.6 shows the final profiles of the tumors without (a) and with surgery (b) at $t=300$ days. Figure 11.7a, b also shows a time-course of the densities of the tumors at the secondary sites ($S_i, i = 1,\ldots,16$). As in the one-dimensional case shown in Fig. 11.5, we came to the following conclusions: (1) the secondary tumors ($S_i, i = 1,2$) near

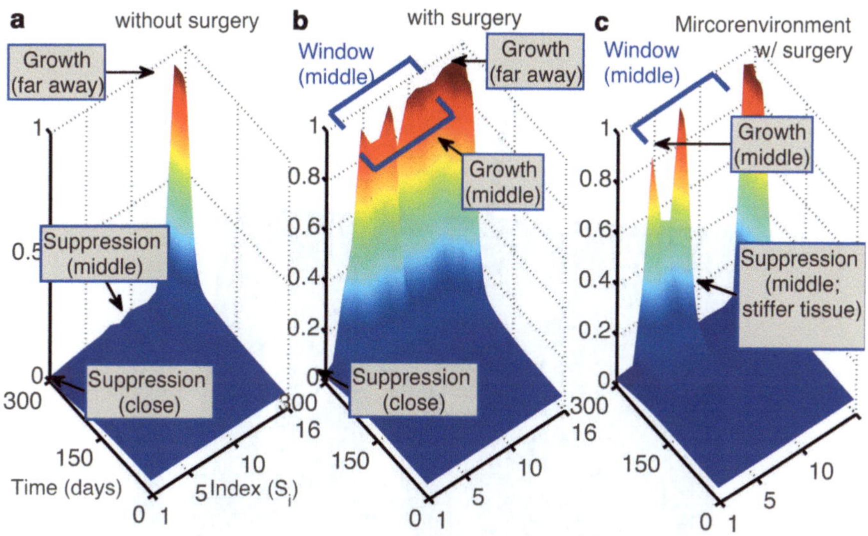

Fig. 11.7 Time evolution of tumor populations at the secondary sites (S_i $(i = 1,\ldots,16)$ on a spiral track in Figs. 11.6 and 11.7) without surgery (**a**), with surgery (**b**), and with surgery in a different microenvironment (**c**). The growth behavior of secondary tumors before/after surgery depends on the distance from the primary tumor site (0.5, 0.5) at the center. The growth behaviors of the secondary tumors in the distance window (those in the *middle range*) after surgery are shown in (**b**), but their growth can be inhibited in the harsh microenvironment (stiffened tissue and slow diffusion of inhibitors) in (**c**)

the primary tumor are closely controlled by the primary tumor and remain dormant with or without surgery; (2) The secondary tumors ($S_i, i = 3,\ldots,13$) in the middle range of distance from the primary tumor remain dormant without surgery but begin to grow after surgery; and (3) The primary tumor cannot inhibit growth of secondary tumors ($S_i, i = 14,\ldots,16$) at the remote sites.

ECM stiffening accounts for changes in the biophysical properties of the tissue [49, 50] and may lead to different efficacies of drug delivery and biochemical control of growth factors and inhibitors in a given system. To study the active, not passive, role of ECM in regulation of growth factors and inhibitors, we tested the effect of space-dependent diffusion coefficients in tumor growth. As shown in Fig. 11.8, we investigated the role of the microenvironment in suppression or growth of secondary tumors $(S_i, i = 3,\ldots,13)$ in a distance window after surgery. We tested whether the different tissue composition in the heterogeneous microenvironment, leading to different diffusion coefficients of growth factors and inhibitors (or random motility of cells), would have any influence on growth of the secondary tumors after surgery in Fig. 11.6. We found that the secondary tumors ($S_i, i = 3,\ldots,13$) remained dormant on the left side of the domain (where smaller diffusivity of LTGFβ was used; $x<0.5$), whereas those ($S_i, i = 4,\ldots,8$) on the right side of the domain (where the normal diffusion coefficient of LTGFβ was used; $x>0.5$) still

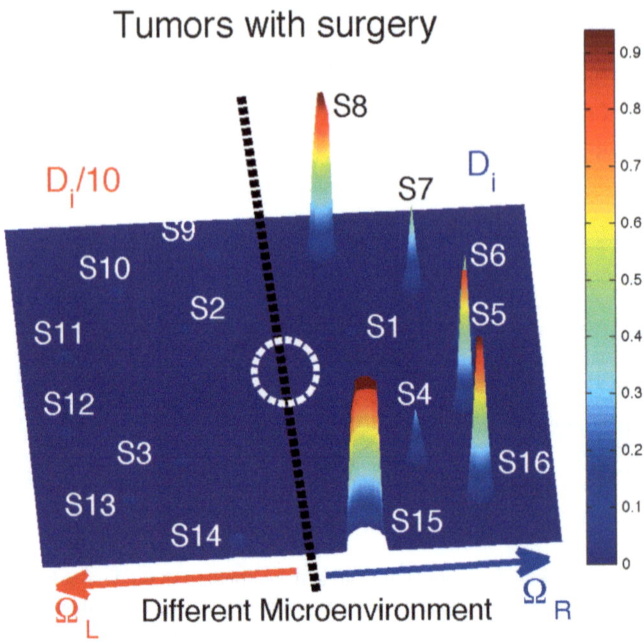

Fig. 11.8 Role of microenvironment (ECM composition) in secondary tumor growth after surgery. A smaller (0.1-fold) diffusion coefficient of the LTGFβ was assumed on the *left half* domain ($\Omega_L = [0,0.5]\times[0,1]$) keeping the same diffusion coefficient on the *right half* domain ($\Omega_R = [0.5,1]\times[0,1]$), as in Fig. 11.6. Growth of the secondary tumors in the distance window on the *left* (Ω_L) was inhibited even after surgery compared to the growth in the control case (growth) on the *right* (Ω_R) (same as in Fig. 11.6)

grow after surgery, as in Fig. 11.6b. This suggests that while LTGFβ, the growth inhibitor, loses its ability to suppress secondary tumors in the distance window after surgery in "normal" or homogeneous tissue, it might still be able to control the growth of the secondary tumors in the middle range in more tough or "stiffened" microenvironment. Tissue stiffening due to ECM deposition from stromal cells, such as fibroblasts, in the cancer microenvironment has been reported in the development of cancers, including breast cancer [49, 50].

Therapeutic Approach

To prevent secondary tumor growth, we propose several therapeutic approaches in the context of the mathematical model. The idea is to achieve better control of the primary tumor's inhibitory control mechanism and to interfere with the system by injecting the necessary molecules for better clinical outcomes. As shown in Figs. 11.9 and 11.10, we investigated the effect of the injection of either LTGFβ

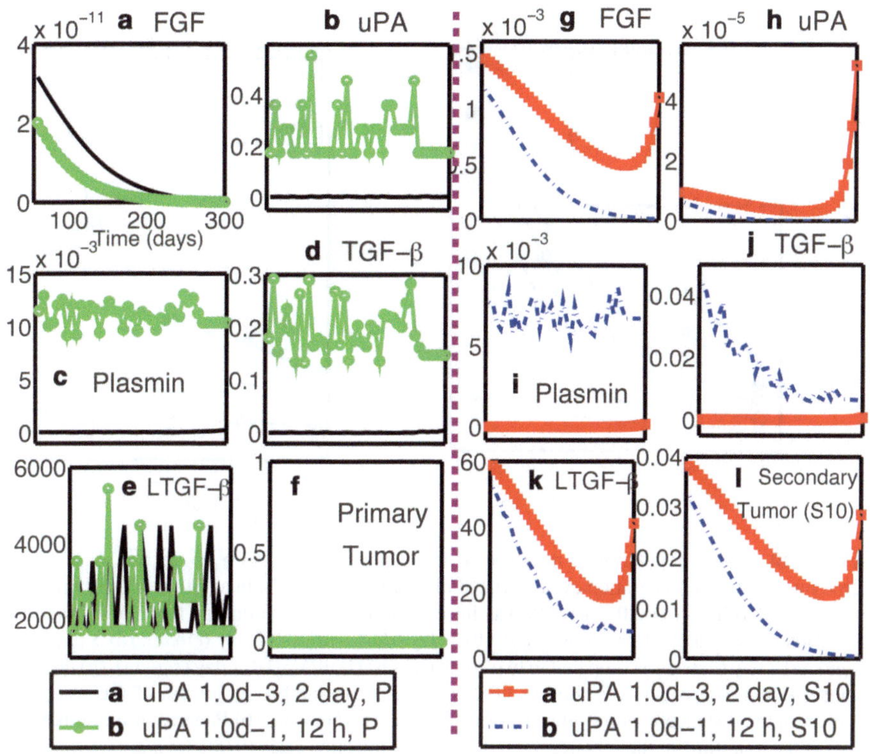

Fig. 11.9 Therapeutic implications. Panel **(a)**, **(b)**, **(c)**, **(d)**, **(e)** and **(f)** and f are time course of concentrations of molecules FGF, uPA, plasmin, TGFβ, and LTGFβ and tumor populations respectively at the primary site (0.5,0.5), and panel g, h, I, j, k and l are time course of the same molecules at the secondary tumor site (S_{10}). Periodic injections of uPA at the low (0.001) and high (0.1) rates at the primary site (0.5,0.5) : administration of uPA every 48h (a) and 12h (b), respectively. Low levels of uPA from periodic injections (black curve in (B); every 48h) at the center of the domain is not enough to prevent the secondary tumor (S_{10}) from growing (red line in (L)) while relatively frequent injections of uPA at the higher rate (0.1) (green line in (B); every 12h) induce the inhibition (blue dotted line in (L)) of tumor growth at the secondary site,. x-axis=Time (days). y-axis=concentration of molecules or tumor population at the primary and secondary site

alone or both LTGFβ and uPA on the inhibition of growth of secondary tumors. These molecules, LTGFβ and uPA, were introduced into the system immediately after surgical removal of a primary tumor in a periodic fashion. The growth behavior of the secondary tumors is determined by the levels of the diffused growth factors and inhibitors from the periodic injections at the primary tumor site. Figure 11.9a–f illustrate the fluctuating levels of almost all molecules (uPA, plasmin, TGFβ, LTGFβ) except FGF at the primary tumor site (0.5, 0.5), where relatively large amounts of uPA (0.1) were administrated every 12 h (green solid lines). Figure 11.9g–l show the corresponding time-course (blue dotted lines) of levels of all molecules (FGF, uPA, plasmin, TGFβ, and LTGFβ) and cell density at one of the secondary tumor sites (S_{10}). In this case, the diffused molecules inhibited the growth

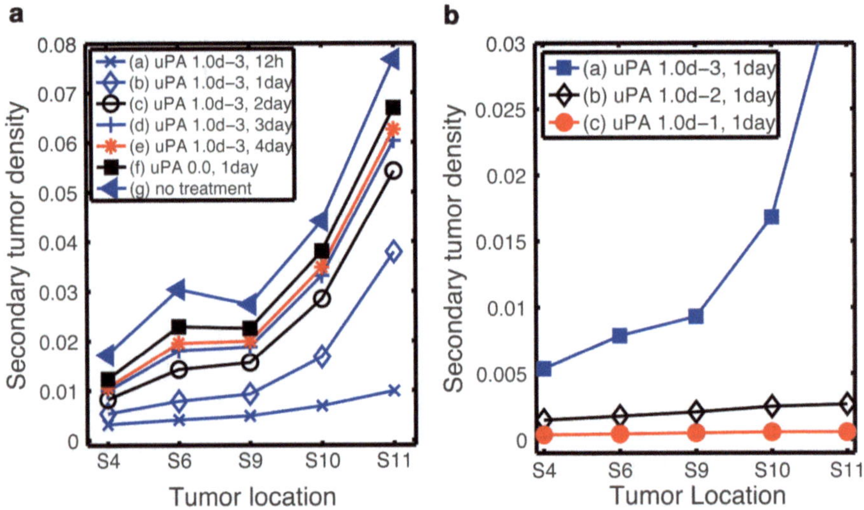

Fig. 11.10 Tumor population at the selected secondary tumor sites ($S_4, S_6, S_9, S_{10}, S_{11}$) at final time for various administration cycles (**a**) and uPA injection rate (**b**) in addition to fixed injection amount of LTGFβ (1,000.0). (**a**) uPA was injected at a fixed injection rate (0.001) with various injection cycles: (a) 12 h (*multiple sign*), (b) 1 day (*diamond*), (c) 2 day (*circle*), (d) 3 day (*plus sign*), (e) 4 day (*asterisk*) compared to two control cases: (f) uPA = 0.0, 1 day (*square*), (g) no treatment (*triangle*). Frequent injection of uPA is necessary to inhibit the growth of the secondary tumors in the distance window after the primary tumor is removed. (**b**) Patterns of tumor population at the final time at the secondary site for various injection rates (0.001 (a; *blue square*), 0.01 (b; *black diamond*), 0.1 (c; *red circle*)) with a fixed administration cycle (1 day). Higher amounts of uPA need to be injected into the system to suppress the secondary tumor growth after surgery

of the secondary tumor (S_{10}), which eventually died out. With the exception of the fluctuating plasmin level, the levels of all other molecules, especially growth factor (FGF), were decreased over time. On the other hand, the injection of smaller amounts of uPA (0.001) over a longer period (2 days) led to growth of the secondary tumor after surgery (Fig. 11.9l). One also notices slower decay of FGF and low levels of other molecules at the primary site, leading to weak control over the secondary tumors in the neighborhood.

Figure 11.10a shows the tumor density at the selected secondary tumor sites ($S_4, S_6, S_9, S_{10}, S_{11}$) at the final time for various injection schedules of uPA, as well as the fixed injection levels of LTGFβ (1,000). The control case is when no treatment [(g) in Fig. 11.10a] was provided. Effects from regular dosage of LTGFβ daily without uPA [(f) in Fig. 11.10a] are too small to suppress the growth of tumors at the secondary sites. As the frequency of the uPA injection was increased, the populations of the secondary tumors ($S_4, S_6, S_9, S_{10}, S_{11}$) at the final time were decreased. When the injection cycle was smaller than 12 h, those injected molecules of LTGFβ and uPA could prevent the secondary tumors from growing. In Fig. 11.10b, a fixed

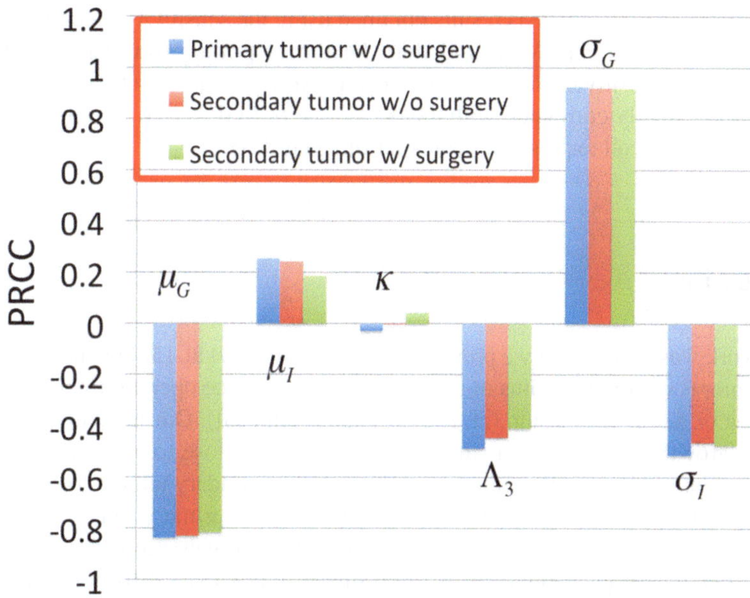

Fig. 11. 11 Sensitivity analysis: partial rank correlation coefficient (PRCC) was calculated for our model using general Latin hypercube sampling (LHS) and methods described in [51]. The reference outputs are tumor populations at the primary and secondary sites at time $t=300$ days with/without surgery at $t=10$ days. The following parameters were taken as the parameter of interest: decay rate of FGF (μ_G), decay rate of LTGFβ (μ_I), Hill-type constant of tumor growth (κ), conversion rate of TGFβ to LTGFβ via plasmin (Λ_3), cell expression of FGF (σ_G), cell expression of LTGFβ (σ_I)

amount of LTGFβ (1,000.0) was administrated in addition to various doses of uPA in the system. Higher doses of uPA (greater than 0.01) were required to suppress the growth of the secondary tumors for a fixed schedule (1 day). These observations lead to the following hypotheses being tested: (1) More frequent injections of uPA in addition to the fixed injection of LTGFβ are necessary to achieve a favorable outcome; and (2) Higher levels of uPA doses must be introduced into the system for a less frequent administration schedule. In a clinical setting with patient-specific data, the model will help us estimate an optimal injection cycle and doses of uPA, as well as the appropriate levels of LTGFβ.

Sensitivity Analysis

The current model includes some parameters for which experimental data is currently unavailable. To determine how sensitive the tumor population is to these parameters, we performed sensitivity analysis of the mathematical model. Figure 11.11

shows the sensitivity analysis results obtained using the sensitivity tools developed by Marino et al. [51]. We found that the tumor population at the final time (day 300) was positively correlated to the cell expression parameter of FGF2 (σ_G), but was negatively correlated with the decay rate of FGF2 (μ_G), conversion rate of TGFβ (Λ_3), and cell expression parameter of LTGFβ (σ_1).

Discussion

This work was motivated by a previous report by Giorgi et al. [52] on the recurrence of a tumor ("within the radius of roughly 5–7 cm from the operation scar") after an excision of a polypoid lesion (cf. Figs. 15 and 16 in [46]). Our model predicts that (1) the growth/shrinkage of a small secondary tumor close to a primary tumor is independent of the growth of the latter; (2) a small secondary tumor close to a large primary tumor should still be under the control of the primary tumor and remain benign due to inhibitors from the latter; and (3) the dormant secondary tumor will grow after the surgical removal of the primary tumor if its diffusion distance is within a certain distance window. However, this notion may be valid only in homogeneous microenvironment. Through two-dimensional simulations, we investigated the role of inhibitors and the microenvironment in growth of the secondary tumors after surgery and explored several possible therapeutic strategies. This model provides a basis for investigating several aspects of therapeutic approaches that have not been explored in the context of re-survival of tumor after the excision of a primary tumor.

It is well established that tumor microenvironment affects tumor growth and metastasis [53–56]. Tumor microenvironment consists of cellular materials, waste products, proteins, ECM, as well as various cell types (e.g., inflammatory cells, epithelial cells, fibroblasts, endothelial cells). These different cell types are involved, by means of growth factors and inhibitors, in a complex crosstalk in which they influence each other's behavior. For example, tumor cells have been shown to recruit fibroblasts/myofibroblasts to promote growth via diffusible molecules, such as EGF and TGFβ, in the early development of breast cancer [48, 53, 54]. The genetic progression of a tumor might be influenced by the microenvironment and host genetics [1]. Understanding the interaction between the tumor and microenvironment might lead to important new therapeutic strategies for inhibiting cancer progression. For example, one could block the host-tumor interaction by targeting stromal cells while simultaneously eliminating the tumor cells [48]. The model predicted that growth behaviors of secondary tumors after surgery of a primary tumor might depend on the microenvironment, for instance, the different diffusion coefficient (D_i) of the LTGFβ inhibitor. Low diffusivity of LTGFβ, which may depend on several factors such as interstitial fluid pressure, tissue composition, and geometry [57], led to the inhibition of secondary tumor growth (see Fig. 11.7). ECM components as well as an increased number of other cells may also contribute to the heterogeneity of tissue, resulting in the different diffusivity of inhibitor (LTGFβ). The current work obviously must be improved to take into account contributions from other cells. Recently, Eikenberry et al. [42] found that metastases may be suppressed by the immune response directed against a primary tumor. While our attempt to understand the role of microenvironment in secondary tumor growth is limited to diffusion of molecules in different microenvironments, our results in this direction serve as a starting point to investigate the specific role of active players of microenvironment, such as stromal cells, in secondary tumor growth.

Surgical removal of a tumor is typically followed by additional treatments such as chemotherapy or radiation therapy. To improve outcomes of surgery, several follow-up treatment options involving inhibitors were suggested in our model. Focusing on an inhibitor-injection method, the model predicted that the best results were produced when both LTGFβ and uPA were injected into the system together, rather than injecting only one molecule, which also suggested a critical cycle time (τ_c) to optimize the expense and favorable outcome. Further experiments must be done to validate our model. As one treatment option, many drugs have been developed to block certain popular mutated pathways. For instance, an antibody against the product of Her2/Neu, a secondary EGF receptor that is often over-expressed in breast cancer, was shown to be effective in inducing G1 cell-cycle arrest [58], which leads to favorable preliminary clinical results in breast cancer patients [59]. To get the whole picture of complex governing dynamics, one may benefit from a multi-scale modeling approach [48], whereby different sub-models in different time and space scales are used altogether to integrate the system at the different levels without losing too much information. The present model serves as only one possible explanation for tumor dormancy and distance-dependent recurrence of secondary tumors after surgery. The complex geometry of organs and altered genetic networks of secondary tumor cells in a given microenvironment could also affect the balance between growth and suppression of secondary tumors at a different level. We will address these issues in future work.

Materials and Methods

Mathematical Model

Tumor cell density changes can be affected by cell migration and the birth/death rate of tumor cells, i.e., tumor cell density change = cell migration + cell proliferation−cell death. Then, using the notion of reinforced (or biased) random walks [60–62], the governing equation for the tumor cell density (η) is given as follows:

$$\frac{\partial \eta}{\partial t} = -\nabla \cdot \left\{ D_\eta \nabla \left[\eta \ln \left(\frac{\eta}{\tau_1(g)} \right) \right] \right\} + \frac{\lambda g}{K+g} \eta (1-\eta/\eta_0) - \mu_\eta \eta \qquad (11.1)$$

Interactions between two tumor masses are mediated by secreted proteins in a biochemical network shown in Fig. 11.1. The reactions between the players in Fig. 11.1 can be summarized by a system of partial differential equations of FGF (g), uPA (c), plasmin (p_m), TGFβ (i_a), LTGFβ (i_i):

$$\begin{aligned}
\frac{\partial g}{\partial t} &= D_g \Delta g + \frac{\sigma_g - \lambda_1 g}{1 + v_e i_a + g / K_m^1} \frac{\eta}{\eta_0} - \mu_g g \\
\frac{\partial c}{\partial t} &= D_c \Delta c + \frac{\lambda_1 g}{1 + v_e i_a + g / K_m^1} \frac{\eta}{\eta_0} - \mu_c c \\
\frac{\partial p_m}{\partial t} &= D_p \Delta p_m + \lambda_2^p c - \mu_p p_m \\
\frac{\partial i_a}{\partial t} &= D_a \Delta i_a + \lambda_3 i_i p_m - \mu_a i_a \\
\frac{\partial i_i}{\partial t} &= D_i \Delta i_i + \frac{\sigma_i}{1 + v_e i_a + g / K_m^1} \frac{\eta}{\eta_0} - \lambda_3 i_i p_m - \mu_i i_i
\end{aligned} \tag{11.2}$$

See Table 11.1 for the definitions and values of parameters [diffusion coefficients ($D_\eta, D_g, D_c, D_p, D_a, D_i$), decay rates ($\mu_\eta, \mu_g, \mu_c, \mu_p, \mu_a, \mu_i$), and other kinetic parameters ($\lambda, K, \sigma_g, \lambda_1, v_e, \eta_0, K_m^1, \lambda_2^p, \lambda_3, \sigma_i$)] in (11.1) and (11.2) above.

Model with Therapy

To incorporate the effect of therapeutic treatments, the following form of modified equations of uPA (c) and LTGFβ (i_i) in (11.2) is used:

$$\frac{\partial c}{\partial t} = D_c \Delta c + \frac{\lambda_1 g}{1 + v_e i_a + g / K_m^1} \frac{\eta}{\eta_0} - \mu_c c + \alpha \tag{11.3}$$

$$\frac{\partial i_i}{\partial t} = D_i \Delta i_i + \frac{\sigma_i}{1 + v_e i_a + g / K_m^1} \frac{\eta}{\eta_0} - \lambda_3 i_i p_m - \mu_i i_i + \beta \tag{11.4}$$

where α and β are the sources of uPA and LTGFβ, respectively. The injection of those molecules is administrated immediately after removal of a primary tumor, with the injection period $\tau_c (= t_{j+1} - t_j, j = 1, \ldots, N_d - 1)$ and the duration τ:

$$\alpha = \sum_{j=1}^{N_d} \alpha_0 I_{\Omega_d} I_{[t_j, t_j + \tau]}, \tag{11.5}$$

Table 11.1 Numerical values used in simulations

Var	Definition	Values	Reference
t^*	Time	1.0 h	
v_e	TGFβ-receptor association rate	$1.1e3(\mu M)^{-1}$	[82, 83]
D_g	Diffusion coefficient of FGF2	$7.92(e-3-e-2)cm^2h^{-1}$	[63]
D_c	Diffusion coefficient of plasmin	$6.48(e-1-e-3)cm^2h^{-1}$	[63]
D_p	Diffusion coefficient of uPA	$7.73e-3-7.73e-2cm^2h^{-1}$	[63]
D_i	Diffusion coefficient of LTGFβ	$6.32e-3-6.32e-1cm^2h^{-1}$	[63]
D_a	Diffusion coefficient of TGFβ	$9.43(e-3-e-2)cm^2h^{-1}$	[63]
μ_g	Decay rate of FGF2	$(14.0-28.0)h^{-1}(21.0)$	[69–71]
μ_p	Decay rate of plasmin	$0.102h^{-1}$	[59, 72]
μ_c	Decay rate of uPA	$8.38h^{-1}$	Estimated
μ_i	Decay rate of LTGFβ	$0.385h^{-1}$	[67]
μ_a	Decay rate of TGFβ	$6.93h^{-1}$	[65, 66, 68]
σ_g	Cell expression of FGF2	$9.7e-3\mu Mh^{-1}$	[81]
σ_i	Cell expression of LTGFβ	$2.05\mu Mh^{-1}$	Estimated
$[p_g]$	Concentration of plasminogen in plasma	$1.0\mu M$	[79, 80]
k_1	FGF association rate with receptor	$2.5e2(\mu Mh)^{-1}$	[63, 73]
k_{-1}	FGF dissociation rate with receptor	$2.88h^{-1}$	[63, 73]
k_2	Induction rate of FGF to uPA	$1.59e2h^{-1}$	Estimated [74–76]
K_m^1	$=(k_{-1}+k_2)/k_1$	$=(k_{-1}+k_2)/k_1$	Estimated
k_4	Dissociation rate of uPA-plasminogen complex CPg	$2.68e3h^{-1}$	[77]
K_m^2	$=(k_4+k_{-3})/k_3$	$25.0\mu M$	[77]
λ_3	$=k_6/K_m^3$	$0.745(\mu Mh)^{-1}$	Estimated
η_0	Carrying capacity of tumor cells	$1.0e9cm^{-3}$	Estimated
λ	Growth rate of tumor cells	$6.25e-3h^{-1}$	Estimated
K	Hill function coefficient of FGF in tumor cell growth	$1.68e-2\mu M$	Estimated
μ_η	Death rate of tumor cells	$1.0e-3h^{-1}$	[78]

$$\beta = \sum_{j=1}^{N_d} \beta_0 I_{\Omega_d} I_{[t_j,t_j+\tau]}, \quad (11.6)$$

where α_0 and β_0 are the injection rates of uPA and LTGFβ, respectively, $I(\cdot)$ is the indicator function, t_1 is the time of surgery for primary tumor removal, N_d is the number of injections, Ω_d is the circular region where the injection occurred: $\Omega_d = \{(x,y) \mid x-05)^2 + (y-0.5)^2 < R_i\}$, where R_i is the injection radius.

Parameter Estimation

Diffusion Coefficients

The diffusion coefficient of FGF (D_g) was estimated using Stokes-Einstein relationship in [63]: $D_g = 5.52\times10^{-5}\,\text{cm}^2/\text{min}$ ($3.3\times10^{-3}\,\text{cm}^2/\text{h}$) and $2.2\times10^{-6}\,\text{cm}^2/\text{s}$ ($7.9\times10^{-3}\,\text{cm}^2/\text{h}$) at 4 and 37°C, respectively. We take $D_g = 7.9\times10^{-3}\,\text{cm}^2/\text{h}$. The other diffusion constants D_c, D_p, D_a, D_i were estimated from the Stokes-Einstein relationship based on D_g and the molecular weights of plasmin, uPA, TGF-β_l, and TGFβ. The diffusion coefficient of large proteins is roughly proportional to $M_w^{-1/3}$, where M_w is the molecular weight of the protein. Also see [64] for further discussion on the importance of the molecule shape in estimating the diffusion coefficient using this assumption.

Decay Rates

An active form of TGFβ has the short half-life of 5–11 min [65, 66]; 2–3 min [67], <11 min [68]. We take the half-life of 6 min, leading to $\mu_a = 6.93\text{h}^{-1}$. In contrast, its latent form TGF-β_l had a much longer half-live (>100 min) [67]. We take $\mu_i = 0.385\text{h}^{-1}$ by assuming the half-life of 108 min. Half-lives of FGF2 in plasma were previously found to be in the range of 1.5–3 min ($\mu_g = (14-28)\text{h}^{-1}$) [69–71]. We took the uPA half-life of 5 min, $\mu_c = 8.38\text{h}^{-1}$. The half-life estimates for plasmin might be from the half-life of plasmin/antiplasmin (PAP) complex, 4.5 h, in [72], and the half-life of plasmin, 14.2 h, in [59]. By taking an intermediate value, 6.8 h, we get $\mu_p = 0.102\text{h}^{-1}$.

Reaction Kinetics

1. The rate constants k_1, k_{-1} were taken from [63, 73], whereas k_2 was estimated as follows: The number of amino acids in uPA was divided by 20 due to the overall cellular transcription-translation rate for proteins, 10–20 amino acids per second [74, 75], leading to an assembly time of 0.012 h per mRNA molecule per cell of a single uPA molecule. Thus, 1.666 protein molecules are assembled from a single mRNA molecule in an hour. From the number of mRNA molecules in breast cancer cells [76], we get the production rate, 45(mRNA molecules/cell)×1.6666 = 75.0 (molecules/h/cell). Since there are roughly $10^{12}-10^{13}$ cells/L, there are between $75.0\times10^{6+12}/6\times10^{23} = 1.25\times10^{-4}$ micromoles and 1.25×10^{-3} micromoles of uPA being produced per hour per liter of cells. Assuming the equilibrium status of the enzyme equation, the absence of inhibitor, and the growth factor concentration of $K_m^1/2$ in addition to the uPA concentration, $[\text{C}] = 6.25\times10^{-4}\,\mu\text{M}$ and $\mu_c = 8.38\text{h}^{-1}$, we get $\mu_c[C] = k_2/3 = k_2\delta\eta_0/3 = 5.3\times10^{-3}\,\text{h}^{-1}$ or $k_2 = 1.59\times10^{-2}\,\text{h}^{-1}$.

2. We obtained $k_4, K_m^2 (k_{cat}, K_m)$ for the enzyme reaction $P_g + uPA \leftrightarrow [P_g uPA] \rightarrow P_m + uPA$ from [77]. Using the value in uPA solution, we get $(k_{cat}, K_m) \approx (0.73/\text{s}, 25\mu\text{M})$ (converted to reciprocal hours) for the tabular entries.

Others

(1) The constant λ is roughly 1/32 of an hour using the turnover rate, 32 h, for malignant cells, which is dependent on the tumor cells under consideration. (2) The carrying capacity of the tumor cells (η_0) is based on a cell volume of $10^3\,\mu\text{m}^3$ and the apoptosis rate μ_η is taken from [78]. (3) Using the estimated value 84–130 μg/L of plasminogen in plasma [79, 80] and the molecular weight of plasma, 88 micrograms per micromole, we used a value of 1.0 μM for the concentration of plasminogen in plasma. (4) The value of σ_g was estimated as follows: We found a production rate of $2(1/8) \times 3{,}600 = 900$ FGF molecules/cell/h using 2–5 FGF mRNA molecules/cell [81], a translation rate of 20 amino acids/s, and approximately 160 amino acids/molecule. Using an estimated cell volume of $100\mu\text{M}^3$, we estimate that the micromolarity rate is $[900\ \text{h}^{-1}/(6\times10^{23})\ \text{M}^{-1}] \times 10^6\mu\text{M}/[10^2\mu\ \text{M}^3 \times 10^{-12}\ \text{cm}^3 \times \mu\text{m}^{-3} \times 10^{-3} \times \text{cm}^{-3}] = 0.015\mu\text{M/h}$. We used a slightly smaller value, $9.7\text{e}-3\mu\text{M/h}$ for the simulation. (5) For σ_i, we used an mRNA/cell range of around 3–8 molecules/cell for TGFβ for sensitivity analysis. (6) The equilibrium constant, v_e, for the TGFβ inhibition of FGF receptor signaling was estimated using the results of [82, 83], which showed that one can expect an equilibrium constant in the range $5\times10^{-4}\,\mu\text{M} \le \frac{1}{v_e} \le 5\times10^{-3}\,\mu\text{M}$ when TGFβ is fully bound to its own receptors.

Acknowledgments K.Boushaba would like to thank Dr. Howard Levine and Dr. Marit Nilsen Hamilton for many discussions. Y. Kim was supported in part by the Faculty Research Grant at the University of Michigan-Dearborn.

References

1. Aguirre-Ghiso JA (2007) Models, mechanisms and clinical evidence for cancer dormancy. Nat Rev Cancer 7(11):2834–2846
2. Hanahan D, Weinberg RA (2000) The hallmarks of cancer. Cell 100(1):57–70
3. Chambers AF, Groom AC, MacDonald IC (2002) Dissemination and growth of cancer cells in metastatic sites. Nat Rev Cancer 2(8):563–572
4. Heyn C, Ronald JA, Ramadan SS, Snir JA, Barry AM, MacKenzie LT et al (2006) In vivo MRI of cancer cell fate at the single-cell level in a mouse model of breast cancer metastasis to the brain. Magn Reson Med 56(5):1001–1010
5. Karrison TG, Ferguson DJ, Meier P (1999) Dormancy of mammary carcinoma after mastectomy. J Natl Cancer Inst 91(1):80–85
6. Kovacs AF, Ghahremani MT, Stefenelli U, Bitter K (2003) Postoperative chemotherapy with cisplatin and 5-fluorouracil in cancer of the oral cavity and the fluoropharynx-long-term results. J Chemother 15(5):495–502
7. Leaf C (2004) Why we're losing the war on cancer (and how to win it). Fortune 149:76–97

8. Sporn MB (1996) The war on cancer. Lancet 347:1377–1381
9. Glotzman J, Mikula M, Andreas E, Schulte-Hermann R, Foisner R, Beug H et al (2004) Molecular aspects of epithelial cell plasticity; implications for local tumor invasion and metastasis. Mutat Res 566:9–20
10. Boudreau N, Bissell MJ (1998) Extracellular matrix signaling: integration of form and function in normal and malignant cells. Curr Opin Cell Biol 10(5):640–646
11. Wicha MS (2006) Cancer stem cells and metastasis: lethal seeds. Clin Cancer Res 12(19): 5606–5607
12. Aguirre-Ghiso JA, Ossowski L, Rosenbaum SK (2004) Green fluorescent protein tagging of extracellular signal-regulated kinase and p38 pathways reveals novel dynamics of pathway activation during primary and metastatic growth. Cancer Res 64(20):7336–7345
13. Weaver VM, Petersen OW, Wang F, Larabell CA, Briand P, Damsky C et al (1997) Reversion of the malignant phenotype of human breast cells in three-dimensional culture and in vivo by integrin blocking antibodies. J Cell Biol 137(1):231–245
14. Godlewski J, Nowicki M, Bronisz A, Palatini GNJ, Lay MD et al (2010) MircroRNA-451 regulates LKB1/AMPK signaling and allows adaptation to metabolic stress in glioma cells. Mol Cell 37:620–632
15. Kim Y, Roh S, Lawler S, Friedman A (2011) miR451 and AMPK mutual antagonism in glioma cell migration and proliferation: a mathematical model. PLoS One 6(12):e28293
16. Aguirre-Ghiso J, Kovalski K, Ossowski L (1999) Tumor dormancy induced by downregulation of urokinase receptor in human carcinoma involves integrin and MAPK signaling. J Cell Biol 147(1):89–104
17. Liu D, Aguirre-Ghiso J, Estrada Y, Ossowski L (2002) EGFR is a transducer of the urokinase receptor initiated signal that is required for in vivo growth of a human carcinoma. Cancer Cell 1(5):445–457
18. Aguirre-Ghiso JA (2002) Inhibition of FAK signaling activated by urokinase receptor induces dormancy in human carcinoma cells in vivo. Oncogene 21(16):2513–2524
19. Coffey JC, Wang JH, Smith MJ, Bouchier-Hayes D, Cotter TG, Redmond HP (2003) Excisional surgery for cancer cure: therapy at a cost. Lancet Oncol 4(12):760–768
20. Oliver RT (1995) Does surgery disseminate or accelerate cancer? Lancet 346(8989):1506–1507
21. Demicheli R, Valagussa P, Bonadonna G (2001) Does surgery modify growth kinetics of breast cancer micrometastases? Br J Cancer 85(4):490–492
22. Braunschweiger PG, Schiffer LM, Betancourt S (1982) Tumor cell proliferation and sequential chemotherapy after partial tumor resection in C3H/HeJ mammary tumors. Breast Cancer Res Treat 2(4):323–329
23. Fisher B, Gunduz N, Coyle J, Rudock C, Saffer E (1989) Presence of a growth-stimulating factor in serum following primary tumor removal in mice. Cancer Res 49(8):1996–2001
24. Fisher B, Saffer E, Rudock C, Coyle J, Gunduz N (1989) Effect of local or systemic treatment prior to primary tumor removal on the production and response to a serum growth-stimulating factor in mice. Cancer Res 49(8):2002–2004
25. Gunduz N, Fisher B, Saffer EA (1979) Effect of surgical removal on the growth and kinetics of residual tumor. Cancer Res 39(10):3861–3865
26. Holmgren L, O'Reilly MS, Folkman J (1995) Dormancy of micrometastases: balanced proliferation and apoptosis in the presence of angiogenesis suppression. Nat Med 1(2):149–153
27. Coffey JC, Wang JH, Cotter TG, Redmond HP (2003) Cytoreductive surgery enhances tumorogenicity by downregulating mitochondrial apoptosis. Ann Surg Oncol 10:S24
28. O'Reilly MS, Boehm T, Shing Y, Fukai N, Vasios G, Lane WS et al (1997) Endostatin: an endogenous inhibitor of angiogenesis and tumor growth. Cell 88(2):277–285
29. Maniwa Y, Kanki M, Okita Y (2000) Importance of the control of lung recurrence soon after surgery of pulmonary metastases. Am J Surg 179(2):122–125
30. O'Reilly MS, Holmgren L, Shing Y, Chen C, Rosenthal RA, Moses M et al (1994) Angiostatin: a novel angiogenesis inhibitor that mediates the suppression of metastases by a Lewis lung carcinoma. Cell 79(2):315–328

31. Brivio F, Lissoni P, Alderi G, Barni S, Lavorato F, Fumagalli L (1996) Preoperative interleukin-2 subcutaneous immunotherapy may prolong the survival time in advanced colorectal cancer patients. Oncology 53(4):263–268
32. Brivio F, Lissoni P, Perego MS, Lissoni A, Fumagalli L (2001) Abrogation of surgery-induced IL-6 hypersecretion by presurgical immunotherapy with IL-2 and its importance in the prevention of postoperative complications. J Biol Regul Homeost Agents 15(4):370–374
33. Cao Y, O'Reilly MS, Marshall B, Flynn E, Ji RW, Folkman J (1998) Expression of angiostatin cDNA in a murine fibrosarcoma suppresses primary tumor growth and produces long-term dormancy of metastases. J Clin Invest 101(5):1055–1063
34. MacLean GD, Longenecker BM (1994) New possibilities for cancer therapy with advances in cancer immunology. Can J Oncol 4(2):249–254
35. Mandelboim O, Feldman M, Eisenbach L (1992) H-2K double transfectants of tumor cells as antimetastatic cellular vaccines in heterozygous recipients. Implications for the T cell repertoire. J Immunol 148(11):3666–3673
36. Coffey JC, Doyle M, O'Mahony L et al (2001) Probiotics confer protection against perioperative metastatic tumour growth. Ann Surg Oncol 89:643
37. Costa ML, Redmond HP, Bouchier-Hayes DJ (2001) Taurolidine improves survival by abrogating the accelerated development and proliferation of solid tumors and development of organ metastases from circulating tumor cells released. J Surg Res 101(2):111–119
38. Zetter BR (1998) Angiogenesis and tumor metastasis, review. Ann Rev Med 49:407–422
39. Finn OJ (2006) Human tumor antigens, immunosurveillance, and cancer vaccines. Immunol Res 36(1–3):73–82
40. Weinhold KJ, Miller DA, Wheelock EF (1979) The tumor dormant state. Comparison of L5178Y cells used to establish dormancy with those that emerge after its termination. J Exp Med 149(3):745–757
41. Zou W (2005) Immunosuppressive networks in the tumour environment and their therapeutic relevance. Nat Rev Cancer 5(4):263–274
42. Eikenberry S, Thalhauser C, Kuang Y (2009) Tumor-immune interaction, surgical treatment, and cancer recurrence in a mathematical model of melanoma. PLoS Comput Biol 5:e1000362
43. Weinhold KJ, Goldstein LT, Wheelock EF (1979) The tumor dormant state. Quantitation of L5178Y cells and host immune responses during the establishment and course of dormancy in syngeneic DBA/2 mice. J Exp Med 149(3):732–744
44. Matsuzawa A, Takeda Y, Narita M, Ozawa H (1991) Survival of leukemic cells in a dormant state following cyclophosphamide-induced cure of strongly immunogenic mouse leukemia (DL811). Int J Cancer 49(2):303–309
45. Willimsky G, Blankenstein T (2005) Sporadic immunogenic tumours avoid destruction by inducing T-cell tolerance. Nature 437(7055):141–146
46. Boushaba K, Levine HA, Nilsen-Hamilton M (2006) A mathematical model for the regulation of tumor dormancy based on enzyme kinetics. Bull Math Biol 68(7):1495–1526
47. Kim Y, Boushaba K (2011) An enzyme kinetics model of tumor dormancy, regulation of secondary metastases. DCDS-S 4(6):1465–1498
48. Kim Y, Stolarska M, Othmer HG (2011) The role of the microenvironment in tumor growth and invasion. Prog Biophys Mol Biol 106:353–379
49. Krouskop TA, Wheeler TM, Kallel F, Garra BS, Hall T (1998) Elastic moduli of breast and prostate tissues under compression. Ultrason Imaging 20:260–274
50. Paszek MJ, Weaver VM (2004) The tension mounts: mechanics meets morphogenesis and malignancy. J Mammary Gland Biol Neoplasia 9:325–342
51. Marino S, Hogue IB, Ray CJ, Kirschner DE (2008) A methodology for performing global uncertainty and sensitivity analysis in systems biology. J Theor Biol 254(1):178–196
52. Giorgi VD, Massai D, Gerlini G, Mannone F, Quercioli E, Carli P (2003) Immediate local and regional recurrence after the excision of a polypoid melanoma: tumor dormancy or tumor activation? Dermatol Surg 29:664–667

53. Cheng J, Weiner L (2003) Tumors and their microenvironments: tilling the soil commentary re: A.M. Scott et al., A phase I dose-escalation study of sibrotuzumab in patients with advanced or metastatic fibroblast activation protein-positive cancer. Clin Cancer Res 9:1590–1595
54. Sadlonova A, Novak Z, Johnson MR, Bowe DB, Gault SR, Page GP et al (2005) Breast fibroblasts modulate epithelial cell proliferation in three-dimensional in vitro co-culture. Breast Cancer Res 7(1):R46–R59
55. Samoszuk M, Tan J, Chorn G (2005) Clonogenic growth of human breast cancer cells co-cultured in direct contact with serum-activated fibroblasts. Breast Cancer Res 7:R274–R283
56. Yashiro M, Ikeda K, Tendo M, Ishikawa T, Hirakawa K (2005) Effect of organ-specific fibroblasts on proliferation and differentiation of breast cancer cells. Breast Cancer Res Treat 90(3):307–313
57. Kim Y, Stolarska M, Othmer HG (2007) A hybrid model for tumor spheroid growth in vitro I: theoretical development and early results. Math Meth Appl Sci 17:1773–1798
58. Demicheli R (2001) Tumour dormancy: findings and hypotheses from clinical research on breast cancer. Semin Cancer Biol 11(4):297–306
59. Takeda Y, Nakabayashi M (1974) Physicochemical and biological properties of human and canine plasmins. J Clin Invest 53:154–162
60. Davis B (1990) Reinforced random walks. Probab Theory Related Fields 84:203–229
61. Othmer HG, Stevens A (1997) Aggregation, blow up and collapse: the ABC's of taxis and reinforced random walks. SIAM J Appl Math 57:1044–1081
62. Levine HA, Tucker AL, Nilsen-Hamilton M (2002) A mathematical model for the role of cell signaling and transduction in the initiation of angiogenesis. Growth Factors 20:155–175
63. Filion RJ, Popel AS (2004) A reaction–diffusion model of basic fibroblast growth factor integrations with cell surface receptors. Ann Biochem Eng 32:645–663
64. He L, Niemeyer B (2003) A novel correlation for protein diffusion coefficients based on molecular weight and radius of gyration. Biotechnol Prog 19:544–548
65. Beck LS, DeGuzman WPL, Lee YX, Siegel MW, Amento EP (1993) One systemic administration of transforming growth factor-beta 1 reverses age- or glucocorticoid-impaired wound healing. J Clin Invest 92:2841–2849
66. Coffey RJ, Kost LJ, Lyons RM, Moses HL, LaRusso NF (1987) Hepatic processing of transforming growth factor/3 in the rat. J Clin Invest 80:750–757
67. Wakefield LM, Winokur TS, Hollands RS, Christopherson K, Levinson AD, Sporn MB (1990) Recombinant latent transforming growth factor beta 1 has a longer plasma half-life in rats than active transforming growth factor beta 1, and a different tissue distribution. J Clin Invest 86:1976–7684
68. Zioncheck SA, Chen TF, Richardson L, Mora-Worms M, Lucas C, Lewis D et al (1994) Pharmacokinetics and tissue distribution of recombinant human transforming growth factor beta 1 after topical and intravenous administration in male rats. Pharm Res 11(2):213–220
69. Lazarous DF, Shou M, Stiber JA, Dadhania DM, Thirumurti V, Hodge E et al (1997) Pharmacodynamics of basic fibroblast growth factor: route of administration determines myocardial and systemic distribution. Cardiovasc Res 36(1):78–85
70. Edelman ER, Nugent MA, Karnovsky MJ (1993) Perivascular and intravenous administration of basic fibroblast growth factor: vascular and solid organ deposition. Proc Natl Acad Sci 90(4):1513–1517
71. Whalen GF, Shing Y, Folkman J (1989) The fate of intravenously administered bFGF and the effect of heparin. Growth Factors 1:157–164
72. Chandler WL, Alessi MC, Aillaud MF, Vague P, Juhan-Vague I (2000) Formation, inhibition and clearance of plasmin in vivo. Haemostasis 30(4):204–218
73. Nugent MA, Edelman ER (1992) Kinetics of basic fibroblast growth factor binding to its receptor and heparan sulfate proteoglycan: a mechanism for cooperativity. Biochemistry 31:8876–8883
74. Anderson SG, Buckingham RH, Kurland CG (1983) Does codon composition influence ribosome function? EMBO 3:91–94

75. Pavlov MY, Ehernberg M (1996) Rate of translation of natural mRNAs in an optimized in vitro system. Arch Biochem Biophys 328:9–16
76. Castello R, Estelles A, Vazquez C, Falco C, Espana F, Almenar SM et al (2002) Quantitative real-time reverse transcription-PCR assay for urokinase plasminogen activator, plasminogen activator inhibitor type 1, and tissue metalloproteinase inhibitor type 1 gene expressions in primary breast cancer. Clin Chem 48:1288–1295
77. Ellis V, Behrendt N, Dano K (1991) Plasminogen activation by receptor-bound urokinase. a kinetic study with both cell-associated and isolated receptor. J Biol Chem 266(19):12752–12758
78. Levine HA, Pamuk S, Sleeman BD, Nilsen-Hamilton M (2001) Mathematical modeling of capillary formation and development in tumor angiogenesis: penetration into the stroma. Bull Math Biol 63:801–863
79. Lijnen HR, Carmeliet P, Bouche A, Moons L, Ploplis VA, Plow E et al (1996) Restoration of thrombolytic potential in plasminogen-deficient mice by bolus administration of plasminogen. Blood 88(3):870–876
80. Ploplis VA, Carmeliet P, Vazirzadeh S, Vlaenderen IV, Moons L, Plow EF et al (1995) Effects of disruption of the plasminogen gene on thrombosis, growth, and health in mice. Circulation 92:2585–2593
81. Mansbridge JN, Liu K, Pinney RE, Patch R, Ratcliffe A, Naugnton GK (1999) Growth factors secreted by fibroblasts: role in healing diabetic foot ulcers. Diabetes Obes Metab 1:265–279
82. Crescenzo GD, Pham PL, Durocher Y, O'Connor-McCourt MD (2003) Transforming growth factor-beta (TGF-beta) binding to the extracellular domain of the type II TGF-beta receptor: receptor capture on a biosensor surface using a new coiled-coil capture system demonstrates that avidity contributes significantly to high affinity binding. J Mol Biol 328(5):1173–1183
83. Crescenzo GD, Grothe S, Zwangstra J, Tsang M, O'Connor-McCourt MD (2001) Real-time monitoring of the interactions of transforming growth factor-beta (TGF-beta) isoforms with latency-associated protein and the ectodomains of the TGF-beta Type II and III receptors reveals different kinetic models and stoichiometries of binding. J Biol Chem 276(32): 29632–29643

Chapter 12
Seeing the Invisible: How Mathematical Models Uncover Tumor Dormancy, Reconstruct the Natural History of Cancer, and Assess the Effects of Treatment

Leonid Hanin

Abstract The hypothesis of early metastasis was debated for several decades. Dormant cancer cells and surgery-induced acceleration of metastatic growth were first observed in clinical studies and animal experiments conducted more than a century ago; later, these findings were confirmed in numerous modern studies.

In this primarily methodological work, we discuss critically important, yet largely unobservable, aspects of the natural history of cancer, such as (1) early metastatic dissemination; (2) dormancy of secondary tumors; (3) treatment-related interruption of metastatic dormancy, induction of angiogenesis, and acceleration of the growth of vascular metastases; and (4) the existence of cancer stem cells. The hypothesis of early metastasis was debated for several decades. Dormant cancer cells and surgery-induced acceleration of metastatic growth were first observed in clinical studies and animal experiments conducted more than a century ago; later, these findings were confirmed in numerous modern studies.

We focus on the unique role played by very general mathematical models of the individual natural history of cancer that are entirely mechanistic yet, somewhat paradoxically, essentially free of assumptions about specific nature of the underlying biological processes. These models make it possible to reconstruct in considerable detail the individual natural history of cancer and retrospectively assess the effects of treatment. Thus, the models can be used as a tool for generation and validation of biomedical hypotheses related to carcinogenesis, primary tumor growth, its metastatic dissemination, growth of metastases, and the effects of various treatment modalities. We discuss in detail one such general model and review the conclusions relevant to the aforementioned aspects of cancer progression that were

L. Hanin, Ph.D. (✉)
Department of Mathematics, Idaho State University, 921 S. 8th Avenue, Stop 8085, Pocatello, ID 83209, USA

Biomedical Engineering, Center for Bioinformatics and Computational Genomics, Georgia Institute of Technology, 6015 State Bridge Road, Apt. 84087, Duluth, GA 30097, USA
e-mail: hanin@isu.edu

H. Enderling et al. (eds.), *Systems Biology of Tumor Dormancy*, Advances in Experimental Medicine and Biology 734, DOI 10.1007/978-1-4614-1445-2_12, © Springer Science+Business Media New York 2013 drawn from fitting a parametric version of the model to data on the volumes of bone metastases in one breast cancer patient and 12 prostate cancer patients.

Keywords Angiogenic switch • Breast cancer • Cancer stem cell • Chemotherapy • Metastatic dormancy • Model identifiability • Poisson process • Primary tumor • Prostate cancer • Radiotherapy • Surgery • Treatment-induced acceleration of metastasis

Introduction

The conventional paradigm of the progression of solid cancers is centered on the following two postulates:

1. Cancer progresses sequentially through the stages characterized by the extent of its anatomic spread—local, regional, and distant.
2. Metastases are independently growing secondary tumors that arise during late stages of the disease from malignant cells shed by the primary tumor and seeded at various secondary sites.

These postulates encapsulate the beliefs of the majority of research oncologists and practitioners, and serve as the foundation upon which oncology is taught in medical schools. Necessary–and highly consequential–implications of these postulates, vis-à-vis cancer detection and treatment, are as follows:

1. Early diagnosis leads to better prognosis and increases the probability of cure.
2. Treatment should eliminate all cancer cells, and the earlier and more aggressive the treatment, the better the outcome.
3. If treatment of the primary tumor fails to prevent metastasis, it does not exacerbate the disease.

The first of these tenets has spawned a massive and costly cancer screening effort, especially for breast, prostate, and colon cancers.

To realize that something is fundamentally amiss in the above paradigm of cancer, one need not delve deep into cancer etiology, phenomenology, and epidemiology. Consider the following well-known facts:

(a) A significant fraction of cancer patients are diagnosed and treated at the earliest stages of cancer progression and still go on to develop distant metastases. Question: *How could one explain this without assuming that by the time of primary cancer detection it has already metastasized?*
(b) Cancer patients who present at late stages of the disease very rarely have clinically manifest secondary tumors. Yet quite often metastases quickly surface after the start of treatment. Question: *Couldn't this be the result of treatment?*
(c) Large-scale randomized clinical trials have shown that the benefits of mammography as a means of early breast cancer detection are uncertain at best, which served as a basis for a recent scaling back of the mammography programs.

Moreover, lymph node-positive premenopausal breast cancer patients aged 40–49 who received mammography were found to have a statistically significant excess *mortality* over those in the control group [1–3]. Question: *While mammography clearly increases incidence of breast cancer through early cancer detection and overdiagnosis, how could it increase mortality other than through treatment-related exacerbation of the disease?*

The Emerging New Paradigm of Cancer

The failure of the traditional paradigm of cancer to explain many clinical, experimental, and epidemiological observations calls for a paradigm shift. Some of the pivotal themes around which the emerging new paradigm of cancer is likely to crystallize are briefly outlined below. Because metastasis accounts for about 90% of cancer-related deaths, many of these themes are focused on metastatic progression. Revision of the traditional paradigm of cancer was initiated in the pioneering works of Bernard Fisher and Judah Folkman. More recently, important contributions to the emerging theory, mostly focused on breast cancer, were made by Michael Retsky, Michael Baum, Romano Demicheli, and their colleagues. For an extensive discussion of the *raisons d'être* of the new paradigm of cancer, supporting biomedical evidence and related hypotheses, the reader is referred to [3–7].

Early Metastatic Dissemination

It is well-known that an avascular tumor can only reach a microscopic size of 1–2 mm in diameter [8]. To support further growth, the tumor must induce angiogenesis. After a capillary network is in place and tumor cells gain competence in migration, degradation of the extracellular matrix and intravasation, the tumor may start shedding metastases into the bloodstream. Additionally, tumor cells may spread to lymph nodes as soon as they acquire the capability to penetrate the lymph channels. Thus, the tumor may spread to various organs and tissues long before it becomes clinically detectable. This possibility has been entertained in the medical literature for decades [9–11]; for example, it was estimated in [10] that more than 70% of cancer patients have occult metastases at presentation. However, the hypothesis of early metastasis deals with events that are typically *unobservable*. Thus, assessment of their relative timing has to rely on indirect evidence, such as that discussed in the *Introduction*, as well as clinical experience and intuition. As yet another validation of the notion of early metastasis, it is worth noting a bold prediction by Bernard Fisher [3, 5, 11] that the extent of, local treatment would have only limited effect on long-term survival and cure of breast cancer patients. This prediction was substantiated later in numerous clinical trials. If confirmed unequivocally, the notion of early metastasis would shift therapeutic emphasis from local to systemic treatment regardless of the observed disease stage at diagnosis and would lead to an overhaul of clinical oncology.

Cancer Dormancy

Cancer dormancy represents a major departure from the linearity of cancer progression, one of the cornerstones of the traditional paradigm of cancer. The term "dormant cancer cell" was apparently coined by Hadfield in 1954 [12]. Dormancy may exist in several forms: cancer cells circulating in blood or lymph vessels; solitary cancer cells lodged in various organs and tissues; residual cancer cells remaining after treatment of the primary tumor; and pseudo-stable, non-progressing microscopic primary or secondary tumors.

The earliest reports on circulating cancer cells go back to the nineteenth century [13, 14]. Numerous modern studies of various types of cancer [15–22] have shown that large quantities of tumor cells can be present in blood, bone marrow, and the lymphatic system of cancer patients without clinically manifest primary or secondary tumors. This suggests the presence of occult microscopic tumors that continuously shed cancer cells into the bloodstream and the lymphatics. The introduction of *in vivo* video microscopy enabled direct observation and quantitative study of dormant cancer cells [23–25]. Transmission of dormant metastases to the recipients of organ transplants has also been reported [26, 27]. A balance between proliferation, apoptosis, and quiescence of cancer cells brings about the possibility that primary or secondary cancer can remain subclinical for an extended period of time; for example, breast cancer recurrence was reported to occur after 20 and even more than 25 years of a disease-free period [28].

Cancer dormancy presents both a challenge and opportunity to cancer treatment. On the one hand, it is a time bomb whose therapeutic dismantling would require development of the capability of targeting quiescent cells rather than actively proliferating cells, or maintenance of the dynamic equilibrium between proliferation and death of cancer cells. On the other hand, once the state of cancer dormancy is achieved by therapeutic means, the urgency of conventional aggressive treatments and even their necessity would be considerably reduced.

Angiogenic Switch

One of the principal mechanisms by which a tumor escapes from the state of dormancy is induction of angiogenesis. In this area, Judah Folkman's groundbreaking work since the early 1970s has resulted in two conclusions of fundamental importance (see e.g., [8]):

1. An avascular tumor can only reach a limited microscopic size characterized by the balance between proliferation and apoptosis. A tumor may remain in this state of equilibrium for a protracted period of time, which brings about tumor dormancy. This state is caused by restricted supply of nutrients and oxygen and by the balance between activation and suppression of growth and apoptosis. Another key condition for maintaining this state of equilibrium is the balance of

angiogenesis-promoting and–inhibiting factors and signals. The equilibrium size of an avascular tumor is species- and organ-specific and depends on the tumor's proliferative potential.
2. Breaking the balance between pro- and anti-angiogenic forces requires an activation event such as stress, infection, trauma, surgery, or irradiation. Disruption of the balance in favor of angiogenesis, called an *angiogenic switch*, causes the tumor to advance along the progression pipeline.

Angiogenesis at the secondary site is a critical, rate-limiting step of the multistage process that leads to a detectable metastasis [8]. As a result, only a small fraction of dormant or slowly growing metastases eventually develop the capillary network enabling their further growth.

Interaction Between Primary and Secondary Tumors

One of the critical flaws in the conventional paradigm of cancer is that malignant tumors are viewed as autonomously growing entities driven by their evolving genotype and constrained by the supply of nutrients, oxygen, and growth factors. It is curious, to say the least, that this theory has persisted to this day in spite of extensive experimental evidence of interaction between primary and secondary tumors in animal models that started accumulating as early as the beginning of the twentieth century [29–32]. Numerous later studies confirmed these early findings, although changed their interpretation (see [3–6] and references therein). The most important discovery within this realm of research is that large primary and secondary tumors make the organism more refractory to the inception of other tumors and inhibit growth of smaller tumors.

The mechanism of inter-tumor interaction was hypothesized in [33] and confirmed in a host of other studies, many of which are reviewed in [3, 5]. The key idea is that tumor dynamic depends on the balance between production and disintegration of factors that promote or inhibit growth and angiogenesis. These factors are produced by tumors and their microenvironment as well as systemically. The growth factors are more easily degradable than growth inhibitors and propagate mostly by diffusion, thus acting locally and promoting tumor growth. In contrast, growth and angiogenesis inhibitors are more stable and, when released into the bloodstream, may reach remote primary and secondary tumors and impede their growth and vascularization.

Treatment-Related Acceleration of Metastasis

If a primary tumor suppresses the growth of secondary tumors, then it would come as no surprise that removal of the primary tumor could accelerate the growth of metastases. The study of this striking phenomenon started in the early 1900s [31, 32]. More than a century of investigation produced a wealth of clinical observations, epidemiological analyses, and experimental studies on animal models, which are reviewed at length in [3–6] and can be briefly summarized as follows: *extirpation of primary or secondary tumors triggers accelerated proliferation of dormant or slowly growing metastases and their vascularization.* This important finding was directly confirmed in a number of well-documented clinical case studies involving various types of cancer:

(a) In eight cases of non-seminomatous germ-cell testicular cancer, resection of bulky metastases caused a dramatic exacerbation of the disease, as evidenced by the clinical picture and/or biochemical indicators [34].
(b) Excision of primary melanomas precipitated metastatic spread of the disease in three skin cancer patients [35, 36].
(c) In one case of pancreatic cancer, excision of the primary adenocarcinoma (via the Whipple procedure) caused surfacing of numerous previously undetectable liver metastases within 34 days of the surgery [37].

Surgery-induced acceleration of metastasis provides a compelling rationale to various clinical and epidemiological findings, including those mentioned in the *Introduction*, that have evaded alternative explanation [3]. Of note, the extent of the acceleration of metastatic growth was found to be proportional to the extent of surgery. Surprisingly, even biopsy was reported to result in a measurable increase in the incidence of lung metastases in mice [38].

What is the mechanism of accelerated growth of metastases following resection of the primary tumor? Removal of the primary tumor reduces production of growth inhibitors and pro-apoptosis factors and signals, which accelerates the growth of metastases. Additionally, wound healing processes following surgery are accompanied by a surge in local and systemic production of various growth and angiogenic factors that act synergistically with the decrease in the levels of growth and angiogenesis inhibitors and apoptosis activators.

This mechanism suggests that the same metastasis-enhancing effect will also manifest for *any* surgery or wounding occurring after removal of the primary tumor. Several studies confirmed that this is indeed the case. For example, one study reported a significantly elevated risk of metastatic recurrence of breast cancer for a group of 125 women who underwent delayed large flap breast reconstruction after mastectomy, compared to a group of 182 individually matched patients who received mastectomy alone [39]. Another study involving statistical analysis of 418 patients with advanced cancer of various types and localization showed increased incidence of metastatic relapse for patients who had surgery unrelated to cancer after resection of the primary tumor [40].

The mechanisms described above also seem to be relevant to radiation therapy. This leads to the hypothesis that the effects of radiation on metastasis are similar to those produced by surgery. A dramatic acceleration of metastasis may also result from systemic treatment involving chemotherapy and hormonal therapies; see [7] and the Conclusions from Model-Based Data Analysis Section below for further details. Although the underlying mechanisms in the case of systemic therapy are likely to be different from those associated with surgery, they surely involve selection of the fastest proliferating and most resistant cells occurring for any protracted treatment.

Cancer Stem Cells

According to the commonly accepted theories of carcinogenesis, a primary tumor originates from a single clonogenic cell. How could one explain, then, the observed heterogeneity of tumor cell phenotypes, including differences in clonogenic capacity, oxygenation level, proliferation pattern, metastatic potential, and resistance to various therapies? Three non-mutually exclusive explanations can be offered:

1. These differences can be traced back to spatial heterogeneity of the tumor that manifests through distinct localization of cancer cells within the tumor body, proximity to blood vessels, and other factors of similar nature.
2. These differences are due to evolution of the genotypes of cancer cells combined with selection processes.
3. The tumor is an organ-like entity that evolves from stem cells by means of differentiation and proliferation.

This third possibility has recently gained prominence after discovery of cancer stem cells in acute myeloid leukemia [41]. The hallmarks of cancer stem cells are self-renewal, pluripotency, high clonogenicity, and high metastatic potential. It has also been demonstrated that stem-like cancer cells share some of these properties and carry identifiable cell surface markers in other types of hematologic cancers as well as in some solid cancers, including cancer of the breast [42–44] and prostate [45]. The existence of cancer stem cells is highly consequential for cancer treatment, as targeting and eliminating this potentially tiny subpopulation becomes a critical factor in cancer therapy.

Spontaneous Cancer Regression

Although spontaneous regression of cancer is a well-documented phenomenon [46–48], recorded cases undoubtedly constitute only a minor fraction of its total incidence. As a clue pointing, if only indirectly, to the true scale of this event, we mention an epidemiological analysis that inferred a 22% rate of spontaneous breast cancer cure from a large-scale study involving more than 229,000 women followed from 1992 to 2001 in Norway [49]. Given that all known modes of cancer treatment have a high failure rate and debilitating side effects, it is tempting to speculate that learning how tumor microenvironment and the action of the immune system bring about spontaneous cancer regression may help discover long-awaited treatments leading to radical cure.

General Methodology

The backbone of the new paradigm of cancer outlined above deals with micro-events and processes that are typically *unobservable*, or only partially observable, in an individual patient. The only reliable way to test hypotheses about the likelihood, timing, rates, and relative importance of these events is to relate them to clinical outcomes and variables that become observable many months, years, or even decades after the chain of causes and effects triggered by the initiating micro-events will have run their full course and manifested as a detectable cancer. The only quantitative methodology available for building and analyzing such relationships for an individual patient is mathematical modeling. The observables and selection of mathematical models warrant a more detailed discussion, which appears in the next two sections.

Observables

A cancer patient's observable clinical variables typically fall into two categories: (1) variables that become available at presentation or as a result of treatment of the primary tumor; and (2) those resulting from follow-up tests that can be either pre-scheduled or prompted by local recurrence or metastatic relapse. Variables in the first category include age, cancer stage, primary tumor volume at diagnosis, and various histological, biochemical, and genomic markers. Observables in the second category usually consist of site-specific numbers and volumes of metastases detected through the use of imaging technology. These data represent an invaluable source of information suitable for validation of the emerging new paradigm of cancer.

Mathematical Models

Following in the wake of molecular biology and genomics that aim to uncover the underlying causes and precise mechanisms of carcinogenesis and cancer progression, the vast majority of mathematical models of cancer seek to describe in great detail various biological processes associated with cancer: the formation and accumulation of oncogenic mutations; evolution of cancer genotypes and phenotypes; metabolism of cancer cells; regulation of their cell cycle, proliferation, and apoptosis; activation and inhibition of angiogenesis; transport of nutrients, oxygen, and growth/antigrowth factors; interaction between tumor cells and their environment including the immune and endocrine systems; cancer cell signaling and signal transduction; complex spatial and temporal dynamics of primary tumor growth; invasion and metastasis; and response to various modes of treatment. Most of these models are formulated as large systems of ordinary or partial differential equations.

Although such models are useful for elucidating specific aspects of cancer development, their integration into a comprehensive, all-encompassing model of cancer is well beyond the reach of contemporary science and technology. To make things even worse, these models depend on numerous, typically dozens or hundreds of unknown parameters (including initial and boundary conditions). Many of these parameters are patient-specific, with the implication that the availability of only a very limited number of observables for an individual patient makes parameter estimation impossible. Finally, these models are based on a large number of simplifying assumptions about the structure and mechanisms of the systems they seek to describe. Overcoming these fundamental difficulties calls for a new class of mathematical models.

A starting point is to drop the overly ambitious goal of uncovering the whys and wherefores of cancer and to instead focus on the accurate phenomenological description of *how* it develops. From this perspective, understanding cancer amounts to being able to determine both the timing of critical micro-events associated with cancer initiation and progression, and the rates of growth of the primary tumor and metastases. These micro-events include emergence of the first malignant clonogenic cell, shedding of viable metastases off the primary tumor, seeding of these metastases in various organs and tissues, interruption of their dormancy, induction of angiogenesis, and the start of their progression to detectable secondary tumors (the event that will be referred to in what follows as *inception*). It is important to emphasize that the fewer biological assumptions built into such a model, the greater its utility and universality.

In the next section, we formulate a very general model of cancer progression that is essentially free of any specific biological assumptions. Its only very mild assumptions are dictated by mathematical convenience and tractability and thus serve to facilitate mathematical analysis without restricting the scope of biological possibilities allowed by the model. Yet another important feature of this model is that it leads to an explicit formula for the distribution of the sizes of detectable metastases in a given secondary site, which—after suitable parameterization of the model—enables quantitative inference on parameters descriptive of the individual natural history of cancer and the effects of treatment. Refinement of this model is limited only by the availability of data on the number and volumes of detectable metastases. Thus, this model is well suited for testing the postulates of the new paradigm of cancer.

At first glance, it seems impossible to describe processes as complex as cancer progression by means of a model that is essentially free of biological assumptions. The "trick" here is twofold: first, the model is formulated in terms of several *arbitrary* functions representing the laws of growth of the primary tumor and metastases pre- and post-treatment, accounting for the rate of metastasis shedding and describing the distribution of metastasis latency times; and, second, this very general setting still allows for computing the output of the model, which is the site-specific distribution of the sizes of detectable metastases, in closed form. Thus, the model obviates the step of initial parameterization that would make it rigid and approximate at the outset; rather, the model is parameterized only when its output is fit to the data.

A Universal Mathematical Model of the Individual Natural History of Metastatic Cancer

The model described hereafter accounts for the timing and rates of the following processes associated with carcinogenesis and cancer progression: (1) cancer initiation; (2) primary tumor growth; (3) shedding of metastases by the primary tumor; (4) selection of viable metastases; (5) dormancy of metastases and their inception at various secondary sites; and (6) treatment-related change in the growth rates of primary tumor and metastases. The model accommodates all modes of cancer therapy including short treatments (such as surgery or external beam radiation) regarded as instantaneous events and protracted treatments (such as brachy-, chemo-, and hormonal therapies). Because cancer initiation and progression are impelled through a number of sporadically occurring, low-probability, high-impact discrete random events [50], stochastic models are a necessary choice. Relevant terminology, notation, and assumptions (essentially mathematical in nature) are formulated below.

1. *Disease-free period.* This period begins with the birth of an individual (or start of exposure to a carcinogen) and ends with the emergence of the first malignant clonogenic cell, an event termed *onset of disease*.
2. *Primary tumor dynamics.* The size of the primary tumor (that is, the total number of tumor cells) at any time t counted from the age T of disease onset will be denoted by $\Phi(t)$. We assume that prior to the start of treatment (age V), the growth of the primary tumor is governed by a function Φ_0 and thereafter by a function Φ_1, which acts multiplicatively on the size of the primary tumor at the start of treatment. Function Φ_0 is strictly increasing, continuous, and satisfies the initial condition $\Phi_0(0)=1$. Furthermore, function Φ_1 is assumed continuous but not necessarily increasing. In particular, for a nonrecurrent excised tumor, $\Phi_1=0$. Functions Φ_0 and Φ_1 may depend on one or several parameters. It follows from the above assumptions that

$$\Phi(t)=\begin{cases}\Phi_0(t) & \text{if } 0\le t\le V-T\\ \Phi_1(t-(V-T))\Phi_0(V-T) & \text{if } t>V-T\end{cases}.$$

   We denote the inverse function for Φ_0 by ϕ.
3. *Metastasis formation.* Current knowledge suggests that the process of metastasis shedding off the primary tumor is "purely random" in the sense that the numbers of metastases shed by the primary tumor over two nonoverlapping time intervals are stochastically independent. This indicates that metastasis shedding is governed by a Poisson process. The rate, $\mu(t)$, of this process at time t is proportional to the number, $N(t)$, of metastasis-producing cells present at time t: $\mu(t)=\alpha_0 N(t)$, where $\alpha_0>0$ is the rate of metastasis shedding per cell. Since $N(t)$ is unobservable, we must relate it to the primary tumor size $\Phi(t)$. Specifically, we assume that $N(t)=g(\Phi(t))$, where g is a positive continuous function on $[1,\infty)$, so that

$$\mu(t) = \alpha_0 g(\Phi(t)). \tag{12.1}$$

A natural choice for function g is $g(x) = \alpha_1 x^\theta$ with some constants $\alpha_1 > 0$ and $\theta \geq 0$, in which case

$$\mu(t) = \alpha \Phi^\theta(t), \tag{12.2}$$

where $\alpha = \alpha_0 \alpha_1$. The value $\theta = 1$ (which corresponds to $g(x) = \alpha_1 x$) implies that a constant fraction of tumor cells has metastatic potential. It is known that many solid tumors contain a core of hypoxic, clonogenically sterile cells or even a broth of proteins, whereas actively proliferating clonogenic cells capable of producing metastases are concentrated near the tumor surface; in this case, one would expect $\theta = 2/3$. Finally, in the case $\theta = 0$, we have $g(x) = \alpha_1$, which suggests the existence of a relatively stable, self-renewing subpopulation of metastasis-producing cells within the primary tumor. In this case, $\mu(t) =$ constant and hence the underlying Poisson process is homogeneous.

It is further assumed that metastases shed by the primary tumor give rise to clinically detectable secondary tumors in a given site independently of each other with the same probability q. Therefore, inception of metastases in the site in question is governed by a Poisson process with rate $\nu(t) = q\mu(t)$, where $\mu(t)$ is given by formula (12.1). Each viable metastasis spends some random *latency time* between detachment from the primary tumor and inception in the secondary site. We assume that latency times for viable metastases bound for a given site are independent and identically distributed with some probability density function (pdf) f and the corresponding cumulative distribution function (cdf) F. Then, see e.g., [51], the resulting process of metastasis inception is again a Poisson process with the rate

$$\lambda(t) = \int_0^t \nu(s) f(t-s) ds.$$

4. *Timeline of the natural history of metastatic cancer and observables.* Suppose that the observed primary tumor size at age V is S. Then the patient's age T at disease onset is given by the formula

$$T = V - \varphi(S). \tag{12.3}$$

We assume that local or systemic treatment was given (or started) at age V, and that at age $W > V$, a certain number, n, of metastases were detected in the same secondary site with the observed volumes $X_1, X_2, \ldots, X_n$, where $X_1 < X_2 < \ldots < X_n$. Thus, $0 < T < V < W$.
5. *Growth of metastases.* Prior to the start of treatment, the growth of the size of any viable metastasis in a given secondary site is governed by a function Ψ_0. After (the start of) treatment, the size of the metastasis grows according to a potentially different function Ψ_1, which acts multiplicatively on the size of the

metastasis at the start of treatment. We assume for simplicity that actively growing metastases originate from a single cell: $\Psi_0(0)=1$. Additionally, we assume that functions Ψ_0, Ψ_1 are strictly increasing and differentiable, and that $\Psi_1(0)=1$. Functions Ψ_0, Ψ_1 may depend on one or several parameters. It follows from our assumptions that the size of a viable metastasis at time t from inception is

$$\Psi(t) = \begin{cases} \Psi_1(t) & \text{if } 0 \le t \le W-V \\ \Psi_0(t-(W-V))\Psi_1(W-V) & \text{if } W-V < t \le W-T \end{cases}. \tag{12.4}$$

This function is strictly increasing, continuous, and piecewise differentiable, and satisfies the initial condition $\Psi(0)=1$. We denote by M the maximum size a metastasis can reach given the model assumptions. It follows from (12.4) that

$$M = \Psi_0(V-T)\Psi_1(W-V). \tag{12.5}$$

6. *Secondary metastasis*. Secondary metastasis (that is, formation of "metastasis of metastasis") to a given site, both from other sites and from within, is assumed negligible.
7. *Metastasis detection*. The volume of a metastasis becomes measurable when it reaches some threshold value m. This value and the accuracy of volume measurement are determined by the sensitivity and resolution of the imaging technology used.
8. *Effects of treatment*. Because the rate of secondary metastasis is assumed negligible, formation of new metastases is stopped at the time of resection of a nonrecurrent primary tumor. Any mode of local or systemic treatment is assumed to affect metastases after their inception only through the rate of their growth (and not through the duration of their latency times).

To estimate model parameters, one must fit the model to a sufficiently rich set of observables. For the above model, this is essentially the set of volumes of metastases detected and measured in a given secondary site for an individual patient. An explicit form of the model-based distribution of the sizes of such metastases is computed as follows.

Theorem. The sizes $X_1<X_2<\ldots<X_n$ of metastases in a given secondary site that are detectable at age W are equidistributed, given their number n, with the vector of order statistics for a random sample of size n drawn from the distribution with the following pdf:

$$p(x) = \omega(W-T-\psi(x))\psi'(x), \quad m \le x \le M, \tag{12.6}$$

and $p(x)=0$ for $x \notin [m,M]$, where the tumor onset time T is given by (12.3), ψ is the inverse function for function Ψ defined in (12.4), M is specified in (12.5), and

$$\omega(t) = \frac{\int_0^{\min\{t, V-T\}} g(\Phi(s)) f(t-s) \mathrm{d}s}{\int_0^{\min\{W-T-\psi(m), V-T\}} g(\Phi(s)) F(W-T-\psi(m)-s) \mathrm{d}s}, \tag{12.7}$$
$$0 \le t \le W - T - \psi(m).$$

The proof of the Theorem is obtained by an appropriate modification of the argument in [52] developed for the case $g(x)=\alpha_1 x^\theta$. Notice that the distribution $p(x)$ given by formulas (12.6) and (12.7) is free of parameters α_0, α_1, q, and sample size n. Observe also that if the process of metastasis shedding is homogeneous (i.e., $g(x)=\alpha_1=\text{const}$), then the distribution $p(x)$ is independent of the laws of primary tumor dynamics before and after the start of treatment. Setting in (12.6) and (12.7) $m=1$ cell leads to the distribution of the sizes of *all* (both occult and detectable) metastases in a given site. Finally, the site-specific total number of viable metastases at age $t>T$ is Poisson distributed with parameter (expected value)

$$q\alpha \int_0^{t-T} g(\Phi(s)) F(W-T-s) \mathrm{d}s \tag{12.8}$$

Because parameters q and $\alpha=\alpha_0\alpha_1$ do not appear in formulas (12.6) and (12.7), knowledge of the volumes of detectable metastases is *insufficient* for estimation of the dynamics of their *number*.

Due to the non-stationarity of the metastasis shedding process, volumes of detectable metastases cannot be thought of as resulting from independent replications of the same random experiment. Therefore, the site-specific sizes (or volumes) of metastases measured at a given time do not form a random sample from a probability distribution. However, according to the above Theorem, the distribution of any rearrangement-invariant statistic based on observations $X_1, X_2, \ldots, X_n$ would be identical to the distribution of the same statistic based on a random sample of size n drawn from the pdf $p(x)$. Likelihood is one such statistic, along with the sample mean and variance. In fact, the joint likelihood of the observations $X_1, X_2, \ldots, X_n$, where $X_1<X_2<\ldots<X_n$, is given by the formula

$$\mathrm{L}(X_1, X_2, \ldots, X_n) = n! \prod_{i=1}^{n} p(X_i)$$

and hence has the same form (apart from the factor n!) it would take should the observations $X_1, X_2, \ldots, X_n$ form a random sample from the distribution with pdf $p(x)$. This makes it possible to estimate identifiable parameters of a suitably parameterized model described above using the method of maximum likelihood.

It should be emphasized again that the main model assumption of some biological import is that metastasis shedding by the primary tumor is a Poisson process whose rate $\mu(t)$ depends only on the current size of the primary tumor [see formula (12.1)]. This assumption was introduced in [53], where μ was assumed to be proportional to the size of the primary tumor, i.e., $g(x)=\alpha_1 x$. A more general functional form, $g(x)=\alpha_1 x^{\theta}$, was explored in [52], and the resulting model was used to reconstruct the natural history of the disease in one case of metastatic breast cancer; however, the model did not account for an effect of surgery on the rate of growth of metastases. An extension of the model accounting for such an effect was designed in [54], which was applied to clinical data on the same patient in [55]. Importantly, this led to a considerable improvement in the model fit to the volumes of detected bone metastases. Furthermore, the model was extended in [7] to incorporate the possibility of recurrence of the excised primary tumor as well as the case of chemotherapy, which may change the dynamics of the primary tumor growth but does not necessarily lead to its essentially instantaneous elimination. In the same work, this extended model was applied to a cohort of 12 patients with metastatic prostate cancer. Finally, the model formulated above altogether removes the reliance on any specific parametric form of the rate of metastasis shedding by the primary tumor.

Model Parameterization

Suppose the primary tumor grows exponentially at a constant rate $\beta_0>0$ before treatment, and at a rate of β_1 after the start of treatment, so that: $\Phi_0(t)=\exp\{\beta_0 t\}$, $0\le t\le V-T$, where time t is counted from the age T of tumor onset, and $\Phi_1(t)=\exp\{\beta_1 t\}$, where time t is measured from the start of treatment (age V). Note that the rate β_1 can be negative. We will also assume that before and after the start of treatment all viable metastases in the site of interest grow exponentially with rates γ_0, $\gamma_1>0$, so that $\Psi_0(t)=\exp\{\gamma_0 t\}$ and $\Psi_1(t)=\exp\{\gamma_1 t\}$, where time t is counted from the inception of metastasis and the start of treatment, respectively. Observe that, in the case of protracted treatment, γ_1 represents the *average* rate of growth of metastases in a given site over the entire period from the start of treatment to the time of metastasis surveying. Finally, metastasis latency times are assumed to be exponentially distributed with pdf $f(t)=\rho^{-1}e^{-t/\rho}$, where t is the time elapsed from the shedding of a viable metastasis and ρ is the expected duration of latency.

The resulting parametric model describes the individual natural history of cancer and the effects of treatment through the following six parameters: β_0, β_1, θ, γ_0, γ_1, and ρ. For mathematical convenience, we re-parameterize the model using an alternative set of six parameters:

$$A=\exp\{\gamma_1(W-V)\},\ M=\exp\{\gamma_0(V-T)+\gamma_1(W-V)\},$$
$$a_0=\frac{\beta_0\theta}{\gamma_0},\ a_1=\frac{\beta_1\theta}{\gamma_1},\ b_0=\frac{1}{\gamma_0\rho},\ b_1=\frac{1}{\gamma_1\rho}. \tag{12.9}$$

Note that parameter M was defined earlier in (12.5) and that $0<A<M$. In terms of these new parameters, the biological parameters of the model are computed as follows:

$$\gamma_1 = \frac{\ln A}{W-V}, \gamma_0 = \frac{b_1 \ln A}{b_0(W-V)}, \rho = \frac{W-V}{b_1 \ln A}, \tag{12.10}$$

$$\beta_0 = \frac{b_1 \ln A \ln S}{b_0(W-V)\ln(M/A)}, \beta_1 = \frac{a_1}{a_0}\frac{\ln A \ln S}{(W-V)\ln(M/A)}, \theta = a_0 \frac{\ln(M/A)}{\ln S}. \tag{12.11}$$

Also, for the onset time of the primary tumor given by formula (12.3) we have

$$T = V - \frac{b_0(W-V)}{b_1 \ln A}\ln\frac{M}{A}. \tag{12.12}$$

The pdf $p(x)$ underlying the distribution of the site-specific volumes of detectable metastases given in the above Theorem has the following form [7, 52, 54].

Case of Resected Nonrecurrent Primary Tumor

1. If $A \le m$ then

$$p(x) = (C_1 x)^{-1}\left[\left(\frac{M}{x}\right)^{a_0} - \left(\frac{x}{M}\right)^{b_0}\right], \quad m \le x \le M, \tag{12.13}$$

where

$$C_1 = a_0^{-1}\left[\left(\frac{M}{m}\right)^{a_0} - 1\right] - b_0^{-1}\left[1 - \left(\frac{m}{M}\right)^{b_0}\right]. \tag{12.14}$$

2. If $A > m$ then

$$p(x) = \begin{cases} \dfrac{b_1}{b_0}(C_2 x)^{-1}\left[\left(\dfrac{M}{A}\right)^{a_0} - \left(\dfrac{A}{M}\right)^{b_0}\right]\left(\dfrac{x}{A}\right)^{b_1}, & m \le x < A \\ (C_2 x)^{-1}\left[\left(\dfrac{M}{x}\right)^{a_0} - \left(\dfrac{x}{M}\right)^{b_0}\right], & A \le x \le M \end{cases}. \tag{12.15}$$

where

$$\begin{aligned} C_2 = {} & b_0^{-1}\left[\left(\frac{M}{A}\right)^{a_0} - \left(\frac{A}{M}\right)^{b_0}\right]\left[1 - \left(\frac{m}{A}\right)^{b_1}\right] \\ & + a_0^{-1}\left[\left(\frac{M}{A}\right)^{a_0} - 1\right] - b_0^{-1}\left[1 - \left(\frac{A}{M}\right)^{b_0}\right]. \end{aligned} \tag{12.16}$$

Case of Non-resected or Resected Recurrent Tumor

1. If $A \le m$ then (12.13) and (12.14) apply.
2. If $A > m$ then

$$p(x) = (C_2 x)^{-1} \frac{b_1}{b_0} \left\{ \begin{matrix} \left[\left(\frac{M}{A}\right)^{a_0} - \left(\frac{A}{M}\right)^{b_0} \right] \left(\frac{x}{A}\right)^{b_1} + \frac{b_1}{b_0} \frac{a_0 + b_0}{a_1 + b_1} \left(\frac{M}{A}\right)^{a_0} \\ \left[\left(\frac{A}{x}\right)^{a_1} - \left(\frac{x}{A}\right)^{b_1} \right] \end{matrix} \right\}$$

$$m \le x < A (C_2 x)^{-1} \left[\left(\frac{M}{x}\right)^{a_0} - \left(\frac{x}{M}\right)^{b_0} \right], A \le x \le M. \tag{12.17}$$

where

$$C_2 = a_0^{-1} \left[\left(\frac{M}{A}\right)^{a_0} - 1 \right] + b_0^{-1} \left\{ \left(\frac{M}{A}\right)^{a_0} \left[1 - \left(\frac{m}{A}\right)^{b_1} \right] + \left(\frac{A}{M}\right)^{b_0} \left(\frac{m}{A}\right)^{b_1} - 1 \right\}$$

$$+ \left(\frac{b_1}{b_0}\right)^2 \frac{a_0 + b_0}{a_1 + b_1} \left(\frac{M}{A}\right)^{a_0} \left\{ a_1^{-1} \left[\left(\frac{A}{m}\right)^{a_1} - 1 \right] - b_1^{-1} \left[1 - \left(\frac{m}{A}\right)^{b_1} \right] \right\}. \tag{12.18}$$

Model (12.13), (12.14), (12.15), (12.16) will be called the *Surgery model*, whereas model (12.13), (12.14), (12.17), (12.18) will be termed the *Full model*. The Surgery model depends on 5 parameters (A, M, a_0, b_0, b_1) and can be obtained as a limiting case of the Full model by letting $a_1 \to -\infty$ (or equivalently, $\beta_1 \to -\infty$). In the case $A > m$, all six parameters (A, M, a_0, a_1, b_0, b_1) of the Full model, as well as parameters A, M, a_0, b_0, b_1 of the Surgery model, are identifiable [52, 54] (for an extensive discussion of the identifiability of stochastic models, see [56]). If $W > V$ (that is, when surveying of metastases occurs after the start of treatment), biological parameters β_0, β_1, θ, γ_0, γ_1, ρ (or β_0, θ, γ_0, γ_1, ρ for the Surgery model) are determined uniquely by the alternative parameters. For a more detailed analysis of the two models and their limiting forms, see [7, 52, 54, 55].

Notice that the function $xp(x)$ depends only on various *ratios* of the quantities x, m, A, and M expressed as sizes, i.e., the numbers of cells. Therefore, the likelihood-based estimates of parameters A, M, a_0, a_1, b_0, b_1 would be the same if the above sizes were replaced by the corresponding *volumes*. This also implies that estimates of these parameters are independent of the volume, c, of a single cancer cell. Observe, however, that the same is true only for the biological parameters γ_0, γ_1, ρ, as well as the age T at tumor onset; see formulas (12.10) and (12.12). Parameters β_0, β_1, and θ depend on c through the size, S, of the primary tumor for which only the volume measurement is available; see formulas (12.11). Measurement of the primary tumor's volume typically involves a significant error due to uncertainty in determination of the tumor boundary. However, neither this error nor the possible deviation of the cell volume from the "standard" value $c = 10^{-9}$ cm^3 assumed in the studies [7, 52, 55] have a significant impact on parameter estimation, because the primary tumor size S appears in formulas (12.11) under the sign of logarithm.

The model of cancer progression described above was validated using the following two criteria:

1. The model with optimal parameters provided an excellent fit to the empirical distribution of detectable bone metastases for all patients analyzed, as confirmed by visual inspection and computing L^2-distance between the theoretical and empirical cdfs. Moreover, the fit was reasonably good even when the 5-parametric Surgery model degenerated into 2- or 3-parametric models for two additional breast cancer patients [55].
2. The Full model applied to surgery patients proved to be sensitive enough to correctly predict that they had surgery [7].

Conclusions from Model-Based Data Analysis

Identifiable model parameters produce, through formulas (12.10), (12.11), and (12.12), estimates of the most important temporal and rate characteristics that describe the natural history of metastatic cancer and the effects of treatment. The model was applied to the following cases, which are described in more detail in [6, 7, 52, 55]:

1. A breast cancer patient who received surgery and adjuvant hormonal treatment with tamoxifen and subsequently developed $n = 31$ detectable bone metastases and six detectable metastases of other localization 8 years post-surgery
2. Twelve prostate cancer patients treated with surgery alone (one case), a combination of surgery and adjuvant radiotherapy (one case), and systemic treatment consisting of various combinations and time courses of chemotherapy and hormonal therapy (all 12 cases). The number of bone metastases detected in these patients ranged from10 to 58

Results of data analysis for all these cases are briefly summarized below, with an eye toward validating the new paradigm of cancer outlined above; see also [6, 7, 52, 55].

1. *Metastasis shedding and cancer stem cells.* Within the parametric version of the model, the rate of metastasis shedding depends critically on the parameter θ; see (12.2). The estimate of θ was found to be uniformly small for all patients analyzed, which means that the process of metastasis shedding was essentially homogeneous. As discussed above, this suggests the tumor contains a small, self-renewing subpopulation of relatively constant size consisting of cells with high metastatic potential. This may serve as indirect evidence for the existence of breast and prostate cancer stem cells.
2. *Onset of metastasis.* For all patients, metastatic dissemination occurred soon after the onset of the disease and much earlier than the appearance of a clinically

detectable primary tumor. In fact, according to the model, the time between the onset of disease and the inception of the first metastasis never exceeded 2.5 years. At that time, the primary tumor was microscopic and definitely undetectable. Shedding of the first viable metastasis occurred even earlier. Thus, for all patients analyzed, their disease was essentially systemic at the outset.

3. *Metastasis dormancy.* The model-based estimate of the mean latency time for bone metastases ranged from a few days to as long as 16 years for prostate cancer patients, and was about 80 years for the breast cancer patient. This implies that, in patients with long metastasis latency times, many metastases were still occult at the time of surveying. Thus, our model and data analysis support the notion of metastasis dormancy, and show that latency times depend on the type of cancer and display significant individual variation. For the breast cancer patient, availability of the volume data for six additional metastases at various secondary sites other than bones made it possible to qualitatively assess the site-specificity of the latency time by estimating it for the pooled set of volumes of all detected secondary lesions. This resulted in a 15-fold drop in the estimated mean latency time, which suggests that latency of breast cancer metastases is markedly site-specific [55]. This also sheds some light on the relative duration of the two components of metastasis latency: (1) free circulation of metastatic cells and (2) dormancy of metastases in the form of solitary cancer cells or microscopic quiescent or slowly growing avascular clumps at various secondary sites. Because the first component is likely to be the same for all sites, the site-specificity of the total latency time leads to the conclusion that the dormancy of breast cancer metastases lodged at secondary sites is prevalent over free circulation.
4. *Timing of metastasis inception.* Observe that parameter A represents the size at the time of surveying of a hypothetical metastasis that was incepted at the start of treatment; see the first formula in (12.9). Comparison of the estimated values of this parameter with the observed volumes of metastases shows that, in all patients, inception of all or most of the detected metastases occurred prior to the start of treatment. Additionally, these early metastases had the largest volumes at surveying.
5. *Treatment-induced acceleration of metastatic growth.* The effect of treatment on metastatic growth is characterized by the ratio γ_1/γ_0 of the rate of growth of metastases after the start of treatment to their pretreatment growth rate. For the breast cancer patient, this ratio was 32 notwithstanding the fact that after surgery the patient was put on tamoxifen, which suppresses metastatic growth and has anti-angiogenic activity. For the prostate cancer patients, the metastasis enhancement ratio ranged from 3.5 to 504, with a median of 27. Thus, resection of the primary tumor, systemic treatment (chemotherapy combined with hormonal therapy), and possibly irradiation of the primary tumor all led to a dramatic exacerbation of the disease.
6. *Inhibitory effect of the primary tumor on the growth of metastases.* Another way to interpret the treatment-related boost of the growth of metastases discussed in Conclusion 5 is to say that the primary tumor *in situ* has a strong inhibitory effect on the growth of metastases. This conclusion is directly, and independently, confirmed by the results of model-based data analysis. In fact, for all patients analyzed, the pre-treatment rate of growth of metastases was smaller than the rate of growth of the primary tumor by an order of magnitude.

7. *Effects of treatment of the primary tumor*. As stated in Conclusion 1, the rate of metastasis shedding depends only weakly on the size of the primary tumor. Therefore, systemic treatment of the primary tumor had only a limited effect on the rate of metastasis shedding and, according to formula (12.8), also on the total number of viable secondary tumors. Additionally, since inception of most metastases occurred before the start of treatment (see Conclusion 4), resection and systemic treatment of the primary tumor had only a minor effect on the number of metastases relevant to patient survival. However, such treatments had a significant amplifying effect on the rate of their growth; see Conclusion 5. Taken collectively, these findings suggest that "local treatment" *per se* does not exist, as any intervention aimed at the primary tumor affects metastatic progression.
8. *Angiogenic switch*. Reconstruction of all 13 individual natural histories of cancer based on the estimates of model parameters suggests that, at the start of treatment (age V), all metastases were microscopic with a diameter far less than the limiting value of 1-2 mm characteristic of avascular tumors. However, by the time of metastasis surveying (age W), the detected secondary tumors reached considerable sizes unattainable for avascular tumors. This suggests that most likely treatment of the primary tumor triggered the angiogenic switch.

Discussion

The model of the individual natural history of cancer described above encompasses all the important stages of cancer progression, from the emergence of the first malignant cell to the surfacing of detectable secondary tumors in various organs and tissues. It enables estimation of the timing and rate of unobservable critical micro-events and processes that collectively constitute cancer progression, as well as assessment of the effects of treatment; in this sense, the model allows us to "see the invisible." The model is universal, based on very minimal assumptions, passes the tests of self-consistency and identifiability, is mathematically tractable and computationally feasible, and it provided an excellent fit to the observed site-specific volumes of detectable metastases. Thus, the model is suitable for testing various biomedical hypotheses about the individual natural history of cancer and the effects of treatment. When applied to clinical data for specific patients and fit to the observables, the model decidedly confirmed all the main hypotheses constituting the new paradigm of cancer delineated in the The Emerging New Paradigm of Cancer.

The model can be extended in the following directions as more data become available:

1. It could incorporate more realistic laws of growth of the primary tumor and metastases than the exponential law.
2. The rate of growth of metastases before and after the start of treatment can be represented as the difference between the rates of proliferation and apoptosis. Estimates of these two rates would provide an answer to the important question

as to what causes treatment-induced acceleration of the growth of metastases: their enhanced proliferation or inhibition of apoptosis, or both.

3. The model could account for the effects of treatment on site-specific metastasis latency times.
4. The model could accommodate a multistage representation of metastatic latency, where durations of the stages of free circulation, dormancy in a secondary site, slow avascular growth, and angiogenesis induction have their own distributions.
5. The model could take into account secondary metastasis.

The most notable finding revealed by the model-based reconstruction of the natural history of the disease in one breast cancer and 12 prostate cancer patients is a dramatic metastasis-accelerating effect of surgery, chemotherapy combined with hormonal treatment, and possibly radiation. Therefore, these aggressive cancer therapies need a critical re-evaluation. In particular, they should be recommended only if the net of their benefits and risks outweighs the risks associated with watchful waiting.

Finally, the results of model-based analysis lend support to the notion of metastatic dormancy. This important and well-documented phenomenon may suggest a new, more conservative therapeutic strategy: maintaining the dormant state of avascular micrometastases while controlling the size of the primary tumor.

References

1. Retsky M, Demicheli R, Hrushesky W (2003) Breast cancer screening: controversies and future directions. Curr Opin Obstet Gynecol 15:1–8
2. Retsky M, Demicheli R, Hrushesky W (2001) Premenopausal status accelerates relapse in node positive breast cancer: hypothesis links angiogenesis, screening controversy. Breast Cancer Res Treat 65:217–224
3. Retsky M, Demicheli R, Hrushesky W, Baum M, Gukas I (2010) Surgery triggers outgrowth of latent distant disease in breast cancer: an inconvenient truth? Cancers 2:305–337
4. Baum M, Chaplain M, Anderson A, Douek M, Vaidya JS (1999) Does breast cancer exist in a state of chaos? Eur J Cancer 35:886–891
5. Demicheli R, Retsky M, Hrushesky WJM, Baum M, Gukas ID (2008) The effects of surgery on tumor growth: a century of investigations. Ann Oncol 19:1821–1828
6. Hanin L (2010) Why victory in the war on cancer remains elusive: biomedical hypotheses and mathematical models. Cancers 3:340–367
7. Hanin L, Zaider M (2011) Effects of surgery and chemotherapy on metastatic progression of prostate cancer: evidence from the natural history of the disease reconstructed through mathematical modeling. Cancers 3:3632–3660
8. Folkman J, Watson K, Ingber D, Hanahan D (1989) Induction of angiogenesis during the transition from hyperplasia to neoplasia. Nature 339:58–61
9. Douglas JRS (1971) Significance of the size distribution of bloodborne metastases. Cancer 27:379–390
10. Barbour A, Gotley DC (2003) Current concepts of tumour metastasis. Ann Acad Med Singapore 32:176–184
11. Fisher B (1980) Laboratory and clinical research in breast cancer: a personal adventure. The David A. Karnofsky memorial lecture. Cancer Res 40:3863–3874
12. Hadfield G (1954) The dormant cancer cell. Br Med J 2:607–610

13. Ashworth TR (1869) A case of cancer in which cells similar to those in the tumour were seen in the blood after death. Aust Med J 14:146–147
14. Goldmann EE (1897) Anatomische Untersuchungen über die Verbreitungswege bösartiger Geschwülstle. Beitr Z Klin Chir 18:595
15. Fodstad O, Faye R, Hoifodt HK, Skovlund E, Aamdal S (2001) Immunobead-based detection and characterization of circulating tumor cells in melanoma patients. Recent Results Cancer Res 158:40–50
16. Pantel K, Otte M (2001) Occult micrometastases: enrichment, identification and characterization of single disseminated tumour cells. Semin Cancer Biol 11:327–337
17. Jiao X, Krasna MJ (2002) Clinical significance of micrometastasis in lung and esophageal cancer: a new paradigm in thoracic oncology. Ann Thorac Surg 74:278–284
18. Sugio K, Kase S, Sakada T, Yamazaki K, Yamaguchi M, Ondo K, Yano T (2002) Micrometastasis in the bone marrow of patients with lung cancer associated with a reduced expression of E-cadherin and beta-catenin: risk assessment by immunohistochemistry. Surgery 131:S226–S231
19. Ellis WJ, Pfitzenmaier J, Colli J, Arfman E, Lange PH, Vessella RL (2003) Detection and isolation of prostate cancer cells from peripheral blood and bone marrow. Urology 61:277–281
20. Pfitzenmaier J, Vessella RL, Ellis WJ, Lange PH (2003) Detection, isolation and study of disseminated prostate cancer cells in the peripheral blood and bone marrow. In: Pantel K (ed) Micrometastasis. Kluwer Academic Publishers, Norwell, MA, pp 87–116, Chapter 5
21. Meng S, Tripathy D, Frenkel EP, Shete S, Naftalis EZ, Huth JF, Beitsch PD, Leitch M, Hoover S, Euhus D, Haley B, Morrison L, Fleming TP, Herlyn D, Terstappen LWMM, Fehm T, Tucker TF, Lane N, Wang J, Uhr JW (2004) Circulating tumour cells in patients with breast cancer dormancy. Clin Cancer Res 10:8152–8162
22. Marches R, Scheuermann R, Uhr J (2006) Cancer dormancy: from mice to man. Cell Cycle 5:1772–1778
23. Chambers AF, Macdonald IF, Schmidt E, Koop S, Morris VL, Khokha R, Groom AC (1995) Steps in tumor metastasis: new concepts from intravital videomicroscopy. Cancer Metastasis Rev 14:279–301
24. Luzzi KJ, MacDonald IC, Schmidt EE, Kerkvliet N, Morris VL, Chambers AF, Groom AC (1998) Multistep nature of metastatic inefficiency. Dormancy of solitary cells after successful extravasation and limited survival of early micrometastases. Am J Pathol 153:865–873
25. Naumov GN, MacDonald IC, Weinmeister PM, Kerkvliet N, Nadkarni KV, Wilson SM, Morris VL, Groom AC, Chambers AF (2002) Persistence of solitary mammary carcinoma cells in a secondary site: a possible contributor to dormancy. Cancer Res 62:2162–2168
26. Jonas S, Bechstein WO, Lemmens H-P, Neuhaus R, Thalmann U, Neuhaus P (1996) Liver graft-transmitted glioblastoma multiforme. A case report and experience with 13 multiorgan donors suffering from primary cerebral neoplasia. Transpl Int 9:426–429
27. Loh E, Couch FG, Hendricksen C, Farid L, Kelly PF, Acker MA, Tomaszewski JE, Malkowicz SB, Weber BL (1997) Development of donor-derived prostate cancer in a recipient following orthotopic heart transplantation. JAMA 277:133–137
28. Karrison TG, Ferguson DJ, Meier P (1999) Dormancy of mammary carcinoma after mastectomy. Natl Cancer Inst 19:80–85
29. Ehrlich P, Apolant H (1905) Beobachtungen über maligne Mäusetumoren. Berl Klin Wochenschr 42:871–874
30. Bashford E, Murray JA, Cramer W (1907) The natural and induced resistance of mice to the growth of cancer. Proc R Soc Lond 79:164–187
31. Marie P, Clunet J (1910) Fréquences des métastases viscérales chez les souris cancéreuses après ablation chirurgicale de leur tumeur. Bull Assoc Franç L'Étude Cancér 3:19–23
32. Tyzzer EE (1913) Factors in the production and growth of tumor metastases. J Med Res 23:309–332
33. Prehn RT (1993) Two competing influences that may explain concomitant tumor resistance. Cancer Res 53:3266–3269

34. Lange PH, Hekmat K, Bosl G, Kennedy BJ, Fraley EE (1980) Accelerated growth of testicular cancer after cytoreductive surgery. Cancer 45:1498–1506
35. De Giorgi V, Massi D, Gerlini G, Mannone F, Quercioli E, Carli P (2003) Immediate local and regional recurrence after the excision of a polypoid melanoma: tumor dormancy or tumor activation? Dermatol Surg 29:664–667
36. Tseng WW, Doyle JA, Maguiness S, Horvai AE, Kashani-Sabet M, Leong SPL (2009) Giant cutaneous melanomas: evidence for primary tumour induced dormancy in metastatic sites? BMJ Case Rep. doi:10.1136/bcr.07.2009.2073
37. Deylgat B, Van Rooy F, Vansteenkiste F, Devriendt D, George C (2011) Postsurgery activation of dormant liver micrometastasis: a case report and review of literature. J Gastrointest Cancer 42:1–4
38. Hoover HC, Ketcham AS (1975) Techniques for inhibiting tumor metastases. Cancer 35:5–14
39. Isern AE, Manjer J, Malina J, Loman N, Mårtensson T, Bofin A, Hagen AI, Tengrup I, Landberg G, Ringberg A (2011) Risk of recurrence following delayed large flap reconstruction after mastectomy for breast cancer. Br J Surg 98:659–666
40. Maida V, Ennis M, Kuziemsky C, Corban J (2009) Wounds and survival in cancer patients. Eur J Cancer 45:3237–3244
41. Bonnet D, Dick JE (1997) Human acute myeloid leukemia is organized as a hierarchy that originates from a primitive hematopoietic cell. Nat Med 3:730–737
42. Al-Hajj M, Wicha MS, Benito-Hernandez A, Morrison SJ, Clarke MF (2003) Prospective identification of tumorigenic breast cancer cells. Proc Natl Acad Sci USA 100(7):3983–3988
43. Dick JE (2003) Breast cancer stem cells revealed. Proc Natl Acad Sci USA 100(7): 3547–3549
44. Kai K, Arima Y, Kamiya T, Saya H (2010) Breast cancer stem cells. Breast Cancer 17:80–85
45. Collins AT, Berry PA, Hyde C, Stower MJ, Maitland NJ (2005) Prospective identification of tumorigenic prostate cancer stem cells. Cancer Res 65:10946–10951
46. Boyd W (1966) The spontaneous regression of cancer. Thomas, Springfield, IL
47. Everson TC, Cole WH (2006) Spontaneous regression of cancer. Saunders, Philadelphia
48. Cole WH (1976) Spontaneous regression of cancer and the importance of finding its cause. Natl Cancer Inst Monogr 44:5–9
49. Zahl P-H, Mæhlen J, Welch HG (2008) The natural history of invasive breast cancer detected by screening mammography. Arch Intern Med 168:2311–2316
50. Kendal WS (2005) Chance mechanisms affecting the burden of metastases. BMC Cancer 5:138–146
51. Hanin LG, Yakovlev AY (1996) A nonidentifiability aspect of the two-stage model of carcinogenesis. Risk Anal 16:711–715
52. Hanin LG, Rose J, Zaider M (2006) A stochastic model for the sizes of detectable metastases. J Theor Biol 243:407–417
53. Bartoszyński R, Edler L, Hanin L, Kopp-Schneider A, Pavlova L, Tsodikov A, Zorin A, Yakovlev A (2001) Modeling cancer detection: tumor size as a source of information on unobservable stages of carcinogenesis. Math Biosci 171:113–142
54. Hanin LG (2008) Distribution of the sizes of metastases: mathematical and biomedical considerations. In: Tan WY, Hanin LG (eds) Handbook of cancer models with applications. World Scientific, Singapore, pp 141–169
55. Hanin LG, Korosteleva O (2010) Does extirpation of the primary breast tumor give boost to growth of metastases? Evidence revealed by mathematical modeling. Math Biosci 223:133–141
56. Hanin LG (2002) Identification problem for stochastic models with application to carcinogenesis, cancer detection and radiation biology. Disc Dyn Nat Soc 7:177–189

Index

H. Enderling et al. (eds.), *Systems Biology of Tumor Dormancy*, Advances in Experimental Medicine and Biology 734,
DOI 10.1007/978-1-4614-1445-2, © Springer Science+Business Media New York 2013

MIX
Papier aus verantwortungsvollen Quellen
Paper from responsible sources
FSC® C105338

If you have any concerns about our products,
you can contact us on
ProductSafety@springernature.com

In case Publisher is established outside the EU,
the EU authorized representative is:
Springer Nature Customer Service Center GmbH
Europaplatz 3, 69115 Heidelberg, Germany

Printed by Libri Plureos GmbH
in Hamburg, Germany